MW00439273

"The truth shall make you free."
(John 8:32)

CHRISTIANITY WITHOUT FAIRY TALES

When Science and Religion Merge

JIM RIGAS

ISBN Number: 1-57087-654-1

Library of Congress Control Number: 2004104527

Pnevma Publications
Distributed by Pathway Book Service
(800) 345-6665 www.jimrigas.com

Professional Press
Chapel Hill, NC 27515-4371

Manufactured in the United States of America
04 05 06 07 08 10 9 8 7 6 5 4 3 2 1

ACKNOWLEDGMENT

Grateful acknowledgment is made for permission to reprint previously published material:

Excerpts from THE AMPLIFIED BIBLE (marked AMP in the text), Old Testament copyright © 1965, 1987 by the Zondervan Corporation, The Amplified New Testament copyright © 1958, 1987 by the Lockmann Foundation. Used by permission.

Excerpts from the *Contemporary English Version* (marked CEV in the text), copyright © 1995, 1999 by the American Bible Society. Used by permission.

Excerpts from THE JERUSALEM BIBLE (marked JER in the text), copyright © 1966 by Darton, Longman & Todd, Ltd. and Doubleday, a division of Random House, Inc. Reprinted by permission.

Excerpts from THE KING JAMES VERSION (printed in the text without source designation), are not copyrighted material but in the public domain. Printed with thanks to the long-dead translators.

Excerpts from The Gospel of Thomas, The Gospel of the Hebrews, and The Gospel of Philip, from THE OTHER BIBLE, edited by Willis Barnstone, copyright © 1985, by Harper Collins. Used by permission.

Excerpts from The Gospel of Mary, p. 525, 10:1-10 pp. 526-7, 18:5-15 from the NAG HAMMADI LIBRARY IN ENGLISH, 3RD, COMPLETELY REVISED EDITION by JAMES M. ROBINSON, GENERAL EDITOR. Copyright © 1978, 1988 by E. J. Brill, Leiden, The Netherlands. Reprinted by permission of HarperCollins Publishers, Inc.

To the memory of my mother,
who taught me that prayers work

PREFACE

Dr. Michael Goulder was a brilliant British scholar. For thirty years an ordained Anglican priest, he spent the last fifteen of them teaching religion courses to his fellow clergy. But in 1981, after having been considered for the position of bishop, he suddenly left the priesthood to become "a non-aggressive atheist," as he put it. He explained that he had lost his belief in God and that he could no longer accept the convictions of all those past generations that peopled the pages of the Bible. The growth of science, history, and archaeology had disproved most of his earlier beliefs. He asked: "Can one still believe in God, if one doubts that he gave Moses the law on top of Mount Sinai?" His answer: "No!"[1]

Dr. Goulder was not the first and, probably, not the last believer to be overwhelmed by the outpouring of new scientific evidence that seems to invalidate one's religious beliefs. Modern archaeologists tell us that there was no wall around Jericho for Joshua to topple with his trumpets. What if there was also no burning bush from which God spoke to Moses? Suppose Jesus did not have a virgin birth? Do these things matter? Do they disprove the existence of God, or even of Christianity? My answer is No! And that is what this book hopefully presents. I believe that scientific truth and faith can go hand in hand, provided that we can keep an equally open mind while examining both the limitations of science and the assumptions of our faith.

---

[1] Abstracted from www.users.global.net.co.uk/~slocks/goulder.html

Every book should have a purpose and an intended readership. I first thought of writing this book when I noticed that the older high school kids in our church, though willing to help around, were not really interested in what we were prepared to teach them. I figured that they were over the Fairy Tale stage. They were ready for steak, but we were still offering them mashed potatoes. Was there a way to teach them the truth about their religion without destroying their spiritual beliefs? Teach them what is in the Bible and the Creed, and yet alert them to the completely human origin of the writings? Explain that even though the Bible is not a historical document, it still contains much truth?

This book also addresses the needs of adult education classes and Lenten study groups which many churches offer. I believe that in addition to refreshing our knowledge of the Bible, it can provide an excellent starting point for an examination and discussion of our beliefs. Single individuals, although missing the excitement of the arguments and sometimes heated discussions that take place in groups, should also be able to learn from this book and to enjoy it. Nevertheless, I can safely assure you of one thing: you will not agree with everything you read in here.

Finally, there are the many individuals who have left the Church because they could not believe its story, but who still wonder about God and life. These people may well have the most to gain by reading this book. I hope I can convince them that not only there is much truth hidden in the Scriptures, but that the world around us reveals the existence of a large, unseen but active spiritual life.

If the Bible and science contradict each other, which one is correct? And if science is right about something and the Scripture is wrong, does this mean that we should throw away the Bible? Many years ago, as a teenager, I asked my father how the Bible could be correct when certain passages seemed so obviously wrong. "Believe what you agree with, and ignore the rest," was his answer. This is not the position that I will be advocating here. Although the Bible does indeed include many things that most people will agree cannot be literally true, I don't think that we should throw away all of the Church's teachings; similarly, we should not blindly follow the pronouncements of science either. Although scientists always present their theories with great conviction, they often replace them later with different ones, which they then present with equal conviction.

Some points regarding the presentation: I have tried to write the book in a logical order from beginning to end. But it does not need to be read that way. Each reader has his own needs and interests, and can read most chapters in any order he wishes. If a chapter bores you, skip it; likewise, the notes can be treated as supplemental material. At the end of each chapter I list a few questions that could generate interesting discussions. Some even question my own positions.

Unless indicated otherwise, all Bible quotations are from the King James Version, with modernized pronouns and occasional words. The bibliography is usually the most ignored part of a book. Most people who look at it do so only to evaluate how well the author has (or pretends to have) researched the material. The scope of this book, however, is large and I have been forced to discuss most subjects somewhat briefly. Hopefully, readers will be sufficiently interested to read a little more about some of them. I hope that the annotated bibliography that I present will help guide their selection according to their interest and knowledge.

A word about the nomenclature that I will be using: "Hebrews" are the followers of the God Jehovah, starting with Abraham; "Israelites" are the members of the ten Hebrew tribes that settled in the independent northern kingdom of Israel, located between the cities of Dan at the north and Bethel at the south (see the map at the end of the book); "Judahites" are the members of the two Hebrew tribes that settled in the independent southern kingdom of Judah around Jerusalem; "Judeans" are the Hebrews who lived in the area now called Palestine from 500 BCE on; "Jews" refers to the Hebrews of the Diaspora living outside of Palestine, who amounted to about seven million people by the time of Jesus. In addition, I am using the term "fundamentalist Christians," or just simply "Fundamentalists" to refer to those people who believe in the inerrancy of the Bible.[2]

I would like to thank all those who were kind enough to read the manuscript and suggest changes. Most important among them were my daughter Maia, a professional editor, who made countless corrections to my writing and suggested changes that rendered the subject matter more understandable and accessible to the reader; Mr. Ed Pabst who offered much constructive criticism regarding the Old Testament section; and Dr. Theodore Rigas for his ideas that improved the readability of some chapters. Any remaining errors, of course, are mine. I would also be interested in hearing your comments and questions. Please address them to RegardingGod@aol.com. Perhaps I can include them in another edition of this book, in the event, of course, that this book proves sufficiently successful to have another edition.

---

[2] Conservative Protestants met annually at Niagara, Ontario, between 1883 and 1897 to protest evolutionary theories and liberal interpretations of the Bible. In 1910 the General Assembly of the Presbyterian Church agreed on five basic points of Fundamentalism: "the verbal inerrancy of Scripture, the divinity of Jesus Christ, his virgin birth, a substitutionary theory of atonement, and the physical resurrection and bodily return of Christ" (www.victorious.org/chur21.htm). For more information see www.wikipedia.org/wiki/Fundamentalist_Christianity.

CHRONOLOGY OF IMPORTANT EVENTS (D. = DEATH DATE)

BCE

13.7 bill.	Big Bang occurs. (First proto-stars formed 200 mill. later.)
5 bill.	Sun forms.
4.6 bill.	Earth forms. (First unicellular blue algae formed 200 mill. later.)
600 mill.	First multicellular organisms appear.
3761	Creation day according to Jewish calendar
1850	Abraham arrives at Canaan.
1640-1540	The Hyksos dynasty of Bedouin nomads governs Egypt.
1480-1250?	Start of Exodus (precise date in question)
1353-1336	Akhetaten, pharaoh who briefly imposed monotheism
1300-1025?	Period of Judges
1213-1203	Egyptian foray into Canaan. "Israel is laid waste."
1025-1010	Saul, king of the Hebrews
1010-1002	Ishbaal, son of Saul, king of Israel
1010-961	David, king of Judah, and then of the United Kingdom
961-922	Solomon, David's son, last king of the United Kingdom
724	Northern kingdom of Israel falls to the Assyrians.
621	Hilkiah "discovers" Deuteronomy text.

587	Babylonians conquer Jerusalem, deport most inhabitants.
538	Cyrus permits Judahites to return to Israel.
429-347	Plato (Greek philosopher)
333	Alexander the Great conquers Persia.
168-165	Maccabean revolt
44	Caesar's assassination
37-4	Herod the Great, king of Judea
4	Probable birth date of Jesus
4-39CE	Herod Antipas, Tetrarch of Galilee

CE

26-36	Pontius Pilate, prefect of Judea
30	Crucifixion and resurrection of Jesus
36or37	Martyrdom of Stephen
37-41	Emperor Caligula
41-44	Agrippa I, king of Judea
47-57	Missionary travels of Paul
48	Council of Jerusalem
54-68	Emperor Nero (persecutions in Rome in 64)
60-130	Papias, bishop in Asia Minor (quoted by Eusebius)
62	Stoning of James in Jerusalem
62-64	Presumed death of Paul
66-74	Jewish revolt
67d.	Presumed date of Peter's death
70	Destruction of Jerusalem
91-101	Pope Clement (4th pope)
98-117	Emperor Trajan (widespread persecutions)
110d.	Ignatius, bishop of Antioch (first Church Father)
110-165	Justin Martyr, converted philosopher
117-138	Emperor Hadrian's rule
130-165	Valentinus, gnostic theologian (taught in Rome)
130-200	Irenaeus, bishop of Lyons, Gaul
132-135	Second Jewish War. Jews expelled from Jerusalem.
156d.	Polycarp, bishop of Smyrna

160d.	Marcion, gnostic theologian
160-240	Tertullian (theologian, turned Montanist in 215)
185-251	Origen
193-211	Emperor Septimus Severus (persecutes Christians)
284-305	Emperor Diocletian (most extensive Christian persecutions)
313	Emperor Constantine establishes toleration of Christianity.
325	Council of Nicaea
330-397	Ambrose
340d.	Eusebius (Church historian)
342-420	Jerome
354-430	Augustine, bishop of Hippo (Africa)
395	Emperor Theodosius prohibits pagan rituals and Olympics.
451	Council of Chalcedon

Contents

PART I

GOD THE CREATOR—THE OLD TESTAMENT

Introduction to Part I

Jews, Christians, and Muslims all believe in the same God, the God of Abraham. Their views of Jesus, however, differ: Moslems believe that he was one of God's prophets, just as Moses was the first and Muhammad was the last; Jewish beliefs vary from ignoring him completely to assuming that he was one of the latter-day holy people, or perhaps even a prophet. The main disagreement among the three faiths, however, stems from the Christian insistence that Jesus is God, something which the other two religions consider to be polytheism, and thus strongly disagree with. Nevertheless, all three creeds view the Old Testament story as an important historical part of their faith. This is where God is defined.

In the next ten chapters we will consider this Jewish God—the God who created the world and everything in it in six days—but we will also examine why scientists maintain that this process required in fact twelve billion years to complete. We will discuss the idea that the first humans whom God created broke his command and thus somehow rendered all humanity sinful, and mortal. That he adopted the Hebrew people as his children, and shepherded them out of Egypt and through the desert for forty years. But then we will discuss the lack of any archaeological finds to substantiate such a journey and what this implies, and who the Hebrews really were. We will examine the Bible's claim that during their travel through the desert, God gave the Hebrews his law, but also that it now appears that this law was actually written down centuries after the event by various writers who had their own purposes to satisfy. We will then talk about these writers, and how their work has survived in the Bible we hold in our hands today.

Having thrown some doubt on the reality of the specific God described in the Bible, we will next face the question of God's existence. Does he exist? If so, what is he like? How can we be sure? And if he really exists, why does such a God allow bad things to happen?

We will close with the final ideas that entered Hebrew thinking as the Christian era approached: the possibility that humans possess souls that survive the body's death in this world, and the existence of an independent evil entity, called Satan, which opposes God and strives to place such souls under its own control.

CHAPTER 1

THE FIRST FIVE BOOKS OF THE BIBLE

Torah, meaning instruction, is the Hebrew name for the first five books of the Bible. Christians call them the Pentateuch (the five volumes). These books describe the period from the creation of the world, to the arrival of the Hebrews at Canaan, the land promised to them by God. The scope of the story is truly immense, giving rise to movies and stories of epic proportions, C.B. DeMille's *The Ten Commandments* being one of the most famous. Yet the main importance of these five books lies not in the story, but in the Judeo-Christian insistence that they represent the true and exact words of God, as copied down by Moses; that they contain all the required rules to guide human actions.

Of the five books, the first, Genesis, is by far the most exciting. God creates the world and everything in it, and then gets angry and almost destroys all the animal life in a great flood. But he allows Noah and his family to survive, and soon the world is populated again. The book ends with the stories of the Hebrew patriarchs, and how the Hebrew people eventually settled in Egypt. The second book, Exodus, tells how God helped his people escape slavery in Egypt, and how he gave them the Law in the desert. The last three describe God's Law in greater detail, and chronicle the Hebrews' entrance into Canaan, the Promised Land. This chapter presents a short summary of these books, but does not add any new information so the knowledgeable reader can probably omit it.

GENESIS

The word, in Greek, means birth. In the beginning the universe was created, and then all the life on earth: plants, fish, birds, animals and finally mankind, Adam and Eve. It was all done in six days, something that we will examine closer in chapter 3. At first Adam and Eve lived in Paradise, the famous Garden of Eden, without any cares or obligations. But they soon sinned by breaking God's command, as a result of which they were evicted from the garden and told they would have to work for their livelihood.

Their first son, Cain, became a farmer and their second son, Abel, a shepherd. On one occasion when they offered sacrifice to the Lord, each one from the fruits of his labor. Abel's offering was accepted but for some unstated reason, Cain's was refused. Thereupon the furious Cain killed Abel with a stone, making this the first of the countless religion-related killings in the history of mankind. God's punishment was swift: the land would no longer respond to Cain's efforts, and he thus became a restless wanderer. Later, Adam and Eve had other children, and so did Cain.

But in a few generations an extraordinary thing happened: the angels sinned; some of them at any rate. Finding human women beautiful, these angels married them and begot giant children. We don't know exactly how big these giants were, but we do know that they were evil, and that most of mankind also became evil with them. So God caused a flood, and drowned everything on the world except for Noah, his wife, and his three sons and their wives. They, and representatives of all animal life, entered the ark, a ship built according to God's specifications, and rode out the weather. Fortunately for the master of this menagerie, all life had originally been created vegetarian, something that greatly simplified the provisioning problem, not to mention the discipline; only after the flood subsided did God cancel this vegetarian imperative.

Ten generations after Noah, Abraham was born, the first of the three patriarchs. By this time Egypt and the Middle East were well populated. You might think that there had not been sufficient time for this to happen, but when people live 400 years,[1] and there is no television to distract them, the birth rate soars. God told Abraham to move to the land that he would show him, and so Abraham started on a great migration from Ur by the Persian Gulf, to Haran at the north end of Syria,[2] to Canaan in the Middle East, to Egypt, and finally back to Hebron in the Middle East. There God promised to him and to his descendants all the land as far as his eye could see. The Koran writes that Abraham even traveled down to Mecca to lay the foundation of Kaaba, Islam's holiest shrine.

God loved Abraham greatly, and maybe for this reason tested his faith many times. When Abraham was approaching a hundred years of age and his wife was past childbearing age, God promised him that his descendants would become more numerous than the stars. Sure enough, he eventually had two sons: Ishmael, the ancestor of the Arab people, from his wife's slave Hagar, and Isaac, the father of the Hebrews, from his wife Sarah.[3] There was yet another test of Abraham's faith. God told him to sacrifice his own son Isaac. But just as the always-faithful Abraham was ready to kill his son, an angel intervened and saved Isaac who grew up to become the second, and perhaps the least noteworthy of the three patriarchs.

Isaac had two sons: Esau, an accomplished hunter, and Jacob, a man of letters who was his mother's favorite. When Isaac became old and blind and was approaching death, the time came to give his blessing to his sons. But Jacob and his mother fooled Isaac so that he bestowed on Jacob, the younger son, the blessing and birthright that should have gone to Esau, the older. When Esau came for his blessing, Isaac realized the deception but it was too late. "'Father,' Esau asked, 'don't you have more than one blessing? You can surely give me a blessing too!' Then Esau started crying again" (Gen. 27:38 CEV). Again, as we saw earlier with Cain and Abel, there appears to be a scarcity of blessings in the Bible; if one gets blessed, the other does not.[4]

The life of the third patriarch, Jacob, was almost as exciting as Abraham's had been. Although he fooled his father in order to steal his brother's blessing and birthright, his father-in-law fooled him in turn, substituting at the wedding his older daughter for the younger, whom Jacob had chosen. So after Jacob had worked seven years as payment for one daughter's dowry, he had to work another seven for his chosen one's dowry. He got even, however, by tricking his father-in-law so that he got most of the flocks when he left. On top of that, his younger wife stole her father's family idols. On the way home, Jacob spent a night wrestling successfully with God or one of his angels.[5] At the end the angel blessed him and renamed him Israel, meaning he who has contended with the divine. From here on in the Bible, the Hebrews are called Israelites, and Israel/Jacob is the father of the new nation. (In order to avoid confusion, however, I will not follow this practice. Between 922 and 724 BCE the nation of Israel included only ten of the original twelve tribes, and competed against the independent nation of Judah formed by the tribes of Judah and Benjamin.)

With the help of his two wives and their two servant girls, Jacob/Israel had twelve sons, and at least one daughter. Joseph, the second-youngest son was his favorite, and the non-Bible-reading public has learned about his exploits through the musical *Joseph and the Amazing Technicolor Dreamcoat*. He ended up in Egypt as the pharaoh's highest aide and, in due time, brought the rest of the family from Canaan to live there with him. Because his eldest brother, Reuben, had gone to bed with one of their father's concubines, Reuben lost his double birthright, which Jacob then bestowed to Joseph's two sons, whom he adopted. Notice that this results in thirteen tribes, not the twelve normally mentioned in the Bible. The descendants of Levi, however, were assigned to the priesthood so they were not counted during the census, nor did they receive a section of the Promised Land when it was finally apportioned among the other tribes.

Exodus

The second book of the Torah chronicles a period of hardship for the Hebrews. A few centuries[6] after Joseph things have changed. The Hebrews were no longer the honored guests of the Egyptians, but instead their slaves. God's promises, however, were coming true: the pharaoh "said to his people, 'Look, the children of the Israelites have become more and mightier than ourselves!'" (Ex. 1:9). So in an effort to reduce their numbers, the Egyptians started oppressing the Hebrews: not only were the men assigned to hard work building cities and pyramids, but they were also kept away from their wives so as to hold down the birth rate; as a final resort, midwives were instructed to drown all male children when they were born. One Hebrew mother, however, tried to save her son by placing him in a papyrus basket and floating it down the river Nile. The pharaoh's daughter found him, adopted him, and named him Moses.

When Moses grew up, he killed an Egyptian who was striking a Hebrew and he was forced to flee into the desert. There he became a shepherd, married the daughter of a Midian priest, and had a son. One day, when he was in the desert with his flock, he noticed a bush burning without being consumed. He approached to take a better look and God spoke to him from the bush. "I have seen the affliction of my people in Egypt...and I will send you to the pharaoh to bring [them out].... Tell the Israelites I AM has sent you" (Ex. 3:7,10,14).

Moses, with his brother Aaron as spokesman, returned to Egypt and tried to convince the pharaoh to release the Israelites, but without success. To change the pharaoh's mind, God sent to the Egyptians ten plagues: the Nile water turned red, frogs came, then gnats, flies, pestilence, boils, hail, locusts, darkness, and finally an angel of death who killed all the firstborn males of the Egyptians in the land (both people and animals). The Hebrews, however, remained safe because the night before the last plague, Moses had given them instructions to mark the lintels of their houses so that the angel of the Lord would pass over them. He also taught them how to celebrate the feast of Passover then and every year thereafter. Throughout all these trials the pharaoh remained unmoved because "God had hardened his heart." But after all the Egyptian firstborn died, he relented, and told the Hebrews to leave.[7]

The Hebrews left Egypt with all their possessions and flocks. The Lord guided their way, as a pillar of smoke in the day and as a pillar of fire at night. They did not follow the road by the Philistine lands, where the Egyptian fortifications were, but traveled through the desert. As they approached the Red Sea, however, the pharaoh changed his mind and hastened after them with his elite troops. The Hebrews thought they would surely die and started complaining to Moses, but he raised his staff and stretched out his hands, at which time God parted the waters of the sea. The Hebrews were thus able to cross on the dry bottom to the other side. But when the Egyptians reached the sea and tried to cross it the same way, Moses stretched his arms over the waters again and they returned, filling the dry channel and drowning the pharaoh and his army.

No sooner were they safe in the desert than the Hebrews started complaining to Moses about lack of food. So every dawn God rained manna from the sky, which served as bread. The people were instructed to collect only what they would need on that day, except on the sixth day when they were to collect a double portion, for none would come on the Sabbath, for "everyone had to stay home and rest on the Sabbath" (Ex. 16:29 CEV). When the Hebrews next complained about lack of water, God told Moses to strike a rock with his staff and water flowed out. The people finally reached the foothills of Mount Sinai where they camped, and here God gave them the Law.

This was not as simple a procedure as you might have thought, because all communications were routed through Moses. Even the first time when God addressed the Hebrews directly from the mountain, the people were too scared to listen and withdrew, asking Moses to hear and tell them what God had said. Three more times God spoke privately to Moses on the mountain. In fact, Moses had to climb the mountain seven times before God completed all his instructions, and the orchestration of the entire procedure appears somewhat incoherent, to say the least. On Moses' second trip up the mountain, God tells him to make sure that the Hebrews don't get close to the mountain, and if they do they should be stoned to death. Later he calls Moses up again and gives him the same injunction, but Moses reminded God, "You said that before."

In the movie *The Ten Commandments*, there is the famous scene where Charlton Heston as Moses comes down from the mountain holding the tablets with the law and finds the Hebrews dancing around the golden calf that Aaron had made for them to be their new God.[8] In a fit of rage, Moses smashes the tablets, and after he orders the killing of some of the apostates, he returns to the mountain again with new tablets to be inscribed—and here is an opportunity to win a bet from your friends: the tablets were inscribed on *both sides* (Exod. 32:15),[9] not on just one side as they are always portrayed.

By this time the Hebrews' lack of faith had made God so angry that he told Moses that he would not continue to travel with them for fear he would get so angry with the people that he would exterminate them. After much cajoling by Moses, however, he relented and gave instructions about the Dwelling that should be built for him, where he would stay while he was with them. (Later on in the Bible, the Dwelling is called the Ark of the Covenant.) The Hebrews did as instructed, and he stayed with them, and guided them through the many years of their travel through the desert.

LEVITICUS

The third book of the Torah is named after the tribe of Levi, the priests of Israel. The first half describes in exquisite detail the priestly duties and the various types of sacrifice that became the basic structure of the Hebrew religion. We normally hurry through this chapter since it is hugely boring and generally does not apply to Judaism as practiced today.

But an interesting event is related in this book. A little after Aaron and his sons became ordained, the two older sons approached the altar to offer sacrifice but, instead, they themselves became incinerated by God's fire. Although Moses tells Aaron not to grieve, no good explanation is given for what happened or why it happened.[10] Moses, however, now tells Aaron that the priests must never drink wine or strong drink. Were the sons drunk when they tried to offer sacrifice? Nobody knows for sure, although this is what some later writers deduced. Next, the book continues with the various purity laws related to both food and people. Fortunately for the reader, since God now lived among the Hebrews, we don't have to deal with Moses' trips up and down the mountain and other complicated arrangements. Depending on the material covered, each section starts simply with either "The Lord said to Moses," or "The Lord said to Moses and Aaron."[11]

NUMBERS

A year after the Israelites left Egypt, God told them take a census of the male members of each tribe over twenty years old and fit to fight. The Bible comes up with the huge number of 603,550, not counting the Levites whose job was to carry the Dwelling. If we were to add the women and the children we would arrive at a total of perhaps two million people, plus their animals. That such a huge throng could successfully survive wandering in the desert for forty years can only be attributed to the efficient air delivery system of manna and the water-bearing rock they continuously encountered. Unless of course it all did not take place as described.

After giving them some additional rules regarding purity and priestly duties, God led the Hebrews from the Mount Sinai area into the desert. When they approached Canaan, God told Moses to send out some scouts from each tribe to reconnoiter the area and its inhabitants. The scouts returned in forty days with mixed reports: the land was indeed as rich as God had promised, but the inhabitants were fierce giants and could not be beaten in battle. Only Joshua and Caleb disagreed, saying that with God's help they could certainly persevere. But the people continued with their grumbling, wanting to return to Egypt. This angered the Lord: "You sinful people have complained against me too many times!...You have insulted me and none of you men who are over twenty years old will enter the land that I solemnly promised to give you as your own—only Caleb and Joshua will go on....Your children will wander around in this desert forty years, suffering because of your sins, until all of you are dead" (Num 14:26,29,30,33 CEV). This shamed the Hebrews, who decided to attack the Canaanites, despite Moses' objections and God's orders. But without the Ark of the Covenant with them, they were soundly defeated by the Canaanites.[12]

Then Korah, one of the Levites, and 250 of his followers rebelled against Moses, accusing him of nepotism for selecting his brother Aaron as the chief priest. But in a confrontation in front of all the tribes, God caused the earth to open up and swallow them.

Soon the people ran out of water and started complaining. Moses and Aaron could not control them and asked for God's help. He told Moses to hit a rock with his staff, whereupon water gushed out. If this sounds familiar, check the story in Exodus a few pages before. Nevertheless God became angry with the two: "Because you refused to believe in my power, these people did not respect me" (Num. 20:12 CEV), and told them that neither one would enter the Promised Land. What bothered him, apparently, was that Moses hit the rock twice, when once would have been sufficient. Soon after this, Aaron died and his son took over the duties of the chief priest, but Moses continued to lead the people for the forty years in the desert.

The rest of Numbers is divided between descriptions of battles with various people the Hebrews encounter and more rules regarding Levites and sacrifices. One unusual thing does happen, however. One time, when the Israelites were again complaining that they had left Egypt only to die in the desert, God punished them by sending them poisonous snakes. The people immediately repented and asked Moses to intervene with God. He instructed Moses to make a bronze replica of a snake and have it held aloft on a pole. If a Hebrew got bitten, he needed only to look up at the snake to recover. The more mathematically inclined readers can calculate, if they wish, the necessary height of the pole for its top to be visible by all of the two million travelers around it. For the rest of us, the important point is that Moses seems to have built a representation of a living thing that had magical powers—an idol—something directly forbidden by the third commandment. In fact the Bible itself relates that many centuries later, Hezekiah, a king of Judah, "smashed the bronze snake Moses had made [because] [t]he people had named it Nehushtan and had been offering sacrifices to it" (2 Kings 18:4 CEV).

DEUTERONOMY

The last book of the Torah is named the "second law." It should really be called the restatement of the law, because it is not another law, but a recapitulation and expansion of some of the laws put forth in the earlier books. Forty days before the end of the forty-year period spent by the Hebrews in the desert, and while they were camped on the east side of the Jordan, Moses addressed the people. He summarized what had happened during the last forty years, adding descriptions of the recent battles, and then restated some of the social laws that had been promulgated in Exodus and Numbers. The important new element that is introduced here, however, is the covenant he makes with the Hebrews: if they follow God's laws, they will be blessed; if they do not, they will be cursed.

For a Christian, perhaps the most important verses in Deuteronomy relate to a minor statement made about the treatment of a criminal's corpse, who after his execution was to be hung on a tree (or possibly impaled on a stake) as a warning to others: "He that is hanged is accursed of God" (Deut. 21:23). Some modern writers believe that twelve hundred years later, these words are what influenced Saul (later called Paul) to persecute so violently the first Christians, the followers of someone who, according to this Bible verse, was accursed.

At the beginning of his speech, Moses tells how he had asked God to change his mind and allow him to enter into the Promised Land. It is ironic that although so many times in the past Moses had successfully persuaded God to change his mind about destroying the Hebrews as punishment for their various misdeeds, he was singularly unsuccessful in pleading his own case.

GOD WAS RIGHT

When we try to think objectively, however, we realize that the decision to keep the Hebrews away from Canaan for forty years was logical, and not as arbitrary as it might appear. The invasion and capture of the land that God had showed them would require fierce fighting, bravery, and faith. These were not qualities possessed by the Hebrews who had left Egypt. These people were not soldiers but slaves, descendants of generations of slaves. They didn't even have the courage to wrest their own freedom from the Egyptians, but had to be coaxed by Moses, again and again. God had fought all their battles on their behalf, and even Moses acted only as a conduit, not as a leader. Moses lifted his arms; God parted the waters of the Red Sea so that the people could cross. Moses lifted his arms again, and the waters returned to drown the Egyptians.[13] And forty years later, they would need a general able to physically lead them, one who could make his own decisions and act upon them. This is not a description that would apply to the octogenarian Moses. But even in the past Moses had never exhibited any evidence of independent leadership. Time and again when the people complained about something, he had turned to God and complained in his turn:

> I am your servant, Lord, so why are you doing this to me? What have I done to deserve this? You made me responsible for all these people, but they are not my children. You told me to nurse them along and carry them to the land you promised their ancestors. They keep whining for meat, but where can I get meat for them? This job is too much for me. How can I take care of all these people by myself? If this is the way you're going to treat me, just kill me now and end my miserable life! (Num. 11:11-15 CEV)

So when God told them that none of their soldiers would enter the Promised Land, he was not punishing them; he was telling them instead that they were not good enough for the job yet. They would only be ready when a new generation had grown up in the desert, prepared from birth for the coming fight. The favorite childhood game in the desert had probably been "Kill the Canaanites."

SOME QUESTIONS TO CONSIDER

1. According to the Bible, after Cain killed Abel, there were only three people left in the world. Where did Cain find a wife?

2. How could God tell Moses to build an idol for protection against the snakes, when he himself had forbidden them? (We will find the answer in the last chapter of the book.)

3. Was the God of Genesis taking sides in the competition for land use between farmers and shepherds? Not only did he refuse farmer Cain's offering, but also by turning him into a nomad, he effectively forced him to become a herdsman.

4. Amihai Hadad, an ardent 22-year-old Jewish nationalist, sitting inside his makeshift home in an illegal settlement in the West Bank, said: "I'm here because God told me 'This is your land, I have given it to you.' Nothing belongs to the Arabs here" (Chicago Tribune, June 4, 2003). Is he right or wrong? Why?

5. God was supposed to have dictated to Moses the laws of the *Torah*. The Hebrew story from the Exodus chapter onwards was history that Moses lived. The patriarchs' stories could have come down to Moses' day by word of mouth. But where did the stories of Adam and Noah originate? (See also similar question in next chapter.)

SELECTED REFERENCES

Armstrong, Karen. *A History of God: The 4000-Year Quest of Judaism, Christianity, and Islam.*

Blenkisopp, Joseph. *The Pentateuch: An Introduction to the First Five Books of the Bible.*

Davis, Kenneth C. *Don't Know Much About the Bible.*

Friedman, Richard Elliott. *The Hidden Face of God.*

Hoffman, Lawrence A. *Covenant of Blood: Circumcision and Gender in Rabbinic Judaism.*

Kugel, James L. *The Bible As it Was.*

Miles, Jack. *God: A Biography.*

Schwartz, Regina M. *The Curse of Cain: The Violent Legacy of Monotheism.*

CHAPTER 2

IS THE BIBLE HISTORY?

We classify books as fiction or non-fiction. In what group does the Bible belong? "It is not fiction," you will say. "It is not a novel." Yet when we look at it, we see a story that covers millennia in history, encompassing the rise and fall of countless dynasties. There are fierce wars of conquest and desperate wars to protect one's homeland. There are passionate love affairs, incest, court intrigues, beautiful women, brave kings, and powerful magicians. It takes place in a faraway land of smoking mountains and endless deserts, of rolling hills and winding rivers, inhabited by colorful people with strange customs. And lording over it all is this incomprehensible, egocentric god, ready to create or destroy on a whim, who is just as likely to fight for his people as to deliver them into the hands of their enemies. It has been rightly described as the greatest story ever told.

"It is not a story," you may respond, "but history. It tells of things that took place a long time ago." In that case, is it a true or a false history? Who was the historian? How did he know what to write about? How can we tell? Although the Bible itself maintains that the first five books, the Torah, were written by Moses, we will see that this is not true. Indeed the internal discrepancies in the story indicate the existence of more than one writer, perhaps as many as four. Although this chapter discusses only the first five books of the Bible, we will see later that most of the other books were also composed by non-witnesses to the events who were promoting their own agendas.

Fact, Fiction, or In-between?

Is the Bible historical? Does it describe actual events that occurred, conversations that took place? Did Moses really speak with a burning bush? Jewish and Christian religions in general say yes, although some are more unequivocal than others. To a Southern Baptist, the Bible is a historical document containing the exact words of God. If it says that God made the world in six days, then this is precisely what happened, and a day means twenty-four hours. Some more liberal denominations agree with the general correctness of the Bible's contents, but qualify the specifics: God may have created the world in six days, but the exact length of a day is open to interpretation. After all, according to Peter, "With the Lord a day can mean a thousand years, and a thousand years is like a day" (2 Peter 3:8 JER).

Is there a meaningful difference between believing that the Bible contains the exact words of God, and believing that it was written by divinely inspired men? The answer is a resounding yes: in the first case there exists no possibility for error; in the second, people may have written what they believed to be true, but their accuracy depends on the quality of their inspiration. Let us not forget that occasionally even the popes have made divinely inspired pronouncements that were later recanted.

Who wrote the Torah? The Bible clearly tells us that it was Moses. "Yahweh said to Moses, 'Put these words in writing, for they are the terms of the covenant I am making with you and Israel'" (Exod. 34:2 JER). And again,

> Moses wrote down all these laws and teachings in a book, then he went to the Levites who carried the sacred chest and said: 'This is the book of God's Law. Keep it beside the sacred chest that holds the agreement the Lord your God made with Israel. This book is proof that you know what the Lord wants you to do.'" (Deut. 31:24-26 CEV)

But even if we assume that Moses wrote down exactly what God had told him, the five books of the Torah contain descriptions of many other happenings that Moses presumably witnessed personally and recorded on his own. And what about Genesis? The last four chapters of the Torah contain numerous conversations between Moses and God, and lengthy specifications of religious rules, but nowhere do they say that God spoke to Moses about the events described in the Genesis chapter. God tells Moses that the Israelites are his chosen people but not that he created them. There is no mention of Adam and Eve, the original sin that caused man's fall, not even of Noah. Although a few times God becomes angry at the Israelites and tells Moses that he will destroy them all and raise another nation from Moses' seed, he never says that he had done this before. From the things covered in the book of Genesis, he only mentions the three patriarchs; he tells Moses that they knew him by another name, El Shaddai (God the Almighty)[1] not Yahweh (Lord), and that he had promised to them that he would give the land of Canaan to their descendants. God never spoke about anything that occurred before Abraham.

Discrepancies in the Bible

For two thousand years or so, nobody questioned Moses' authorship of the Torah. It had been assumed true by all early writers, including the Jewish historian Josephus[2] and the early Christian Church historian Eusebius.[3] Yet there exist many discrepancies in the various stories.

Consider the Creation narrative. The familiar story where God creates the world in six days is followed immediately by another, somewhat different Creation story: First, God creates man before the vegetation; then he creates the Garden of Eden and assigns man to cultivate it; then he creates the animals so that man will not be alone, and only after man does not find a suitable partner among the animals does God create Eve out of one of Adam's ribs. But if Eve was an afterthought, what about procreation? Was there going to be only one human being forever? In the first story, man and woman are created simultaneously; in the second, it appears as if the woman had originally been thought to be superfluous. She is not even told directly about the prohibition regarding the forbidden fruit. Interestingly, the Church teaches a story that combines the six creation days from the first version with the belated creation of Eve from the second, but ignores the duplications.

Or look at the flood narrative. God first tells Noah: "Of all living thing of all flesh you shall bring two of every sort in the ark, to keep them alive with you; they shall be male and female" (Gen. 6:19 AMP). But only a few verses later he tells him: "Of every clean beast you shall take with you seven pairs, the male and his mate, and of beasts that are not clean a pair of each kind, the male and his mate; Also of the birds of the air seven pairs, the male and the female" (Gen. 7:2-3 AMP). What is it, one pair or seven pairs?

According to Exodus, Moses himself wrote the Ten Commandments on the second set of tablets (Ex. 34:28), but in Deuteronomy he relates that God had written them himself, just as he had done the first time (Deut. 10:4). In the chapter of Numbers, God tells Moses, "Send men to reconnoiter the land of Canaan" (Num. 13:2), but in Deuteronomy, when Moses recalls the event to the assembled Hebrews, he tells them, "Then you came up to me everyone of you and said 'We will send men ahead and they will reconnoiter the land'" (Deut. 1:22).

There is even a variation about the reason Moses was not allowed to enter the Promised Land. In Numbers it occurs because Moses cannot control the people who complain about the lack of water; in Deuteronomy he is punished, together with the others, after the scout incident.

The Importance of the Discrepancies

Perhaps there are some who do not think that such minor discrepancies are very serious, yet they do bring up questions regarding the absolute truth of the Bible as a historical document. As any lawyer can tell you, the fastest way to discredit somebody's testimony is to find discrepancies in it. A witness who first says that he was watching a movie at a particular time and date and later says that he was at home asleep is either lying or confused. In either case his testimony cannot be trusted. If fundamentalist Christians and Orthodox Jews are correct in saying that the Bibles their groups acknowledge contain the exact, literal words of God and we find that these words contain obvious discrepancies, then either God makes mistakes or there must exist some other explanation.

Indeed, in the two centuries before the birth of Christ, numerous Jewish thinkers and philosophers tried to reconcile the obvious problems posed by appearance of inconsistencies in the Bible writings. They did this by relating the stories anew and inserting details and explanations that had not been there originally. These writings are now included among what we call the *apocrypha*: stories which have not been included in the final approved version of the Bible. Such explanations, however, are interpretations of the Bible and conceptually are no different than Peter's saying that one day means a thousand years.

As an example, there is an interesting explanation of the two creation stories, which may first have appeared some thousand years ago and is probably based on Sumerian mythology: Adam really had two wives; the first, Lilith, was created simultaneously with him, was independent and uncontrollable, and eventually flew away and left him.[4] When Adam asked for a replacement he got Eve, a properly subservient wife, just as the mores of those days required.

MORE THAN ONE WRITER?

Nobody questioned Moses' authorship of the Torah until the eleventh century, when a Jewish physician in Spain noticed that a list of Edomite kings in Genesis included some who had lived after Moses;[5] his contemporaries laughed at him and named him "the blunderer." In the fourteenth century, a scholar in Damascus suggested that a later prophet had added the contested words to Moses' writings. In the fifteenth century, the bishop of Avilla pointed out that the account of Moses' death could not have been written by Moses himself, but perhaps by Joshua. In the sixteenth century, two Jesuit scholars thought that other writers had expanded on Moses' writings. In the seventeenth century, a French Calvinist wrote that, based on some obvious anachronisms, Moses had not written the first five books of the Bible; his book was burned. Similar fate met the writings of Spinoza in the Netherlands, of Richard Simon in France, and of Hampden in England.

By the eighteenth century, people noticed that, as we saw earlier, many stories in the Bible were written down as doublets, or even triplets, each one usually referring to God by a different name (Elohim vs. Jehovah). Some recent researchers claim to identify as many as six different writers of various parts of the Torah.[6] Although there may be disagreement about the exact authorship of each verse, there is general agreement now that many authors were involved in the writing. Who they were, and when and why they wrote what they did are things that we will discuss in a later chapter.

The important point here is that the Bible was not written directly by God, and so it is not inerrant.[7] In some denominations it is now safe to question the accuracy of anything you read and to ask for proof. But not everywhere. In order to teach at Wheaton College near Chicago, for instance, you must first sign a document that you agree with the literal interpretation of the Creation story as it is described in Genesis.[8] And this brings us to our next subject: creation as related in the Bible compared with the current theories developed by the scientists.

SOME QUESTIONS TO CONSIDER

1. The Hebrews believed that in addition to the written Torah, there was an oral version that had been given by God to Moses, and by Moses to Joshua, which survived intact to the days of the Pharisees. Why could the Genesis story not have been part of this oral history, passed down by Moses?

2. In the ancient days, oral transmission was considered to be a very accurate way of passing stories to the next generation. People were meticulously trained to do exactly that. The Zoroastrians even forbade the writing down of their scriptures. Why should we assume that oral transmission would result in alterations of the content?

3. Almost every ancient civilization around the world has a myth about a prehistoric flood. Doesn't this provide justification for the biblical version?

Selected References

Friedman, Richard Elliott. *Who Wrote the Bible?*

CHAPTER 3

CREATION: SIX DAYS OR A BIG BANG?

Some of the matters touched upon in the first chapter's overview of the Torah are of sufficient importance to require their own discussion. The first of them is the Creation story. Nothing characterizes fundamentalist Christians better than their insistence that God made the world in six days of twenty-four hours each and that we are living in the sixth millennium after that event. In their eyes, the Bible is inerrant, so if it says that God made the world in six days, it means exactly that. We will see, however, that the Bible presents two separate creation stories differing in many important points, including the creation of mankind. Obviously they cannot both be correct.

Scientists tell us another story. According to them, the universe was created almost fourteen billion years ago, and depending on how one defines "man," mankind has existed for the last thirty thousand to two hundred thousand years. We are going to look at their theories but, unfortunately, they are neither easy to explain, nor easy to understand. I do hope to be able to show, however, that their ideas leave gaping holes. Scientists can account, for instance, for only four percent of the matter and energy in the universe; they have absolutely no idea what the remainder is or how it came to be. Similarly, in order to explain why matter possesses mass, one of its most important properties, they are forced to postulate the existence of an as yet undiscovered viscosity field.

Even worse, it appears that our universe could not have resulted in the conditions that permitted the development of man had it not been for the exact values possessed by half a dozen physical parameters. It almost appears as if creation was planned, or at least guided. I can only conclude that both sides have failed to prove their point, but read on and make up your own mind.

The First Creation Story

First things first, and the first line of the first chapter of the first book of the Bible starts with the description of the Creation story. Let us skim through it:

> In the beginning God created the heavens and the earth.
> The earth was barren with no form of life.
> It was under a roaring ocean covered with darkness
> But the Spirit of God was moving over the water.
>
> **The First Day**
>
> God said, "I command light to shine!"...He separated light from darkness and named the light "Day" and the darkness "Night."...
>
> **The Second Day**
>
> God said, "I command a dome to separate the water above it from the water below it." And that's what happened. God made the dome and named it "Sky."...
>
> **The Third Day**
>
> God said, "I command the water under the sky to come together in one place, so there will be dry ground." And that's what happened. God named the dry ground "Land," and he named the water "Ocean."...
>
> God said, "I command the earth to produce all kinds of plants, including fruit trees and grain."...

The Fourth Day

God said, "I command lights to appear in the sky and to separate day from night...." God made two powerful lights, the brighter one to rule the day and the other to rule the night. He also made the stars....

The Fifth Day

God said, "I command the ocean to be full of living creatures, and I command birds to fly above the earth."...Then he gave the living creatures his blessing—he told the ocean creatures to live everywhere in the ocean and the birds to live everywhere on earth....[1]

The Sixth Day

God said, "I command the earth to give life to all kinds of tame animals, wild animals, and reptiles."...

God said, "Now we will make humans, and they will be like us. We will let them rule the fish, the birds, and all other living creatures."

God created man in the image of himself,
in the image of God he created him,
male and female he created them.[2]

God gave them his blessing and said:

Have a lot of children! Fill the earth with people and bring it under your control. Rule over the fish in the ocean, the birds in the sky, and every animal on earth.

I have provided all kinds of fruit and grain for you to eat. And I have given the green plants as food for everything else that breathes. These will be food for animals, both wild and tame, and for birds....

The Seventh Day

> By the seventh day God had finished his work, and so he rested. God blessed the seventh day and made it special because on that day he rested from his work.
>
> That's how God created the heavens and the earth. (Gen. 1:1-2:4 CEV)

And so ends the first creation story, the one "Creationists" usually quote when they say that God made the world in six days.

Two things immediately jump out. First, the earth is the beginning and center of creation; second, God's actions are contrary to the currently known laws of physics. We know for instance that the earth is one of nine planets revolving around the sun; that our sun is a rather smallish example of the billion suns that belong to our galaxy; and that our galaxy is only one of the billions of galaxies existing in the universe, all interacting through an all-pervasive gravitational field. It is a little hard to imagine that God first made the earth floating in the middle of nothing and then, four days later, he made the rest of the universe and the gravitational laws.

There are other things also that are hard to comprehend. Light and dark are properties, not independent entities. Yet God first made light and three days later he created the source of light. He created sunlight-dependent plants before he created the sun. By definition, an all-powerful entity can do anything it wants, including creating order from chaos through a simple command, but let us examine the world after he finished creating it.

Wherever we look we see order, and order is the result of rules. In a house with teenagers, for instance, one rule might be "Don't put the dirty cake dish in the sock drawer." An even better rule is "Put the dirty cake dish in the dishwasher" (in my house, add, "after rinsing it"). Three thousand years ago, the writers of the Creation story didn't have the scientific knowledge that we now have, so they could only observe the results of the creation, not the rules that made them possible. On the second day, for example, God was supposed to have separated the waters above the earth from those on it. In the story he did it by a simple command; in practice it must have involved the implementation of laws for evaporation, condensation, heat transfer, gas dynamics, and so on. Only in this manner could the results of his creation last throughout the seventh day, when he stopped giving orders.

Another interesting thing to notice is that "Creation" did not begin on the "First Day." Instead, "In the beginning he created the heavens and the earth," and in the six days that followed he refashioned the materials at hand to shape the world, as we know it. So the six days do not include the Big Bang of the physicists, which we will discuss later.

So, did God create the world or not? The world obviously exists, so we can assume that it was created. But was it created by God? Since we have not examined yet who or what God is, we cannot answer this question. One thing that we can definitely say, however, is that the world was not created in the exact manner described in the Bible.[3]

Although most people don't take the Creation story seriously, almost all accept the statement that God created mankind in his image, something that generates unending theological arguments. What exactly does it mean? Does God have a body like we do? No, since we all agree that he is a spiritual being and does not resemble the old bearded man painted on the ceilings of some churches. So perhaps it is mankind's spiritual side that resembles God's. This concept is a very important one, especially among Christians, and it appears in many discussions relating to man's future life. Yet one could rightly ask why we accept that one statement in the Creation story when we reject the rest. I cannot answer that. One explanation you often hear is that "it is not God who created man in his image, but man who created God in his." But wouldn't that take "Godness" away from God?

The Second Creation Story

Immediately after the first Creation story, the Bible tells a second story.

> When the Lord God made the heavens and the earth,
> no grass or plants were growing anywhere. God had

not sent any rain, and there was no one to work the land. But streams came up from the ground and watered the earth.

The Lord God took a handful of soil and made a man.[4] God breathed life into the man, and the man started breathing. The Lord made a garden in a place called Eden, which was in the east, and he put the man there.

The Lord God placed all kinds of beautiful trees and fruit trees in the garden. Two other trees were in the middle of the garden. One of the trees gave life—the other gave the power to know the difference between right and wrong....

The Lord God put the man in the Garden of Eden to take care of it and to look after it. But the Lord told him, "You may eat fruit from any tree in the garden, except the one that has the power to let you know the difference between right and wrong. If you eat any fruit from that tree, you will die before the day is over!"

The Lord God said, "It isn't good for the man to live alone. I need to make a suitable partner for him." So the Lord took some soil and made animals and birds. He brought them to the man to see what names he would give each of them. Then the man named the tame animals and the birds and the wild animals. That's how they got their names.

None of these was the right kind of partner for the man. So the Lord God made him fall into a deep sleep, and he took out one of the man's ribs. Then after closing the man's side, the Lord made a woman out of the rib. (Gen. 2:4-22 CEV)

The second story differs from the first in many ways:

- Nothing is created except the man and the animals; the earth, the sky, and everything else already exist.
- God creates man before the animals in order to take care of the garden.
- God himself creates the animals, instead of ordering the earth to bring them forth.
- Only after none of the animals prove to be a proper partner for man does God create the woman.
- There are two special trees in the middle of the garden and the fruits of one are forbidden to man, because if he eats them he will learn the difference between right and wrong (now why do you suppose this is so bad?).

This story sets the scene for what will eventually result: man's fall into sin. Logically, most of it makes very little sense. God appears to think that man will find a suitable partner among the animals.[5] Only when this does not occur, almost as an afterthought, God creates a woman. He gives man an order to follow, yet at that time man did not know right from wrong; how was he expected to know that it was right to follow orders?

This second Creation story has been used by the early Christian Fathers to justify the downgrading and oppression of women. The fact that the woman was made later than the man and from a part of his body, was used as proof that she was obviously inferior to him and was created to satisfy his needs. That she later succumbed to the snake, ate the apple and convinced the man to do the same, was used to paint her as the tool of the devil, responsible for man's perdition.

The existence of two separate creation stories poses all sorts of problems to those who believe in the Bible in a verbatim manner. As we will see later, however, the duplication was caused by multiple writers whose works were included side by side in the Bible, instead of being smoothly merged into one.

THE BIG BANG CREATION THEORY

Scientists tell us that our universe was created out of one humongous explosion 13.7 billion or so years ago. How do they know that? They first discovered that the galaxies are moving away from each other at speeds proportional to their distance from us. Next, they measured these speeds and positions and calculated backward to discover that all galaxies had originated from the same point in space.[6] It appears that all the matter and energy in our universe had been concentrated once in a sphere the size of a grapefruit before it burst violently outwards. This explosion occurred 10^{-32} seconds after the start of creation, that is one trillionth, of a trillionth, of a millionth, of a hundredth of a second after the first something occurred in the original nothing that existed before the beginning.

The inside of this grapefruit-size volume was so hot and dense that matter could not exist in it, only energy. Obviously it could not remain confined in this small space, so it exploded outwards, an explosion that is still continuing. And as this energy expanded out, it created new space where previously there had been none. The best way to visualize this is by imagining that the universe is the surface of a balloon; as the balloon is blown up, its surface increases, and all points on it move apart at the same rate. The fireball's expansion resulted in its cooling from the original 10^{28} K to lower temperatures where some of the energy could be changed into matter[7] first quarks, then after 3.5 minutes into protons, neutrons, and electrons, and finally after 300,000 years, into atoms.

At that point in time, three quarters of the matter was hydrogen, and the rest was helium with traces of lithium. Although the expanding cloud of matter was very homogeneous, it did contain small discontinuities of higher density. These locations possessed increased gravity and they attracted additional matter from their surroundings, causing the eventual coalescence of stars and galaxies. It required about 200 million years or so to form these proto-stars (meaning first stars). The higher gravity in the center of the stars started pulling the star matter closer together, compressing it and raising it to temperatures in the millions of degrees, eventually starting nuclear reactions. In the smaller stars, hydrogen sustained fusion and became converted into helium; in the larger ones, the process continued, forming successively, but also simultaneously, carbon, oxygen, silicon and finally iron. This last element is stable and cannot enter into any further reactions.

When a star eventually runs out of fuel, it contracts sharply towards its center until the resulting pressure causes it to explode in what we call a supernova. It is during such explosions that the heavier elements are produced and then blown out into the surrounding space. Gravitational forces then coalesce the dust from these explosions, producing a second generation of stars. Unlike the first stars that consisted only of hydrogen and helium, these newer stars have a full complement of chemical elements; our own sun belongs to this second generation.

The scientists are very happy and confident about their star-building theory. They have written equations that describe the formations of the various particles and have performed experiments that they claim prove many of their deductions. There is, however, one unanswered question marring all this certainty. They have calculated that in order for the universe to behave as they say it does, there must exist ten times more mass in it than can be seen. Since we cannot see this material, they have called it dark matter and have concluded that it must consist of some entirely new substance.

Is this really important? Well, suppose you are an auditor examining the books of a homebuilder. Everything looks neat and balances to the penny. You see no problems and are ready to sign the books and approve them. But then you notice that the company has used ten times more lumber and bricks than were needed for the houses it has built. Now you see a problem! There is a similar situation with the universe creation theory; if it accounts for only ten percent of the created matter, it has obviously left out a lot. Is the theory ten percent right and ninety percent wrong?

Unfortunately, the astronomers were in for worse news. The Big Bang theory had predicted that the expansion rate of the universe should be slowly decreasing due to the gravitational effect of all the matter it contained. By 2002, however, more precise measurements revealed that instead of slowing down, the stars were actually speeding up. It seems that there exists in the universe a mysterious and previously unknown antigravitational force, whose strength is assumed to increase as the universe ages. The new calculations for the matter and energy content of the universe now yield the following components: seventy-three percent unknown antigravity (dark) energy needed to explain the recent acceleration of the stars; twenty-three percent unknown dark matter needed to explain the original agglomeration of the stars and galaxies; and four percent visible matter, explained by the Big Bang theory.[8] It now appears that the scientists' creation theory can only explain four percent of the creation!

Average people like us are not qualified to evaluate what our scientists tell us or to question their information; we are expected to believe them and accept their pronouncements as "gospel truth." But just as we are now questioning the gospel, perhaps we should keep some reservations regarding what scientists tell us, especially since they often disagree among themselves and equally often change their minds when new facts appear.[9]

Plant and Animal Evolution Theory

Let us return to the formation of earth. About five billion years ago, dust particles from a supernova explosion started coalescing to form our Sun. These particles contained all the elements of our periodic table, and atoms of hydrogen, oxygen, carbon and silicon already combined with each other forming the hundreds of organic chemicals that are necessary for the development of life. Most of the dust particles continued to attract each other until they formed our sun, while others built boulder-size planetoids that revolved around it. The chaotic movements of these so-called planetoids caused many of them to crash into each other and join them to eventually form the Earth and the other planets. Other masses, however, remained intact and continued circling the Sun as comets (which consist mostly of gases and water) and denser planetisimals. Within the first billion years after their formation, most of these also disappeared through collisions with the newly created planets until, eventually, the solar system reached its current, reasonably ordered state.

Scientists tell us that our Earth, as we know it today, came into existence about 4.6 billion years ago, but that its surface remained molten until 3.8 billion years ago. Within 100 to 300 million years of that event, single-cell blue-green algae came to life in the seas, as we can tell from fossil remains in Australian and African rocks. It required another three billion years, however—until barely 600 million years ago—before the first multicellular organisms appeared. It was another 200 million years before life emerged from the sea and moved on to dry land, first in the form of plants and then of animals.

In 1859 Charles Darwin published his now-famous book, *On the Origin of Species*. In it he described how animal species evolve over periods of years so as to compete more successfully in their environment.[10] As could be expected in the strongly religious atmosphere of those days, his ideas met immediate and strong opposition. Darwin was challenging the Bible's assertion that God had created everything in its present form. Today, however, his theory is supported by our knowledge of DNA, the molecule that stores all of an organism's genetic information. We know that random, accidental changes in DNA occur throughout an organism's life, and that these cause variations in the characteristics of their descendants. If these variations improve the ability of the organism to survive, compared to that of the standard organism, then according to Darwin the new organism will last longer, leaving behind more descendants than does the old one, and in time will replace it. If the change is not beneficial, however, as is generally the case, the organism will leave proportionally fewer descendants and the change will disappear in time.

Despite the continuing objections of the fundamentalist Christians, the evolution theory is completely accepted today; it is logical and seems to be justified by what we know about the world around us. Nevertheless, it does not automatically refute the existence of a creator. As we saw earlier, if there is a creator, he almost certainly does not micromanage, but governs through rules. Furthermore, for the believer in the Bible, God's direct creation was finished after the sixth day. So appearances of new species can be explained only by assuming that evolution is one of the creator's tools.

But unfortunately things are not that simple. For one thing, the results of natural evolution cannot be predetermined. Changes occur due to random, unpredictable causes. More important, evolution does not explain creation, only the changes to what has already been created. We saw earlier that almost as soon as meteorites stopped bombarding the Earth's surface, single-cell organisms came into being. This of course is not evolution, but creation. Scientists have been able to create similar organisms by simulating the chemicals and conditions that were present on earth 3.5 billion years ago. But we also saw that it required three billion years to advance from unicellular to the next level, multicellular life. But when that happened, all but one of the main divisions of animal life were created, almost explosively, all within a period of five to ten million years called the Cambrian Explosion of Life. It almost appears as if once nature discovered how to make multi cellular life, it tried all possible combinations. Many explanations have been advanced for this unique sudden proliferation, but none sound really satisfactory.

There is something else that puzzles scientists when they look at the history of the universe, and they have named it the Anthropic Principle. The ability of our universe, our solar system, and our Earth to support human life depends on specific, precise numerical values for hundreds of physical parameters in nature. A small deviation in any of them would result in a universe that is inhospitable to human life. It thus appears that our universe was either planned, or a chance event with an infinitesimally small probability of occurrence. The first possibility presumes an intelligent creator; the second is realistic only if an infinite number of universes exist, since something having an infinitely small probability of existence can occur only when there is an infinite number of occurrences. In the latter case, our universe is that "one in a million" in which conditions are just right for human life.

Before the Big Bang

The obvious question at this point is how did that grapefruit-size volume of space that contained all the matter-energy of our present universe come into being? Unfortunately our scientists are in far less agreement about what happened before the Big Bang than about what happened afterwards.[11] The problem is that the theory of relativity is not valid during the first 10^{-43} seconds of the creation process. The laws governing that period will be explained by the unified theory of quantum gravity, if and when it is developed.[12] So now scientists are forced to use the general quantum theory to describe the events that they think occurred at that time.

Quantum theory is a very strange bird indeed and it goes against all natural understanding of what is physically possible. Quantum physics applies to subatomic particles and predicts many strange properties for them. For example, they can simultaneously exist in more than one location: they may exhibit properties of either a physical particle or of a mathematically described wave function, but never both at the same time; a neutrino (an article assumed to have zero mass and zero electrical charge) can move simultaneously along two different paths. What is more important, we are unable to measure precisely and simultaneously both the location and the velocity of a particle; if one is accurate, the other will be in error. These constraints have led scientists, starting with Niels Bohr, to conclude that such particles don't have specific position, velocity, or other properties until an attempt is made to measure them!

Apparently what is true about the position and velocity of a particle applies also to the energy magnitude and time duration of its existence. Because we cannot measure anything in less than 10^{-43} seconds, it is perfectly permissible for a particle or a quantum of energy to appear and disappear during that interval for no apparent reason. This is based on the argument that if we cannot measure it, then it does not exist.[13]

Of course as long as things keep appearing and disappearing, we are not going to start a universe-building process. But if one particle can pop into existence alone, it is also possible for two or more of them to appear simultaneously, and the applicable mathematical formulas then describe an interesting phenomenon: the potential energy due to the mutual gravitational energy of these particles is equal and opposite in sign to the energy of their masses; the two cancel each other. So according to these equations, our entire universe possesses zero total energy! As hard as it is to believe this, the concept is similar to borrowing money from the bank to build a house. At the end of the process you own a house, but you are not any richer than before since you owe the bank what the house is worth. Of course this analogy breaks down when you consider interest and monthly payments.

As I said before, scientists don't know anything about the period that covers the first 10^{-43} seconds of our universe's life, the so-called Planck period. They tell us, however, that there was only one type of particle, and one single unified force that combined gravity, electromagnetism, and the strong and weak nuclear forces. At the end of that period, gravity separated from the other forces and quark particles started forming. At 10^{-35} seconds, the universe had reached the size of a proton and was at a temperature of 10^{28} K. According to quantum theory, the maximum permissible density could have been 1094 gm/cm^3, and at this density we would expect that gravitational attraction would prohibit any further expansion. Permeating everything was a mysterious Higgs field; unfortunately, although the physicists have postulated this field's properties in order to explain their various physical observations and theories, they have not discovered it yet.[14]

The currently most attractive hypothesis of what happened next in the creation process is the inflation theory.[15] It postulates that the strong nuclear force that holds together protons and neutrons did not separate from the weak nuclear force at the theoretically correct time but 10^{-33} seconds later, because of a super-cooling phenomenon that occurred. According to the physicists' equations, this effect (which they call a false vacuum) caused the pressure to reverse the effects of gravity, generating a repulsion between the particles. As a result, the volume doubled a hundred times, once every 10^{-35} seconds, growing from the size of a proton to that of a grapefruit. When the strong nuclear force finally separated at the end of the super-cooled period, a huge amount of energy was suddenly released from the Higgs field and was converted into all the particles and radiation that now exist in the universe. This simultaneous particle formation throughout the entire volume is credited for the uniformity of matter throughout the existing universe.[16]

The inflation theory has been popular for the last ten years, mainly because it explains the universe's density uniformity, but not everyone subscribes to it. Many other theories have been proposed, some of which postulate the existence of as many as eleven space dimensions.

A quick word on the possibility of multiple universes to which I alluded earlier. Some scientists maintain that the creation process took place in what was essentially a black hole, a singularity in space where gravity is so strong that nothing can escape from it. But since our own universe contains innumerable black holes, it is conceivable that each one of these generates its own universe, connected with ours by what the scientists call wormholes. Furthermore it is possible that each of these universes has different basic laws, with different basic parameters.

CONCLUSION

We first looked at the Bible's description of creation and noted that many details were at odds with the laws of our universe, as we understand them now. In addition, it contains numerous contradictions and appears to have been written by at least two people who did not see eye to eye in all things. We don't know the identity of the writers, although in chapter 6 we will discuss where they probably lived, but there is nothing to suggest that they wrote down God's actual words.

We then looked at the scientists' description of creation. Despite the confidence with which they have presented the creation process, we noted that there remain many unexplained items, including the extremely small probability that our world, as we know it, could have come into existence without external manipulation. The scientists' explanations of creation leaves 96 percent of the created matter and energy unexplained and unidentified. Furthermore, the origin of creation is as tenuous as the Bible story. To quote one of the writers in the field: "The fact that a quantum fluctuation has no deterministic cause has already been mentioned. This is one of the mysteries of quantum theory that we have to accept."[17] When I was young and asked awkward religious questions I was often told, "Believe and do not search." (Nowadays we say that one must first have faith and then understanding will come.) I feel that I am being asked now to exchange the mystery of religion for the mystery of science. But we can make another observation about the scientific story. The creation process depends on pre-existing rules, like those of quantum theory, and pre-existing structures such as the Higgs field that we cannot even detect. How did all these come into being? If creation is based on rules, how were the rules created?

So what is more reasonable? Purposeful creation or a statistical fluke among an infinite number of universes? Take your pick. We cannot understand the physical processes that could have given rise to these universes, any more than we can comprehend a purposeful power that could have created them.[18] Besides how did that power originate?

Some Questions to Consider

1. We should not expect that the creation process of our universe could be described in two pages of prose written by someone who lived three millennia ago; thus the fact that it is obviously wrong or simplistic should not affect our opinion regarding the existence or not of a creator. Do you agree?

2. Scientists calculated that all galaxies would have originated from the same point, based on the assumption that they had been traveling at constant speeds since they were formed. How is this conclusion changed by the fact that scientists now admit the existence of both accelerating and decelerating forces about which they know nothing, including their distribution in space?

3. The pre-creation universe of the scientists implies the existence of a non-space filled with mathematical fields that permit self-creation of energy and matter with specific properties. This is more than just a complete nothing. How did it originate? Can scientific theories be considered believable if they do not cover this point?

4. Assume that it were possible to create a computer-simulation program in which one would set forth the physical laws and the game would construct the resulting

universe. What would that tell us about the reality of our existence?

5. What would convince you of the existence or non-existence of a purposeful force that created the universe?

Selected References

Brown, Walter. *In the Beginning: Compelling Evidence for Creation and the Flood.*

Delsemme, Armand. *Our Cosmic Origins: From the Big Bang to the Emergence of Life and Intelligence.*

Greene, Brian. *The Elegant Universe.*

Guth, Alan H. *The Inflationary Universe: The Quest for a New Theory of Cosmic Origins.*

Morris, Richard. *The Universe, The Eleventh Dimension, and Everything: What We Know and How We Know It.*

CHAPTER 4

THE GARDEN OF EDEN—ORIGINAL SIN

The other story from the book of Genesis that requires some more detailed discussion, because it is basic to our theology, is that of Adam and Eve's original sin in the Garden of Eden. They disobey God by eating the forbidden fruit and learn the difference between right and wrong. God punishes them and they become mortal, although it is not really clear that they had ever been immortal. Since their sin passes down through the generations, all mankind is supposed to have inherited their sinfulness until Jesus died on the cross to redeem us. Thereafter, according to Paul, by being baptized in his death we become absolved of the original sin and regain immortality. Since we still die, however, we must conclude that immortality is a delayed event achieved after resurrection from our graves. Immortality is the most basic concept of Christianity and perhaps the main reason for its early popularity.

Except perhaps for the fundamentalist Christians, everybody today believes that the Adam and Eve story is just a myth. Despite this, however, the concepts of the original sin being transmitted down through the generations and of our salvation through Jesus Christ remain unchallenged.

THE BIBLE STORY

There is probably no story in the Old Testament better known than the one where the snake convinces Eve to eat the forbidden fruit from the tree of knowledge of good and evil. The snake asks Eve:

> "Did God tell you not to eat fruit from any tree in the garden?" The woman answered, "God said we could eat fruit from any tree in the garden, except the one in the middle. He told us not to eat fruit from that tree or even to touch it. If we do we will die." "No you won't," the snake replied. "God understands what will happen on the day you eat fruit from that tree. You will see what you have done, and you will know the difference between right and wrong, just as God does" (Gen. 3:1-5 CEV).[1]

Since God gave the command to Adam but not to Eve, for she had not been created yet, she must have been repeating what Adam had told her. So it was Adam who added the additional prohibition of not even touching the fruit, a detail about which later Bible analysts made much ado. In any case, Eve agrees that the fruit looks good, and that the result of learning to tell good from evil is also good; she eats some of the fruit and offers some to Adam, who also eats it. The result is immediate: they lose their innocence and realize that they are naked.

God gets angry. "You will suffer terribly when you give birth," he tells Eve. "You will have to sweat in order to earn a living," he tells Adam. "You were made out of soil and you will once again turn into soil," he adds, thus rendering them mortal. "The Lord said, 'These people now know the difference between right and wrong, just as we do. But they must not be allowed to eat fruit from the tree that lets them live forever'" (Gen. 3:22 CEV). Although it has always been assumed that Adam and Eve lost their immortality because of their sin, this assumption may be wrong. They may have only lost the opportunity to remain immortal by eating the fruit of that second tree.[2] In any case, God makes some clothes for them from animal skins and sends them out of the Garden of Eden.[3]

ANOTHER MYTH

This tale is a continuation of the creation story, which we have already branded as false. Although fundamentalist Christians believe that a divinely inspired writer composed it, there is no evidence of that. Furthermore, it could not have been passed down by word of mouth through the generations; the story was put on paper thousands of years after the event, and after a flood that presumably killed all but eight people on earth. It would be quite naive to expect any degree of accuracy in the retelling. Besides, the fact that men and chimpanzees have at least 97% of their DNA genes in common indicates that they have shared the same creation process; if God picked up clay from the ground to make man, he must have used the same clay for the chimpanzees.

The Garden of Eden story, and that of Abel and Cain that follows it, are just myths whose purpose is to explain the presence of evil in the world, and why a perfect God brought forth such an imperfect creation.[4] Not only is the story illogical, but it does not even present God in a favorable light. In it he creates two innocent, child-like people who cannot tell good from bad, and he does nothing to protect them from the outside world. He gives them an order and then tempts them to disobey it; when they do, he punishes them outrageously severely. It can be compared to a father who buys a TV set and then kicks his children out of the house when he catches them watching it instead of studying.

Or we can look at the story as a fable: a fictional account with a moral. The snake told Eve: "[T]he day you eat [of the apple] your eyes will be opened, and you will be like gods knowing good and evil" (Genesis 3:4). It is this desire of man to become like god, which is the sin that brings disaster. A similar fable tells about the Tower of Babel through which men attempted to reach heaven. That effort also resulted in disaster. Again the moral seems to be that man is only human and should not aspire to godhood. Yet even today we play god whenever through our justice system we condemn someone to death.

From an evolutionary point of view, however, the story makes sense. Mankind starts with an undeveloped brain, living naked in the forest, eating freely from the available fruits and plants. Man is nothing more than an animal, a mammal to be exact. Then the animal-man evolves into the human-man. This requires a larger brain, which not only gives him the ability to tell good from evil, but also turns birthing into a very painful experience. Somewhere along the way he learns how to talk. With more desires to be satisfied, life becomes more complicated; man becomes civilized and has to work for a living, tilling the crops, taking care of the herds, and building houses and a civilization.[5] In the Garden of Eden man was an animal; outside he is a human being. Do you suppose that perhaps God interrupted his seventh-day rest period and sent the snake to tempt man and complete his development?[6]

One thing that had bothered the ancient Bible scholars is that Adam and Eve did not die immediately after eating the fruit. After all, God had told Adam, "Of the tree of knowledge of good and evil you shall not eat, for in the day you eat of it, you will surely die" (Gen. 2:17). And yet they did not die, just as the snake had predicted. But the scholars thought they found an answer in psalm 90 (verse 4): "A thousand years in your [God's] eyes are but a yesterday." So if one of God's days is equal to a thousand of our years, and Adam died when he was 930 years old, then he only lived one of God's days.[7]

THE CHRISTIAN PERSPECTIVE

As with almost everything else in the Book of Genesis, the stories of the Garden of Eden and of Adam's fall are rarely, if at all, referenced in the rest of the Old Testament. But they do play a prominent role in Paul's theology:

> "Adam sinned, and that sin brought death to the world. Now everyone has sinned, and so everyone must die." (Romans 5:12 CEV) "Everyone was going to be punished because Adam sinned. But because of the good thing that Christ has done, God accepts us and gives us the gift of life. Adam disobeyed God and caused many others to be sinners. But Jesus obeyed him and will make many people acceptable to God." (Romans 5:18-19 CEV)

Although both Jews and Muslims share the story of the original sin,[8] it has special importance for Christians, because they alone believe that it has been canceled by Christ's death on the cross. As we will see in chapter 20, the explanation and manner of Christ's death were early stumbling blocks in the spread of the Christian religion.

According to Paul, the devil obtained power over man because of Adam's sin. If Adam had not sinned, the devil would not have had any power over us. But Jesus, being a sinless man, vanquished him. So now, those of us who believe in Jesus are protected from the devil's power. Paul argued that during baptism, when we are submerged under water, we figuratively die with Christ, and when we come up out of water we are resurrected sinless like him, new immortal people beyond the devil's power. Of course we still die in this world, so Paul had to amend his theology and introduce the idea that the dead will rise from their graves.[9] Paul probably assumed that baptized people will not sin any more. Unfortunately this is not true; we still sin and so we are still sinners. The Church resolves this problem by the sacrament of absolution in which a priest, acting as a representative of Christ, forgives our sins.[10] For just a little time between absolution and our next sin, we remain sinless![11]

A few centuries after Paul, the great Catholic theologian Augustine theorized that although God had condemned mankind to eternal damnation because of Adam's sin, this was not related to any apples he might have eaten. According to Augustine, the sin was that instead of contemplating God, mankind had taken pleasure in mere creatures, that is each other, particularly during the sexual act. We will meet Augustine again in chapter 27.

SOME QUESTIONS TO CONSIDER

1. Except for the strongest Fundamentalists, everybody agrees that the Adam and Eve story is a fable. Why then is everybody obsessed with the original sin that had to be expiated?

2. Although Adam and Eve did not die immediately after they ate the fruit, they were separated from God at once.

It has been proposed that this represents death. Does this idea have any merit?

3. God said, "These people know right and wrong just as we do." To whom was he talking?

4. Can you justify God's punishment of Adam and Eve?

5. Was the immortality that Adam and Eve seemingly had the same as the immortality that we have presumably gained through Jesus' death on the cross?

6. If there was no original sin from which Jesus redeemed us, how did he save us, and from what?

SELECTED REFERENCES

Kugel, James L. *The Bible As it Was.*

Friedman, Richard Elliott. *The Hidden Face of God.*

CHAPTER 5

EXODUS—TWO MILLION PEOPLE IN THE DESERT?

Perhaps the single biggest event in the history of the Hebrew people has been the Exodus of their ancestors from Egypt and their forty-year journey through the desert in the company of their God. According to the Bible, it was during these forty years in the desert that the Hebrew religion was born.[1] Yet archaeologists have failed to discover any records or other evidence to substantiate the departure from Egypt of such a great multitude followed by its lengthy sojourn in the desert. This has led many of them to conclude that, if this event did occur, it probably involved only handfuls of escapees who were not missed by the Egyptians. Some experts have gone even further to suggest that, in that case, the Canaanite population might have even welcomed the escapees as priests. Though completely unsubstantiated, such a theory would explain many distinctive features of the Levites, such as their Egyptian names and their lack of land ownership.[2] According to this thinking, the true Israelites were the original dwellers in Canaan—Abraham's people, who were later joined by Moses' followers.

PASSOVER

At sundown on the 14th day of Nissan (which falls during April or late March) every Jewish household gathers around the dinner table to start the eight-day celebration of Passover that commemorates the escape of the Hebrews from Egypt. More accurately, it observes the occurrence of the tenth plague in Egypt, when the angel of the Lord passed over the houses killing every Egyptian firstborn, man and animal.[3] To escape the same fate, Moses told the Hebrews to kill (sacrifice)[4] a blemish-free, one-year-old male lamb and to paint some of its blood on the posts and lintel of their door to their homes. As the Bible tells it, the next morning, shocked by all the deaths, the Pharaoh let his Israelite slaves go.

This remembrance, however, also plays a very important role in the Christian religion. Because Jesus observed the Passover (or, according to John, the preparation day ahead of the Passover) before he died on the cross, Christian Easter is often celebrated during the same week. But more important, Paul and the gospel writers have compared Jesus' salvation of the Christians to the lamb that saved the Israelites in Egypt. He is pictured as the sacrificial lamb for our deliverance, something that is still referenced directly during the Christian mass. But we will discuss this in later chapters; here we will concentrate on the Hebrew story.

Is the Exodus a Historical Event?

We have questioned the validity of almost everything in the Genesis book of the Bible, but the Exodus is such a defining part of Jewish history, that one would assume that it has to be true, at least in a general way. After all, this was the period when Hebrews were given the Law, when their God lived among them in a tent. It is the root of Israel's history.

Besides, there are quite a few archaeological findings to substantiate the events that could have given rise to the Exodus. There is evidence that before 1200 BCE, many Canaanites and people from the Transjordan had moved to Egypt to escape droughts, or were brought in as captured slaves, and that they had worked to build cities and pyramids, just as the Bible says.[5] It is also accepted that four hundred years earlier, starting at about 1674 BCE and lasting for a little over 100 years, Egypt had been governed by the Hyksos dynasty of Bedouin nomads, the so-called Shepherd-Kings. (Does that bring to mind Joseph?) In addition, excavations have found the remains of many cities mentioned both in the Exodus and in the book of Joshua that describes the conquest of Canaan.

Nevertheless, many archaeologists have recently questioned the reality of the Exodus event: they have been unable to find any Egyptian records reporting a massive departure of people from the country, and they have not found in the desert any material or campsite remains that could be dated to 1200 BCE, although they have found artifacts from earlier and later periods.[6]

The huge number of people that the Bible claims escaped Egypt is also suspect. If we add women and children to the reported six hundred thousand fighting men, we arrive at something in excess of two million people. To visualize this horde, assume that they marched fifty people across, in rows six feet apart. The column then would have been forty-five miles long. Add their flocks and the carts carrying their possessions, and we can easily see that the crowd would have occupied such a huge area that it would have been uncontrollable even by professionals. Do you remember reading in the first chapter that, according to Leviticus, Moses had fashioned a bronze serpent for the people to look at when they were bitten by snakes? How could they have done this over such distances?

Although the Bible story may be unrealistic, there could still be some truth in it. Suppose that instead of two million people marching out of the city of Ramses, with all their animals and belongings, only a small band of one or two hundred sneaked out into the desert. It would have been easier for such a small number to escape the pursuing Egyptians,[7] and also to find food and water. We usually think of the Hebrews wandering through the desert for forty years, followed by Miriam's well;[8] but according to the Bible they spent thirty-eight of those years in the oasis of Kadesh-barnea. Certainly this oasis could accommodate two hundred people far easier than two million.

Rethinking the Beginnings of Israel

If only a small group escaped from Egypt, how can we explain the conquest of Canaan, the land that God had promised to the Hebrews and where they eventually settled, or the formation of the nation of Israel? Looking at the first question, those who say that there was no Exodus, obviously say that that there was no conquest either. After all, they also maintain that the so-called conquered towns were unoccupied at the time of their supposed conquest. Those who believe in a small group of escapee-conquerors, however, have a different theory. Remember the movies *The Seven Samurai*, and *The Magnificent Seven*? In each case, a small group of inspired and dedicated professionals proved to be far superior in combat to the more numerous, but easily intimidated villagers.[9]

So the argument is made that when the small band of Hebrews entered Canaan, they encountered only simple agricultural and shepherding communities. The large armies and city fortifications mentioned by the Bible did not exist yet, nor would they for another few centuries. Such a small band of intruders might not have posed a discernible threat to the existing communities and might not have given rise to a military response.

So, assuming that a small band of Hebrews entered Canaan without opposition, how did the great Israelite nation come into being? A monument has been found in Egypt, describing Pharaoh Merneptah's (1213-1203 BCE) campaign into Palestine. The inscription includes the following: "Israel is laid waste; his seed is not."[10] Since this happened before the Hebrews got to Palestine, it must refer to a different tribe-nation by that name. It has been proposed therefore, that after Moses' group conquered or joined this group of people, it assumed their name. (This is another reason for calling the group that left Egypt "Hebrews" instead of "Israelites.")

It has been also suggested that perhaps only the Levites, the Hebrew priestly tribe, participated in the Exodus. Apparently most people in this group had Egyptian-sounding names, and this one tribe is the only one that was not assigned any territory. The Levites were not even included in the count of "the twelve tribes of Israel." Perhaps when the Levites reached Canaan, instead of trying to substitute Yahweh for the local god El, they proposed that the two were one and the same. So the Canaanite god El, who was associated with the patriarchs, was merged with the Hebrew god Yahweh, associated with Moses. Then the Levites assumed the priestly duties for a tithe.[11]

The Bible tells us that during the period of the Judges, which followed the entry of the Hebrews into Canaan, the Israelites formed independent cities, without kings, in the hill country of the area. Until 1025 BCE when King Saul was elected, the people of these cities lived together in a spirit of cooperation without formal government, following the ethical teachings of their religion.

The obvious question, of course, is what was the origin of this Hebrew religion? Did it start with Abraham and the other patriarchs? Was it formed during their stay in Egypt? Was it developed or given to them while they were in the desert?

ORIGINS OF THE HEBREW RELIGION

According to the Bible, the Hebrew story starts with patriarch Abraham. God appears to him, tells him that he will make a great nation of him, and sends him traveling towards the land that he will show him. Again and again he appears to him, promising that his descendants will be more numerous than the stars. By the fifth time, Abraham is almost 100 years old, his wife Sarah approaching 90, and they still have no children. This time God makes a pact with him, requiring that he and all his household should be circumcised, and although Abraham agrees, he cannot help laughing when he hears again the promise of all the descendants to come.[12] When at his sixth appearance God tells Abraham again about the large number of his future descendants, even Sarah starts laughing.[13] Eventually, however, Sarah does have a son, Isaac. But after she dies 27 years later, Abraham remarries and has another six sons, at an even more advanced age, and presumably without any announced help from God.

Interestingly enough, God does not introduce himself to Abraham until the fifth time, when he says that his name is "El Shaddai," where "El" means "god," and the entire name is usually translated as "the almighty god." Note that "almighty" here is not a descriptive term for God, but an identifier for a god. There were numerous gods in Mesopotamia in those days, and Abraham's god identified himself as the mightiest one, the one above all the others. Back then, each household had its own personal god whose job was to look after the interest of the family—remember in the Genesis chapter when Jacob's new wife stole the household idol (god)? There were also gods who reigned over certain geographical localities, and others who were worshipped by entire nations.[14]

At first God appears to be the personal god of Abraham's household.

> Jacob solemnly promised God, "If you go with me and watch over me as I travel, and if you give me food and clothes and bring me safely home again, you will be my God. This rock will be your house and I will give back to you a tenth of everything you give me." (Gen. 28:20-22 CEV)

As the patriarchs built altars to God, he also became the god of various localities: "I am the God you worshipped at Bethel" (Gen. 31:13 CEV). When the Exodus starts, he becomes the God of the Hebrews: "You shall be to me a kingdom of priests, and a holy nation" (Exodus 19:6). Interestingly, God never tells Abraham that he is the only god, only that the Hebrews should not worship any other gods.

The most distinguishing feature of Judaism was developed during the Exodus. It was then that God pronounced his laws—not necessarily the detailed instructions regarding sacrifices and other celebrations,[15] but the ethical rules encapsulated in the Ten Commandments.[16] It is the adherence to these rules, presumably enforced by a powerful divinity, or by the fear of it, that would allow the diverse tribes in Canaan to live together in peace without kings or any formal government. God was their king; it was a true theocracy.[17]

One must assume that something happened to this motley bunch of slaves in the desert to change them from grumbling cowards to an inspired, close-knit group of people capable of conquering or at least converting the residents of Canaan, and in due time building one of the largest kingdoms in the area. Just because they might have amounted to only a couple of hundred souls instead of two million does not mean that they could not have encountered a theophany, a vision of the supernatural. On the other hand, perhaps they were forced by the privations of the land to depend on each other and develop a morality-based system of behavior.

There is a very interesting detail supplied by an archaeologist who tries to refute the Exodus event, but which seems to argue against his thesis: the people who lived on the highlands of Canaan, the eventual country of the Israelites, did not eat pork; no pig remains were found in the excavations. The people in the plain, however, did eat pork.[18] Now pigs eat everything and live anywhere. Why would anyone refuse to eat one of the most convenient animals to raise unless there was a religious ban? And where did that ban come from? Interestingly, the Egyptian priests also did not eat pigs and were also circumcised, another tie between them and the early Israelites.

Obviously nobody knows for sure who was in the desert and what experiences they had there. Maybe they encountered a palpable theophany, or maybe they discovered God in the world around them and within their own soul, just the way many people do today.[19] But who wrote this story anyway, and why?

SOME QUESTIONS TO CONSIDER

1. Can we relate the Hebrews and the Egyptian priests through their common customs of circumcision and abstention from eating pork?

2. In 1356 BCE, Pharaoh Akhenaten tried to impose monotheism in Egypt, going as far as to build a new capital in the desert to move away from the power of the priests. (See a beautiful article about him in the April 2001 issue of the *National Geographic*. Egyptian artifacts relating to his wife Nefertiti and son Tutankhamun represent some of the most valuable Egyptian archeological finds.) Some have suggested that this is where Moses got his monotheistic ideas. What do you think?

3. Is it possible that the Exodus, the most important event in Hebrew religion did not actually occur?

4. The author seems to imply that Abraham was a Canaanite and that the practice of circumcision was first brought to Canaan by Moses' Levites, not by Abraham and the patriarchs. Do you find the argument satisfactory?

5. From what we know, until the last few centuries, every civilization in the world had believed in a god or gods. We have no record of the existence of atheists until the development of modern science (although some Romans did accuse the early Christians of atheism). Do you think that science and religion are incompatible?

SELECTED REFERENCES

Finkelstein, Israel, and Neil Asher Silberman. *The Bible Unearthed: Archaeology's New Vision of Ancient Israel and The Origin of Its Sacred Texts.*

Mendenhall, George E. *Ancient Israel's Faith and History: An Introduction to the Bible in Context.*

CHAPTER 6

WHO WROTE THE OLD TESTAMENT?

The side-by-side appearances in the Bible of contradictory versions of the same stories have led us to conclude that there were multiple authors. But who were these writers? Scholars tell us that no written law of any kind existed during the time of Exodus or the periods of Judges and the United Kingdom. Sometime during the eighth century, however, someone in the Northern Kingdom of Israel, whom we call "E," wrote down a version of the books of Genesis, Exodus, and Numbers. At about the same time, someone in the Southern Kingdom of Judah, whom we call "J," wrote down his version of the same books. Then when the Assyrians conquered Israel, some of its inhabitants took refuge in Judah, bringing with them their "E" Bible, and almost immediately the two versions were combined to generate the "JE" version.

Soon thereafter, during the reign of King Hezekiah, a member of the priestly clan in Judah, whom we call "P," sat down and wrote his own version of the story, adding the entire book of Leviticus with its detailed religious rules. Finally, almost a century later, during king Josiah's reign, someone else (perhaps Jeremiah) wrote the fifth book of the Torah, Deuteronomy. Then Judah fell to the Babylonians. Fifty years later, Persia's King Cyrus conquered Babylon and allowed the exiled Judahites to return to Jerusalem. After another eighty years, the Persian king's scribe

Ezra, a Hebrew, arrived from Babylon with the completed version of the Torah that combined all the earlier versions in the one story we now have. Later he probably finished the job by writing most of the other historical books of the Bible.

From the Exodus to Alexander

It is obvious that the Bible was not created in a historical vacuum. It was affected by the rise and fall of the Hebrew kingdoms, as well as those of the surrounding nations. So before we talk about the Bible's authors in a meaningful way, we should take a brief look at the history of the Hebrew people between the years of 1200 and 350 BCE.

The Bible describes how the Hebrews under Joshua invaded Canaan, killed many of the inhabitants, destroyed the towns, and divided the land, mostly the highlands, among their tribes. As we saw earlier, this story may or may not be true. But however these tribes settled there, there was no organized government to link them together during the two centuries after 1200 BC. They were all independent, having God as their king. On occasion, however, a brave or wise person, called a judge, would rise and pull together some of the tribes to fight their common enemies.[1] Although they fought against all their neighbors, and sometimes even against each other, their most powerful opponents were the Philistines, who had been known earlier as the People of the Sea, and who had settled just to the west of the Hebrew area on the east coast of the Mediterranean.

Despite occasional successful battles by the Hebrews, the Philistines held the upper hand—so much so that the Hebrews were not allowed to practice blacksmithing lest they make weapons for themselves. So the Hebrews reasoned that if they had a king, as all their neighbors had, they might be more successful against their enemies, and they asked Samuel, their last prophet-judge, to get them one. In the beginning both Samuel and God objected. But they eventually agreed, and God directed Samuel to anoint Saul as the first king. Now Saul had not been looking for the job. He had gone to Samuel to get help in locating some donkeys that he had lost and became anointed king instead; wisely, he kept it a secret. Later, however, the Hebrews got together to select the new king by lot,[2] and he was publicly chosen. Even then, however, they had a hard time finding him, for he had hidden among the baggage.

Soon thereafter, the Ammonites attacked. Saul raised an army from among all the tribes, and though they lacked real weapons, he defeated them in the ensuing battle. Despite his earlier reluctance, Saul quickly became accustomed to all the trappings of royal power, and so the military successes of his young protégé, David, scared him. Saul tried to kill David, who escaped and hid in the countryside, becoming a mercenary and eventually fighting on the side of the Philistines. He was even ready to take part in the last battle against Saul when, according to the Bible, the Philistines asked him to stay away.

After Saul's death on the battlefield in 1010 BC,[3] David was made king of the tribe of Judah and Saul's son, Ishbaal,[4] king of all the other Israelite tribes. This marked the first division of the Hebrew nation into two separate kingdoms, Judah on one side, and the other eleven Hebrew tribes on the other. Seven and a half years later, however, after the assassination of Ishbaal, David became the king of all Hebrews. The Bible says that he was the greatest king they had ever had, and he expanded the borders of the kingdom to limits never attained again. He was also God's favorite. God promised him that one of his descendants would always occupy the throne of the Hebrews and that the throne would stand firm forever (2 Samuel 7:16). Of course we know that the throne fell and that David's line ended. How was this explained? It so happens that in Deuteronomy, in his last words to the Hebrews, Moses graphically detailed all the curses that would befall them if they did not follow God's laws (Deut. 28:15-69), and one of them was that they would lose their country and end up back in Egypt offering themselves as slaves. So it has been argued that God's promises recorded in 2 Samuel were contingent on the Hebrews following his laws.[5]

After David's death from old age in 961 BCE, his son Solomon became king. He is known as the wisest of all people, perhaps because when God asked him what he wanted most, he replied "wisdom." Unfortunately, history does not substantiate this reputation. It is true that he inherited a kingdom without any active enemies[6] and that he wisely kept peace by marrying the daughters of every neighboring king, even of Egypt's pharaoh, and thus made all the kings in the surrounding lands his in-laws. He built houses for all his wives, and they moved in with statues of their personal and national gods (the neighborhood must have looked a little like an embassy row). He also went on a great building binge, erecting the opulent first Temple and even more impressive palaces for himself. But he obtained the wood and the expertise to build them from the king of Lebanon, repaying him with twenty Israelite cities in the land of Galilee (1 Kings 9:11). And he conscripted thirty thousand workers from Israel, but not Judah, to work every third month in Lebanon. According to the Bible, his riches were proverbial, but he alienated the northern Hebrew tribes by oppressing them for the benefit of Judah. He was the last person to reign over the United Kingdom. After his death, in 922 BCE, the Hebrews split into two: in the north, the kingdom of Israel consisting of the ten northern tribes, and in the south, the kingdom of Judah, which also included the relatively small tribe of Benjamin.

The kingdom of Israel lasted from 922 BCE until 724 BCE, when it fell to the Assyrians. In order to achieve religious independence from Jerusalem's temple, and thus prevent the drain of all the sacrifice money that would have been paid to Judah in the south, King Jeroboam (922-901 BCE) set up places for sacrifices at Dan, the northern end of Israel, and at Bethel, the southern end.[7] This made it unnecessary for the people to travel all the way to Jerusalem for the feast days and sacrifices.

Ahab (869-850 BCE) was one of Israel's greatest kings, although he was thoroughly reviled in the Bible; he built a great army and an elite chariot corps and conquered many of the neighboring people.[8] Another great king was Jeroboam II (786-746 BCE), who defeated Syria and extended Israel's borders to their greatest extent. During his reign the kingdom achieved unprecedented prosperity and great building projects were carried out. After his death, however, the Assyrians came from the east, conquered the country, and in an act of ethnic cleansing, deported its inhabitants, replacing them with people from other areas of their empire.[9]

The southern kingdom of Judah lasted 140 years longer, until 587 BCE when the Babylonians overran it. It did not really become important until Israel fell and many of its inhabitants escaped to Judah. Then its capital, Jerusalem, underwent an unprecedented population explosion, increasing from an area of about twelve acres to 150, surrounded by a mighty defensive wall.[10] Some of its kings, like Ahaz (735-715 BCE) and Manasseh (687-642 BCE), courted the friendship of the Assyrians and the Babylonians and paid tribute to them; this usually resulted in great prosperity. Others, like Hezekiah (715-687 BCE), rose in independence against the foreigners, almost invariably bringing invasion and destruction to the countryside. Since the Bible evaluates the kings according to their piety, not their political acumen, the successful kings are called bad, and the ones who brought destruction to the country, good. Another burst of glory occurred in Judah when the great king Josiah (640-609 BCE) came to power and tried to emulate King David by attempting to recover his original kingdom. Unfortunately, the Egyptian pharaoh Necho killed the king and claimed control of the territory. Soon thereafter, the Babylonians became active in the area, defeated the Egyptians, and conquered Judah. They took king Jehoiachin (609-598 BCE) and all the aristocracy and priesthood captive to Babylon and installed his uncle, Zedekiah (597-587 BCE), as their vassal king. But when he revolted and tried to form an alliance with some neighbor kings, the Babylonians came back, destroyed all the cities in the countryside, laid siege to Jerusalem, and eventually burned it to the ground. All the king's sons were murdered, and he himself was blinded and taken to Babylon. Thus ended the nation of Judah, and with the eventual death of Jehoiachin, the family tree of David was extinguished.

During the next fifty years or so, all the Judahite nobles and priests were kept captive in Babylon. Some of them were sent to work in the fields, but it appears that many served at the temples and in government positions. A kind of low-level aristocracy with high-level contacts thus developed.[11] In the meantime, the peasants were left alone in the burned-out land around Jerusalem. They continued their lives there, working what was left of their fields and worshipping God with sacrifices at the destroyed temple.

Suddenly, in 539 BCE, the situation in Babylon improved for the Hebrews. A new power arose in the East—Persia—and under Cyrus it conquered Babylon. Although the Persians were in theory monotheists, believing in Ahura Mazda, most of their leaders, and particularly Cyrus, were willing to accept all local deities.[12] In addition, Cyrus believed that he could use people's religion and their priestly administration to help control the population and collect taxes. He thus restored the temples of the conquered nations, and allowed those held captive to return to their lands.[13] Isaiah (actually the second writer by that name, usually called DeuteroIsaiah) probably misinterpreted Cyrus' motives when he referred to him as the Messiah (Isaiah 45:1). At any rate, Cyrus allowed, or ordered, the return of almost 50,000 Hebrews back to Jerusalem, together with his follower Sheshbazzar to govern them.[14]

Cyrus (539-530 BCE) was succeeded by his son Cambyses (530-522 BCE and then by Darius (522-486 BCE) a member of the extended family. Darius developed a unique approach to governing subject nations. He helped them build temples and a strong religious presence and insisted that they codify and publish their laws in a language that could be understood by the population. He might even have been a Zoroastrian monotheist. In any case, Darius sent Zerubbabel to Jerusalem as governor and Joshua as chief priest with orders to rebuild the Temple but not the city walls. This work started in 520 BCE but did not progress very far. Darius also dispatched a huge army to Egypt and required the Judeans to help support it, but he did repay them to some extent by providing help for the Temple work.[15]

Darius's son, Xerxes (486-465 BCE), ascended on the throne upon his father's death. But because there were no new countries to conquer and plunder, he had fewer financial resources available than his father. He was thus forced to raise taxes and reduce support to the various local temples in his empire. This increased the pressure on the Judean population: not only they had to pay higher taxes to the Persians, but they also had to give greater support to the Temple.[16]

Under Artaxerxes I (465-423 BCE), who followed in his father's footsteps, the empire continued its slow disintegration. Egypt allied itself with Greece and rebelled again, forcing Artaxerxes to fortify his borders. He sent Ezra and then Nehemiah to Jerusalem as governors with orders to rebuild the city's walls and bring the population under strict control, and he provided them with substantial funds and unlimited power over the people. Ezra brought with him the Book of the Law, which he read to the Judeans. The king's orders to him were:

> Set magistrates and judges, who may judge all the people who are beyond the river, all those who know the laws of your God, and teach those who do not know the laws. Whoever does not obey the law of your God and the law of the king, let speedy judgment be executed upon him, whether death, or banishment, or confiscation of his goods, or imprisonment. (Ezra 7:25-26)

It is important to note that during the first century of the Hebrews' return to Jerusalem, the Persians had absolute control over the population through the governors and chief priests they appointed, and even more significantly, through the political and religious laws that were published and enforced by the Temple priesthood.

Not much is known about the second hundred years of the Persian Empire's occupation of Judea (435-335 BCE) and how it affected life in Jerusalem. We can assume, however, that as Persia's power diminished, its interference in the religious life in the Temple disappeared. The Judeans then probably fell increasingly under the influence of the Egyptians and Greeks with whom they traded.

MONOTHEISM

It is important to keep in mind that in the pre-Christian era, gods were more than just local deities. They were also symbols of the nations that worshipped them. Just as today we honor a visting dignitary by raising the flag of his nation and playing its anthem, so in the ancient days a host nation would erect a statue of the other nation's god and offer sacrifices. This was important when one wanted to show respect and have friendly relations with another country, and it was mandatory if one was the vassal of that other country.

One of the most important jobs of a god in those days was to help its client nation win wars. So when two nations went to war and one of them won, it was proof that its god was superior to the defeated nation's god; the vanquished were then expected to put up statues of the winner's god and offer sacrifices to it. Since everybody agreed with this logic, it was always done.[17] Furthermore, most gods were not jealous; as long as you worshipped them, they did not mind if you worshipped other gods also.

It is not surprising, therefore, that the politically more successful Hebrew kings allowed the worship of Canaanite and other deities. It was only when monotheists gained the upper hand in the nation that the kings destroyed the worship places of the foreign gods and alienated their neighbors. This first occurred in Israel when Jehu (842-815 BCE) had Ahab's dynasty killed and Baal's places of worship destroyed. Soon thereafter Syria attacked and took control of substantial portions of Israel's area, forcing Jehu to pay homage and ransom. It is probable, however, that in this particular case the two events were unrelated.

In Judah, the two most pious kings had been Hezekiah and Josiah. It was these two who imposed strict monotheism in the nation following the "discovery" of important religious documents. In both cases they destroyed the places of sacrifice to pagan gods and refused to pay tribute to their powerful neighbors, and in both cases the affronted neighbors invaded and destroyed the country. Only later, during the two hundred years of Persian hegemony, were the Judeans free to practice their religion as they wished, for the Persians had a completely hands-off attitude regarding the religions of their vassals.

During this tempestuous period of Hebrew history the writings we now find in the Old Testaments were written. Some were composed in Israel, some in Judah, and some in Babylon during and after the exile. We can now turn our attention to the identity of their authors.

The Old Testament Authors

It is generally agreed that no written religious records existed before the separation of the two Hebrew kingdoms, but we can assume that in earlier years many stories were passed orally down the generations.[18] Then two authors, one in Israel and one in Judah, gathered the stories together, edited them, and wrote them down. This must have happened sometime during the two hundred years' existence of the kingdom of Israel (922-

724 BCE). Since both authors cover the same material, either one of them copied from the other or both used the same original sources. Despite their general similarity, however, there are substantial differences in the way the stories are presented.

The writer in Israel is called "E" because he used Elohim as God's name (the plural of "El," meaning god in the Canaanite language).[19] It is thought that he might have been a Levite priest from Moses' lineage: he shows interest in liturgical ceremonies, and his stories concentrate on Moses, ignoring Aaron. He wrote the story sometimes between 922 and 722 BCE, I am guessing during the reign of Jeroboam II (786-746 BCE). Not only was this a prosperous time that could have given rise to culture and writing, but it is also the period when the first prophetic writings appeared: those of Amos and Hosea, both of who attacked the materialism of the period.

The Judahite author is called "J" because his name for God was Jehovah. In his stories, Aaron is almost always placed at the side of Moses and sometimes plays the more important part. This of course represents the interest of the Judahite priesthood who, in addition to being Levites, were also descendants of Aaron. Nevertheless, there is no reason to believe that J was a priest, and the sensitive treatment of some women in the stories has raised suggestions that the author might have been a woman, perhaps from an aristocratic family.

Soon after the two stories were written, Israel fell to the Assyrians. Some of its inhabitants were deported, but certainly many fled to Judah, probably with some copies of the E text. Now there were two sacred narratives in Judah and they did not agree with each other. Discarding one to keep the other would not solve the problem since there were people who remembered each one. The answer was to combine them, interleaving the two stories wherever they dealt with the same subject. This resulted in the "JE" version.

But almost immediately, another writer in Judah became unhappy with the JE account: it did not stress the importance of worshipping God in only one location, the Temple; it did not insist that only Aaron's descendants could be priests; it did not specify the proper worship procedure in sufficient detail. So this person sat down with a copy of the JE narrative in front of him and wrote another version from scratch. He rewrote the parts in JE, slanting the stories to satisfy his points of view. And he added extended sections describing the manner in which the priests should carry out sacrifices.

This writer probably lived in the time of King Hezekiah (715-687 BCE), since that was when religion was first centralized to restrict worship only at the Temple. Hezekiah responded to the newly discovered writings by destroying the high places where some people worshipped Jehovah and other gods, and also Nehushtan, the bronze serpent that Moses had made in the desert. This third writer has been named "P" since he was almost certainly a priest, but we don't know his identity. Again there were two sacred texts in the country.

A few years later, in 622 BCE, King Josiah was told that a priest named Hilkiah had discovered a hidden scroll in the Temple. When it was read to him, the king tore his clothes in despair since his people had not been following the laws described in this scroll. It was the book of Deuteronomy, which, as we saw in chapter 1, includes powerful curses for those who do not follow its teachings. So the king called all the men of Judah to the Temple, had the scroll read to them, and they all made a covenant with the Lord to follow the commands in the scroll. He then proceeded to destroy all places of sacrifice in Judah other than the Temple, and emboldened by the weakening status of Assyria, attacked the old Israel lands and destroyed Beth-El, its old southern temple.

We call this fourth Hebrew writer the Deuteronomist, or "D" for short. His writings are characterized by the covenant; when the people do what is good in the eyes of the Lord they fare well, when they don't, they get punished. Biblical scholar Richard Elliott Friedman believes that he can identify this person as the prophet Jeremiah, or perhaps his scribe Baruch.[20] But according to Friedman, D did more than write this one book: he essentially rewrote the entire Bible. He started by combining the writings of JE with those of P, turning them into one long story that contained the many doublets and triplets we now find in it. Thus the first story of Creation, where God did his work in seven days came from P, and the second story where Adam and Eve sinned came from E.[21] At the end of the first four books, D added his own book of Deuteronomy. And then he completed the work! He took existing stories, now lost,[22] edited them and wrote the books covering the story of the Hebrew people from their entrance into Canaan, to the time of Josiah: Joshua, Judges, 1 and 2 Samuel, 1 and 2 Kings. And his style is unmistakable: everyone is judged according to whether he does good or bad in the eyes of the Lord.

This authorship explains the archaeologists' objections regarding Joshua's conquests of the Canaanite cities, which according to them did not exist in Joshua's time but did exist in King Josiah's time. It also explains the exuberant praise for Josiah and the high expectations for the results of his actions. Josiah's sudden and ignominious death, followed soon thereafter by the complete destruction of Judah, must have taken the author by complete surprise.

There is one more writer to discuss and then we have covered all the Bible authors. We have already seen that in 458 BCE, King Artaxerxes sent Ezra and Nehemiah to Jerusalem with the "Law of God and the King." The Bible relates how Ezra gathered "the men, women, and children old enough to understand," and read to them "The Law," with the Levite priests interpreting it. The reading included the celebration of the Feast of Booths, which "the children of the Israelites had not done so from the days of Jeshua, son of Nun, until that day" (Nehemiah 8:17).[23] It is obvious that we are dealing here with a new writing, something that had originated in Babylon during the exile. Was this author Ezra? Friedman thinks so. For our purposes, it is sufficient to agree that this was the final version of the Jewish Bible, composed in Babylon during the exile. How much of the previous Hebrew beliefs were changed as a result of Persian influence? We don't know. It is interesting to note, however, that the Zoroastrian Persians followed even stricter purity laws than those prescribed in the Jewish Bible.[24]

The Old Testament also contains the two books of Chronicles. They represent a recapitulation of all the previous books, with a somewhat different point of view. It is thought that they were written sometime around 400 BC, perhaps also by Ezra.

From everything said above, it would appear that the Old Testament is the result of the work of many people who modified stories of past and contemporary (to them) events in order to validate their own beliefs, interests, and points of view. Why should we read it? Why should we attribute any special meaning to its words? Was God somehow involved in its writing? Is there a God?

Some Questions to Consider

1. If the Old Testament was written many years after the events it describes by people who had their own agendas to fulfill, why would we assume that it relates actual events?

2. The Bible attributes the recurring defeats of the Hebrew people at the hands of their pagan neighbors to their punishment by God for paying homage to their neighbors' gods. The historians, however, think that the Hebrews' troubles arose whenever they showed disrespect to their neighbors' gods, prompting these neighbors to attack them. Who is correct?

3. The prophets of Israel and Judah condemned treaties with their neighboring nations as sinful, believing that the treaties violated the exclusivity of God as the sole protector of their nations. Whenever the prophets succeeded in having the treaties abrogated, the nations' neighbors attacked. Would you view this as an example of the results of religious fanaticism? Can you think of other occasions in history where religious fanaticism has caused great human suffering? Does this mean that we should water down our religious beliefs?

4. Can we accept the proposed theories regarding the various authors of the Hebrew Bible and still believe that the writings were God-inspired? Justify your position.

Selected References

Berquist, John L. *Judaism in Persia's Shadow: A Social and Historical Approach.*

Friedman, Richard Elliott. *Who Wrote the Bible?*

Yamauchi, Edwin M. *Persia and the Bible.*

CHAPTER 7

IS THERE A GOD?

"Do you believe in God?" asked my atheist friend. "It depends," I replied, "on what you mean by God." End of discussion. Nobody wants to think about God, only to argue about him or her to illustrate the most recent topic of disagreement. I am proposing below that God is pertinent to us only to the extent that he can affect us: guide us in our lives, answer our prayers, and reward or punish us here or in the next life, if there is one. I further believe that the fact that our prayers are so often answered proves the existence of such an interested and powerful power. People, however, have always preferred to travel the long, fruitless philosophical path.

What Kind of God?

For almost two millennia, countless theologians and philosophers have tried to prove God's existence. That so many people still refuse to believe indicates that their arguments have not been very persuasive. Perhaps this is because all these theologians have the same problem, they assume God's attributes: he is all-knowing, all-powerful, he created the world, he is good, and so on. But these assumptions open the door to all sorts of objections: why do bad things happen to good people? Why did God make mosquitoes? Why, according to the Bible, is he so often surprised by things that take place? Why does he admit to making so many mistakes? And so on.

One of my daughter's friends dated a boy for ten years before marrying him, and then six months later she divorced him. "I didn't realize what he was like," she said. If we can't figure out people who are very much like us, whom we know for years, how can we comprehend a power about which we know nothing? What makes us think, for instance, that God knows everything, that he can do anything, that he is good, and so on? Not only are we trying to prove the existence of God, we are also specifying his characteristics. What exactly do we mean when we say "God"? Can you and I have an intelligent discussion regarding his existence when we don't have the slightest idea what he is? Is it necessary?

I am proposing here that the only important characteristics of God are those that affect us in a meaningful way:

- God is capable of affecting our existence in this or any other life that may exist.
- God is aware of our actions and of our entreaties.
- God is interested in us.

Let's look at them closer. If God is not capable of affecting our lives in any way, then God is completely irrelevant to us. The father of someone in another city, for example, does not concern me; my own father at home does, since his actions do affect me. Similarly, if God is unaware of what we do or say, then we cannot affect God's disposition towards us: our actions, good or bad, are completely immaterial if God doesn't know of them; so are our prayers. Finally, if God is not interested in us, for whatever reason, then our affairs and existence will not influence God's actions, even though God's actions may affect us; this is something akin to our thoughtless destruction of the environment because we are not concerned with it.

You may have noticed that I did not use any personal pronouns in the paragraph above. Unless someone can prove otherwise, God is not a he or a she. God is an it. If God exists, then it is almost certainly a power: a sentient power, capable of interacting with its environment.[1] Maybe this power created the world, or maybe the world created this power. Whatever the case, this issue is completely irrelevant to us except in a purely theoretical way. So having simplified the matter in this way, how can we prove God's existence?

PROVING GOD'S EXISTENCE

From what I said above, it would appear that proving God's existence would be an easy matter. Pray for something, and see whether God delivers. If he does, then he exists; otherwise he does not. Unfortunately things are not so simple. If he does answer our request, how can we prove that it was not chance that brought about what we prayed for? Anything that can possibly occur can occur whether we pray for it or not. Conversely, if our request is not answered, it may just mean that God decided not to grant our request.[2] After all he is God, not a genie that we have released from a bottle who is forced to do our bidding.

Much work has been done in this area, but we need to separate it into two categories: requests that affect ourselves and requests that affect others. Regarding the first area, it is undeniably true that our expectations can influence the reactions of our own body and mind. This is called the placebo effect, and it is the reason why, in clinical tests of a new medicine, the effect of the medicine is always compared to that of a plain sugar pill or equivalent. There is no doubt that our minds can unconsciously change the reactions of our bodies. There are cases of people with multiple personalities where one personality has diabetes, and the other does not;[3] or one is nearsighted, and the other not. So when we pray to God to give us strength for a task or to cure us from a sickness, and we believe that our prayers will be answered, then we ourselves may be the cause of the favorable response to our prayer.[4] In ways that we still don't understand, our thoughts and beliefs do determine the reactions of our body. One could argue that in such cases we can define God as our own unknown and uncontrollable powers. So although praying for ourselves cannot prove God's external existence, it can help us with our personal problems; and the more we believe, the more we are helped.

To prove God's external existence, we must demonstrate that our prayers can affect others, or external events.[5] We can pray for someone to get well, and examine the results: does the sick person get well or does he die? But what would that prove? We know that even without being prayed over, some sick people get well and others die. We also know that eventually we will all die, no matter how many people will pray for us. There have been only two or three clinical studies to test this effect.[6] Hospital patients were divided into two similar groups. Then some people who did not know any of the patients, prayed that those in one group would get well and ignored those in the other group. It appears that there was some statistical evidence that prayer helped a little, even if it was just to reduce the length of the hospital stay. I am personally unconvinced about the validity of such tests, however, mostly because I find it difficult to visualize a person praying for somebody whom he knows only by name, say John Smith. Are we asking God to cure every John Smith in the world, or perhaps just those in a particular hospital? To me, a meaningful prayer is an intense thing involving a clear request, even if it is just a simple "Please help me."

The Power of Prayer

"The effectiveness of prayer is one reason why people continue to believe in supernatural entities."[7] This is undoubtedly the most important thing that I have to say in this book: prayers work! For some reason, young people normally don't notice it. Perhaps when they were children they took it for granted; then as they grew older and prayed for all sorts of things that didn't materialize, they figured it was a hoax and started to just pay lip service to the idea. But upon reaching middle age, many people become increasingly aware of a prayer's effectiveness.

When I am talking of prayer, I don't mean reciting mindlessly some Hail Marys or the Lord's Prayer. What I do mean is concentrating intensely, trying to send someone a mental request. It doesn't matter whom you are addressing: it could be God, Jesus, Mary, your favorite saint, or even the soul of your dead grandmother. The important thing is trying really hard to send someone a message asking for help about something.

Jesus dealt with the subject of prayer a number of times. "Therefore I tell you, whatever things you desire, when you pray believe that you will receive them, and you shall have them" (Mark 11:24). And again:

> So I tell you to ask and you will receive, search and you will find, knock and the door will be opened for you. Everyone who asks will receive, everyone who searches will find, and the door will be opened for everyone who knocks. Which one of you fathers would give your hungry child a snake if the child asked for a fish? Which one of you would give your child a scorpion if the child asked for an egg? As bad as you are, you still know how to give good gifts to your children. But your heavenly Father is even more ready to give the Holy Spirit to anyone who asks. (Luke 11:9-13 CEV)

Do we have to believe that our prayers will be answered for this to happen? According to Jesus, this is a requirement for a miracle to occur, and a prayer is a request for a miracle. He often said, "Your faith has saved you." Conversely, in his hometown where people knew him as the carpenter's son and where the belief level was low, "He did not do many mighty deeds there because of their unbelief" (Matt. 13:58). On the other hand, it happens quite often that an unbelieving person who is at the end of his resources asks for help, from nobody in particular since he does not believe in anything specific, and, miraculously, help arrives. Certainly this applied to my own mother, a baptized but non-believing person, who often spoke of the many occasions when she had received unexpected help in times of need.

How do we acquire faith in prayer? There are three steps: noticing, experiencing, and deciding. "First, you hear about other people's miracles. Then...you raise your awareness and begin to notice that you too are receiving miracles, in your life. Finally...you begin to expect them."[8] We will discuss miracles in chapter 18; here I would like to repeat that you should try to remain aware of what is happening in your life, and you should keep an open mind to new possibilities so that you can recognize the miracles when you encounter them.[9]

Jesus also told two really strange stories. In the first one, someone receives an unexpected visitor after sunset, and not having any food at home, goes to his neighbor's house to borrow some flour; the neighbor is in bed and refuses to open the door, but eventually, when the pounding on the door does not stop, he gets up and gives him the flour so that he can sleep in peace (Luke 11:5-8). In the second story, a poor widow has been wronged, but the judge will not pay any attention to her; she makes such a pest of herself, however, that he finally rules in her favor so that he can find some peace (Luke 18:3-5).

Are we to conclude that our prayers interfere with God's day, and that he grants them only in order to find peace? Is it like the child in the supermarket who picks a toy off the shelves and starts whining to his mother, "Can I have it? Can I have it?" on and on, until his mother finally agrees so that she can finish her shopping in peace? Or is it like the human brain, where millions of neurons keep sending messages to each other, but only those whose strength exceed a certain threshold level get transmitted? Are we perhaps connected to God in some similar fashion?

Do all prayers get answered? Obviously not. If you ask for snow on a hot July afternoon, you will almost certainly be disappointed; unless your faith is strong enough to move mountains, the odds are too much against you. If you pray to win the lottery, again you will be disappointed; think how many others are also praying for the same thing. On the other hand, if you pray to get rich, and you do it with enough belief, then it may well happen. But beware! As the saying goes, be careful what you pray for, for you may get it. The money that makes you rich may well be the insurance payment for the accident that paralyzes you. If you pray really hard about not having to go to work tomorrow, don't be surprised by the phone call that announces your dismissal. It is almost as if our prayer requests are answered literally by a computer. We ask for something and we get exactly that; since we say nothing about our other needs, wants, likes, and dislikes, the computer uses only the information we give it.

The correct way to pray, I believe, is to ask for help with a problem. You may not necessarily want to be rich; you may just want some help to get out of a financial tight spot. So just ask for help with a problem, don't specify the solution. Remember the child in the supermarket asking for that toy? His mother may have bought it to keep him quiet, but it might have been nothing that he could use. He probably picked it off the shelf attracted by the pictures on the box. But if the same child had asked his mother for a toy, not necessarily this toy, his mother would have scanned the shelves and picked something that he would really enjoy. Don't tell God what you want him to do for you, just ask him for his help.

Unfortunately this problem of misdirected solutions has scared me so much that I only pray for something specific when I have already reached the end of my resources. Although I have no doubt that my prayer will be answered, I don't dare to ask for anything minor such as help in losing weight. The fastest way to lose weight is to get sick, and this is what will probably happen; in fact something very similar has actually happened to me. So although a prayer is a powerful tool, it has to be used very carefully. Say thank you often for the little miracles that happen to you every day, and ask for help to strengthen you in your various commitments. Keep the really important prayer for the time where you have reached the end of your rope and cannot deal with the situation by yourself.

Let us assume that you do believe in a power that is both interested in you and capable of helping you. How is this related to the Bible? We have already seen that it was written by all sorts of people who had their own agendas? Where does the real God come in?

About Prophets and Scientists

> Surely the Lord God will do nothing without revealing his secret to his servants the prophets.
>
> The lion has roared, who will not fear? The Lord God has spoken, who can but prophesy? (Amos 3:7-8)

Let us agree that there is a God, that he is interested in us, and that he is willing to help us when we ask. If in addition, he is also capable of showing initiative, then we should expect that he would try to find a way to let us know what behavior is to our advantage and what is not. Hence the concept of inspired writing. Let us look at a story regarding a latter-day prophet.

> He was asleep or in a trance when he heard a voice say: "Read!" He said, "I cannot read." The voice again said, "Read!" He said, "I cannot read." A third time the voice, more terrible, commanded, "Read!" He said, "What can I read?" The voice said:[10]

> Read: In the name of thy Lord Who created.
> Created man from a clot.
> Read: And it is thy Lord the Most Beautiful
> Who taught by the pen,
> Taught man that which he knew not.[11]

It appears that Muhammad, the founder of Islam, did not know how to read or write. Yet the words that flowed out of his mouth created a masterpiece of Arab prose and poetry. "Many of the first believers were convinced by the sheer beauty of the *Qur'an*, which resonated with their deepest aspirations, cutting through all their intellectual preconceptions in the manner of great art, and inspiring them, at a level more profound than the cerebral, to alter their whole way of life."[12]

Is it possible that an uneducated, simple person could produce the *Qur'an*, one of the greatest literary masterpieces of the world, one that apparently defies adequate translation? We are told that while the meaning comes through when translated into another language, the feeling gets lost. Of course, all of the Hebrew prophets maintained that they were speaking God's words. They would often start with "Thus says the Lord." Even Jesus said,

> Everybody who speaks a word against the Son of Man will be forgiven, but he who blasphemes against the Holy Spirit will not be forgiven.

> When they take you before synagogues and magistrates and authorities, do not worry about how to defend yourselves or what to say, because when the time comes, the Holy Spirit will teach you what you should say. (Luke 12: 10-12 JER)

If the authors and prophets in the Bible were all following their own agendas, how can we say that their pronouncements were divinely inspired? Almost certainly some of them were not. There probably were many false prophets making up sayings, just as there were many writers who, being uninspired, filled in the empty spaces in their stories with their own imagination.[13] Moreover, even though many might have been divinely inspired, their stories could still have been corrupted. God speaks, and we hear. But don't we often hear only what we want to hear? And how often do we say things that we really did not mean? How often are we misunderstood? This is why we cannot read the Bible as if it were a history book and interpret it verbatim. It also contains many outright untruths.

It is not only prophets who soak up information from the world around them: musicians, scientists, and many creative people do it all the time. Mozart once said that music flowed into his mind faster than he could write it down. "Beethoven believed that when he created music, he was seeing into the mind of God....Einstein believed in a universal order that could be understood by an intuitive human mind. What is amazing about his outlook is that his intuitions were virtually always correct."[14] Many of the most important theorems in theoretical physics were arrived at almost intuitively by young men who then spent the rest of their lives trying to develop rigorous proofs to substantiate them.[15] The same applies to Freud's work. Along similar lines, "The astrophysicist Chandrasekhar once remarked that when he found some new fact or insight, it appeared to him to be something 'that had always been there and that I had chanced to pick up.' According to this view, the equations that underlie the workings of the universe are in some sense 'out there,' independent of human existence."[16] The great psychiatrist Carl Jung admitted that all his work was based on a still mostly secret, semi-mystical treatise he wrote between 1912 and 1917. Part of it was a short work called *Seven Sermons to the Dead*, which he apparently wrote in three days under the influence of spirits.[17] Similarly, Alfred Wallace arrived at his theory of evolution during a period of intermittent delirium.

Whether God created the world or the world created God may not matter; perhaps the two are largely the same. And since we are part of the world, we also share in this sameness. Some of our more philosophically minded scientists are starting to believe in the existence of morphic fields that permeate the universe and encode the pattern of everything that existed and has ever occurred, aiding in their reoccurrence. According to this thinking we are surrounded by knowledge that is only waiting to be accessed.[18] We will discuss these ideas again in chapter 17 when we also examine memory.

SOME QUESTIONS TO CONSIDER

1. The author proposes that only a god who affects our destiny is relevant. Do you agree?

2. Isn't it unscientific to ascribe responses to our prayers to the actions of an unknown God?

3. Assuming the validity of the author's observation that God often answers requests literally, in an unexpectedly undesirable manner, what can we conclude from it?

4. The proposition that prayers are answered is of such importance that we would expect that much more effort would have been spent to prove it or disprove it. Whys hasn't this been the case?

5. We would expect young people with uncluttered minds, who are not brainwashed yet by the generally accepted beliefs, to originate new ideas, even if they cannot completely justify them yet. Why does the author need to propose metaphysical concepts of ideas floating around the universe and ready to be plucked from thin air?

SELECTED REFERENCES

Hoeller, Stephan A. *The Gnostic Jung and the Seven Sermons to the Dead.*

Tart, Charles T., ed. *Body Mind Spirit: Exploring the Parapsychology of Spirituality.*

Walsch, Neale Donald. *Moments of Grace.*

CHAPTER 8

WHY BAD THINGS HAPPEN

Where was God on 9/11? If God is good and all-powerful, why didn't he stop the evil? These are the types of questions that people ask when unhappy things happen. We can't know for sure whether God is good and all-powerful, but any scientist knows one thing: if God created the world, he had to first create the physical laws to run it. And these physical laws also determine the results of our own actions and those of everyone and everything around us. God cannot interfere without repealing the laws. The only thing that God could do on 9/11 was to try to embrace and soothe the thousands of his children's souls that were suddenly set adrift.

The Age-Old Question

Whenever anyone suffers, dies, or meets with bad luck, we ask ourselves why God allowed it to happen. We expect him to protect us even from the so-called acts of nature, like earthquakes, tornados, volcanic eruptions, and so on. We assume that not only he has the moral responsibility to protect us from all unhappiness, but that he also has the capability to do it. All this is based in our belief that God is good, omnipotent, omniscient, and interested in our welfare. But as we saw earlier, these beliefs are not based on any factual knowledge. I hope and assume that God is good, but I have no way of really proving it. We will discuss the possibility of his being omniscient later, when we get to the subject of free will. At this point I will only claim that he is not omnipotent.

God Is Not Omnipotent

There are two main reasons why God's power to interfere in everyday life is limited: everything in creation must follow existing rules, and sentient life is free to act as it sees fit. Bricks and stones are usually attached to building walls with mortar. But mortar deteriorates with time and lets go of what it is holding. So every now and then, while somebody is walking on the sidewalk next to a tall building, a stone, or even a piece of broken glass, comes loose, hits him on the head and sends him to the morgue. Here is a case where God, assuming he was omniscient and thus knew what was about to happen, could have stepped in to save the person. Perhaps he could have slowed down the mortar deterioration by a couple of seconds, or ordered a strong wind to blow so that the falling object would miss the person. Such actions, however, would change the world from a physically ordered environment to the equivalent of a puppet theater. Physical laws would not depend on rigid rules, but on the effect that their results have on us. Nothing like this has ever been demonstrated to exist. Physical laws always hold true. Occasionally however, an event that has a very small probability of occurrence does happen, and sometimes we call that event a miracle.

For instance, instead of interfering with the natural laws, God could have attracted the person's attention to a window display, or cause him to stumble, or otherwise slow him down sufficiently so as to miss the falling object. Believe it or not, such things do happen. And when they do, the unbeliever marvels at his good luck, and the believer thanks his guardian angel. But no matter how loud our guardian angel shouts, we do not always listen to him.[1] We will talk about angels in chapter 18. Right now it is sufficient to point out that unless God takes detailed control of the environment, thus intervening with the laws of physics, or of us thus taking away our freedom of action, he cannot protect us from harm.

THE STORY OF JOB

Christians and Jews alike believe that God's explanation of why bad things happen to good people can be found in the story of Job, written sometime between 600 and 400 BCE. It appears that there once lived a blameless and upright man named Job, who was blessed with seven sons, three daughters, and much wealth. He lived a proper life and even sacrificed to God to atone for any accidental sins that his sons might have unknowingly committed.

One day God asks Satan whether in all his travels on earth he had noticed Job and seen how blameless and upright he is, fearing God and avoiding all evil. Satan replies that it is easy for Job to be so upright, since all this bounty and God's protection surround him. "Take these away," he says, "and things will quickly change." So God and Satan make a bet, and God allows Satan to assume power over Job. In quick order, Job loses all his wealth and all his children and is eventually reduced to a boil-covered vagrant sitting on ashes.

When his wife and three friends try to point out that God is obviously punishing him for his sins,[2] he indignantly refuses to admit that he had ever sinned and continues praising the Lord. Finally, however, Job loses his patience and cries out indignantly, proclaiming his righteousness and questioning what good has all his piousness brought him. He calls for a judge to hear his case, for the right to see his accuser. God appears and addresses Job out of a storm:

> Why do you talk so much when you know so little?
> Now get ready to face me!
> Can you answer the questions I ask?
> How did I lay the foundation for the earth? Were you there?
>
> Doubtlessly you know who decided its length and width.
> ...Did you ever tell the sun to rise? And did it obey?
> ...Do you know the laws that govern the heavens,
> and can you make them rule the earth? (Job 38:2-33)

On and on the Lord rants and raves at Job telling him how insignificant and ignorant he is relative to himself. Finally Job answers:

> Who am I to answer you?
> I did speak once or twice, but never again. (Job 40:4-5)

But the Lord continues to rage:

> Are you trying to prove that you are innocent
> by accusing me of injustice?
> Do you have a powerful arm
> and a thundering voice that compare with mine?
> ...Show your furious anger!
> Throw down and crush all who are proud and evil.
> Wrap them in grave clothes
> and bury them together in the dusty soil.
> Do this and I will agree
> that you have won this argument. (Job 40:8-14)

Job meekly agrees that he is insignificant when compared to God, and "repents in dust and ashes." God restores his good fortune, gives him a new set of sons and even prettier daughters, double the wealth that he had previously, and lets him live a second life until he is 140.

Notice that God did not really answer any of Job's complaints. He just shouted him down, comparing his own greatness to Job's stupidity. "You are too lowly and stupid to understand, so I will not bother to explain." Many books, and even plays, have been written analyzing this story, but they usually ignore the introductory passage where God and Satan make a bet. Without it, the message of the story is clear: man is not wise enough to understand why bad things happen in the world.[3]

TODAY'S ANSWER

With the great advances of science, we can understand today many more things than in the days of Job. We might guess, for instance, that perhaps Job had located his house and cattle over a radioactive or chemically polluted site. Indeed, exactly such a thing happened during the Seventies at the Love Canal development, outside Niagara Falls, where the houses were built on the site of an old chemical dump. We can try explaining to a grieving young mother that her child was stillborn because of genetic damage to one of the DNA strands, but this will not answer the question "Why to *my* child?"

Some time ago, a family in a nearby suburb came home at midnight after an evening out. The sixteen-year-old daughter decided to drive home her cousin, who had been babysitting. While turning at an intersection, a car with a drunk driver ran through the red light, killing both girls. Was God at fault? Was he responsible for the drunk driver or for the young, inexperienced driver being out late at night, breaking curfew law?

Bad things happen to us because of our own actions, because of the actions of others, or because of chance events that could happen to anybody, and this "anybody" turned out to be us. We cannot blame God for them; he is not our babysitter. Within a year of our birth we want to move, to explore, to go where we have never been, to be independent. It is part of our growing up, but it is also fraught with danger. A parent can try to be watchful, but he can't protect his children from all accidents. The only sure way to prevent a boy from falling off a tree is to keep him from climbing it; but then you are also keeping him from growing and learning how to climb trees.

We tend to blame God for the results of our actions in a universe that is controlled by fixed rules. But in order to protect us from all harm God would have to either keep adjusting the universal laws, something that would soon result in our complete confusion, or to control all our actions, thus taking away our independence.

If we consider Job's story in its entirety, we recognize that Job did not build his house on the site of an old dump. Although Job does not know it, we know that his sufferings resulted from a bet that God had made with Satan while bragging about his servant Job; certainly this was something that Job was capable of understanding, yet God covered it up with a loud, arrogant, and demeaning speech. In the end, however, God tried to make amends by restoring to Job another lifetime to replace the one that he destroyed.

The most unfortunate part of the story, however, is Job's response to God's speech. Yes, you are great, and I am insignificant, he says. I spoke before, but I will not speak again. The concluding remark is missing but you can almost sense it: now go away and leave me alone in my misery. By erecting such a deep chasm between him and us, God rendered himself irrelevant. If he will not answer us, if we can't hope to understand him, then he does not matter. It makes little difference if he is there or not. We will see in chapter 20 that this is one area where the Christian God differs from the Hebrew one.

SOME QUESTIONS TO CONSIDER

1. Doesn't it appear capricious that God will sometimes interfere on our behalf, and sometimes not?

2. The author claims that God cannot interfere in our lives, but he then argues that God-caused miracles occur. Is he not being inconsistent?

3. If God is not really omnipotent, how can he really be God?

4. Could the story of Job have been written without including God's bet with Satan? Would it have been better or worse that way?

5. Having permitted the destruction of Job's life, God bestows upon him a second life to live without outside interference. Does this indicate that God felt he had erred?

6. It is generally accepted that the story of Job is a writing of fiction. Why then, has it been given so much attention over the years?

SELECTED REFERENCE

Miles, Jack. *God: A Biography.*

CHAPTER 9

APOCALYPTIC WRITINGS—PROLOGUE TO CHRISTIANITY

During the last two centuries BCE, Hebrews started to question some of their basic religious beliefs. It had become obvious by then that in their world God not only did not reward the good people and punish the bad ones, as was expected, but he appeared to be doing the exact opposite: the good suffered while the evil prospered. To explain it, there arose during this period a new genre of writings, called apocalyptic, that introduced two new ideas: there is life after death, in which rewards and punishments are meted out, and there exists an evil deity whom God cannot, or will not, keep under control. Although most of these writings never made it into the Bible, they were very important in giving birth to some of Christianity's basic ideas.

From Alexander to the Maccabees

The two hundred years of Persia's domination of the Near East ended in 333 BCE when Alexander the Great and his Greek army defeated the Persian king Darius III. Alexander's technique was effective: he destroyed and burned any city that fought against him, but did no harm to those that opened their doors to him. In a short time he conquered Syria and Judea, where he left intact the religious arrangements that had been approved by Persia.[1] He then invaded Egypt, where he was greeted as a deliverer from the hated Persians and was proclaimed pharaoh. After leading his army all the way to India and back, he died in 323 BCE at the age of 33. Three of his generals promptly murdered his infant son and divided his empire among themselves: Antigonus Gonatas took over the Greek Macedonian homeland; Seleucus Nicator laid claim to the old Babylonian empire, including Syria and Judea; and Ptolemy Soter claimed Egypt, Syria and Judea.

The conflicting claims of the last two generals over Syria and Judea kept them at war for the first hundred years and divided the people of Jerusalem into two political camps. Although Ptolemy kept the upper hand, he had to campaign often in the area. In 302 BCE, his army broke into Jerusalem on a Sabbath, unopposed by the Judeans who did not bear arms on their holy day even for their self-protection.[2]

Ptolemy resettled more than a hundred thousand Judeans in Egypt, drafting many of them into his garrisons, and restructured the administration of Palestine. He replaced the appointed civil governors of the Persian era with a bureaucratic system of administrative officials, each one responsible to a higher-up. He also separated the military matters, assigning them to a general with garrisons in various cities. The removal of the governor probably enhanced the power of the high priest, a hereditary position restricted to the descendants of Zadok, David's original high priest. Also participating in the administrative affairs was an assembly of rich and aristocratic families, which eventually developed into the Great Sanhedrin—the word sanhedrin, derived from the Greek *synedrion,* means council. This was a seventy-one-member central legislative and judicial council which interpreted biblical laws, passed new ones, and governed Judea to whatever extent its various conquerors permitted.[3]

The Ptolemies introduced the tax-collector system that was later adopted by the Romans, which is so often derided in the New Testament. Under this system, the rights to collect taxes in various districts and cities were auctioned annually, first in Alexandria, and later in Rome. The successful bidders contracted to collect for the government the appointed taxes, keeping of course any overcollections for themselves. Although the winners were held personally liable for any failure to meet the specified quota, these lucrative contracts were avidly sought by the major aristocratic families in Judea and gave rise to considerable friction and animosity among them.

The greatest historical importance of Alexander's conquest of the East was the spread of Greek culture throughout the area. Following his orders, his generals married locals, settled down, and introduced the institutions of their homeland. They built Greek cities with temples to the Greek gods, and gymnasiums where, according to Greek custom, young men gathered and participated naked in athletic events. These gymnasiums attracted the local population of elite who sought to join the new aristocracy of the administrative circles.[4] Greek became the new language of administration and commerce.

In Egypt, Ptolemy built his new capital, Alexandria, and founded the new royal library that was intended to eventually contain a copy of every existing book. At about 285 BCE the Hebrew Old Testament Scripture was translated into Greek, in the version now called the Septuagint.[5] As more and more Jews in the Diaspora became Hellenized and forgot the Hebrew language, this Greek version became the standard Scripture for Jews, as well as for Greeks and Romans.

By the beginning of the second century BCE, the Judeans in Jerusalem were divided in pro-Seleucid and pro-Ptolemaic camps, which waged war against each other. In 198 BCE, however, the Seleucid king Antiochus III defeated Ptolemy V and became ruler of Judea. He granted the city many privileges, including the right to live as an autonomous priestly nation, something that in effect made the high priest a civil leader as well as a religious one. During this period, however, the Romans began their own empire building. In the process of spreading their field of influence over all the civilized world, they engaged and defeated Antiochus III and forced him to pay severe reparations.

After his death in 187 BCE, Antiochus III was succeeded first by his oldest son Seleucus IV, and later in 175 BCE by a younger son, Antiochus IV, who replaced the Temple's high priest, with the priest's Greek-loving brother, Jason. This launched one of the most turbulent periods in Judean history. Jason introduced

Greek ways to the community and had a gymnasium built near the Temple, which soon became the social center of the city. "Even our priests gave up worshipping at the altar. They cared nothing about the temple, and they neglected offering sacrifices. And when the signal was given, they hurried off to take part in the games that were against our teachings" (2 Maccabees 4:14 CEV).[6]

Antiochus IV was far different from his benevolent father. Because of a misunderstanding, he attacked Jerusalem and in three days massacred forty thousand men, women, and children and sold an equal number into slavery. He then stole all the gold utensils and other treasures he could find in the Temple (2 Maccabees 5:11-16). A couple of years later, in 167 BCE, he sent a representative with a military force back to the city. They massacred many, tore down the city walls, and built a fortified citadel for the troops. Antiochus IV then decreed that all his subjects, including the Judeans, should henceforth honor and offer sacrifices to the same gods. A statue of Zeus was erected in the sanctuary, and pigs and other animals considered unclean by the Judeans were sacrificed to him. The Judeans in the whole country were forbidden to practice their Hebrew religion or to circumcise their young, and they were forced to follow the pagan Greek practices.

FROM THE MACCABEES TO JESUS

Although many Judeans went along with the new ordinances of Antiochus, some opposed them strongly. Among the first to oppose them were the members of a priestly family, the Hasmoneans. They escaped to the mountains, where they were soon joined by many other discontented religious and freedom-loving people, many of them members of the poor class. After the death of the family patriarch, Mattathias, the leadership went to his youngest son, Judas, who was given the nickname "Maccabee," the Hammerer. At first he concentrated his attacks on those of his fellow Judeans who followed Antiochus's decrees, killing them, destroying the pagan altars and burning their villages. But soon the group became strong enough to attack directly the Seleucid troops and defeat them in battle, aided by the fact that the main Seleucid army was away fighting the Parthians, a remnant of the old Persian empire, on the eastern front.[7]

In 164 BCE, after a bloody battle between Judas's group and the Seleucid forces, the law requiring one uniform religion throughout the country was repealed, and the Judeans were allowed to celebrate again according to their old customs. Judas and his group entered Jerusalem, cleaned and purified the Temple, and rededicated it on the anniversary of the exact date on which it had been profaned, in a ceremony now celebrated as Hanukkah.[8] Soon, however, a much stronger Seleucid army returned, killed Judas in a battle, took control of the country, and started executing the Hasmonean supporters. Judas's brother, Jonathan, took over leadership of the group, which was forced to flee the city.

During the next eighty years there was much unrest in the Seleucid kingdom, as many pretenders vied for the throne. The Hasmoneans took advantage of the situation and extracted increasingly greater amounts of independence for Judea. In 154 BCE Jonathan got himself appointed high priest, despite the fact that he did not belong to the hereditary priestly family. A few years later, after many expensive gifts to both the Seleucid and Ptolemaic kings, he was appointed general and governor of Judea. He proceeded to annex some areas away from Samaria, the old nation of Israel.

When Jonathan was killed in 142 BCE, his brother Simon took over. The Seleucids appointed him high priest and ruler, and ended the collection of taxes in Jerusalem. Militarily, Simon was very successful, expelling the last foreign soldiers from the fortified citadel, expanding the area controlled by Judea, and building fortresses for its defense. The Judeans in turn voted to bestow upon him, and upon all his descendants in perpetuity, the powers of high priest and ruler, and the right to wear gold and purple; in short, in all but name, they made him king. At last, 440 years after the city's fall to Babylon, there was again a king in Jerusalem, although somewhat beholden to both the Seleucids and the Romans. This was to last for almost eighty years, and during that time the kingdom grew until it covered all the area that had originally been conquered by Joshua, from the south end of the Dead Sea to north of the Sea of Galilee, from the Mediterranean coast to east of the Jordan River.

Along with the new royalty, there arose in this period the three main religious parties: the Sadducees, who were from the rich aristocratic group and believed only in the written Torah; the Pharisees, who were mostly teachers and who, in addition to the Torah, believed in the unwritten law, presumably handed down orally by Moses but actually being developed by them through an ongoing interpretation of the Scripture; and the Essenes, who were an apocalyptic sect that judged all things as either black or white, and whose members eventually left the city to escape its evil and pollution. Almost nothing was known about this last group until the discoveries of ancient texts in the caves of Qumran, half a century ago.

The successive generations of the Hasmonean kings became increasingly arrogant and unpopular with their subjects, often using force to keep them under control.[9] Eventually, the Roman army under Pompey arrived in Syria, and when two of Simon's quarreling great-grandsons asked for his help against each other, Pompey invaded Judea and, in 63 BCE, captured Jerusalem. The Romans were to remain in the country until the Muslim era—Jerusalem was conquered by the Muslim army in 638 CE, six years after Muhammad's death.

The Romans divided the country into five areas, and appointed one council (Sanhedrin) to govern each area. In 40 BCE Rome appointed Herod (who had sided with Caesar during his confrontation of Anthony) as king of Judea, but it took him three years and the help of the Roman army before he succeeded in ascending to the throne. He was a very bloody and unscrupulous person; he had ten wives and killed many of them and their sons. But he was a great builder, constructing an entire Roman city, Caesarea, and also the Masada and Herodeum fortresses. In 20 BCE he started work on the final temple, which was still under construction in Jesus' days. Upon his death in 4 BCE, the Romans divided the kingdom into four parts, naming four of Herod's sons as ethnarchs, a somewhat lower title than that of king. As ethnarch of Galilee, they appointed Antipas, whom Luke calls Herod in his gospel. The Judean territory, with Jerusalem as capital, was originally given to Archelaus, but two years later the Romans took direct control of it and assigned its governance to Roman prefects, the most famous of whom was Pontius Pilate between 26 and 36 CE.

WRITINGS

When I discussed the authors of the Old Testament in chapter 6, I probably did not make it clear that I was referring only to the so-called historical books. In addition to these, the Old Testament also contains two other groups of writings: the prophetic books that narrate what the prophets of Israel and Judah said, and the wisdom books (called writings in the Hebrew Bible) that include the psalms, proverbs, and other wisdom-imparting writings.

Once Ezra came to Jerusalem in 458 BCE with a copy of The Law, the first five books, and read it to the people, the period of the prophets ended. God's law was now in writing, and could not be altered or declared anew by inspired men. Except for the additional historical books that we attributed to Ezra, and a brief prophetic work by Joel composed about 400 BCE,[10] nothing more was written for the next 250 years.

But in the last two centuries BCE, two new kinds of writing appeared: the first reiterated the Hebrew history, making changes whenever necessary to explain some obvious discrepancies in the Torah; the other, called apocalyptic (of a revealing nature), described mankind's eventual end, the final judgment of all individuals, and the new world that would replace the old.[11] This last type of writing represented an entirely new concept in Jewish literature and thought. While in Egypt, the Hebrews witnessed and helped their Egyptian masters build pyramids and prepare for their life after death; yet in their literature they never mentioned the possibility of such life after death.[12] It had always been assumed that all of God's rewards and punishments took place in the present world; after death there was nothing, a belief that many Jews still hold today. Now, all of a sudden, the apocalyptic writings make their appearance, describing the eternal rewards for the faithful and the eternal punishment of the sinners. Why?

Looking back at the history of their nation, the Judeans of the second century BCE could clearly recognize the sins of their ancestors and why they incurred God's punishments. But a unique situation arose during this period. It now appeared that the more pious a person was, the greater the evil with which he was rewarded in his life. Judeans were morally required to abide by the rules of their kings, who they believed were appointed by God, and to not rebel against them; yet their kings were acting in an obviously immoral manner. If to follow their orders was immoral, to revolt against them was equally immoral. Some Judeans elected to fight for their beliefs, only to get slaughtered on the Sabbath, when their faith forbade them to protect themselves.[13] It became obvious that the world in which they lived was unjust. But how could God allow the existence of such an unjust world? The solution was to propose the existence of another life, following the present one on earth. It was in this next life where the good would finally get their proper reward, and the evil their just retribution.[14]

None of the apocalyptic writings are normally included in the Hebrew Bible, but some are occasionally included in recent versions of the Christian Bible and called Apocrypha (meaning "from the secret ones" in Greek). Despite their non-inclusion, many of these books were very influential during this period, as well as in the following few centuries. Perhaps the most noteworthy of them are 1 Enoch and The Jubilees; this last one was a new retelling of the entire Bible story, endeavoring to explain away some of the puzzling occurrences in it.

1 AND 2 MACABEES

These two historical books written during the second century BCE discuss the early years of the period when members of the Maccabee family affected the history of Judea: the first, 1 Maccabees, covers the period of 175 to 134 BCE, and was probably written about 100 BCE; the second, 2 Maccabees, covers the far shorter period between 180 and 161 BCE, and was probably written sometime between 124 and 100 BCE. In this book, for the first time in the Bible, we find references to "the resurrection of the just in the last day, the intercession of saints in heaven for people living on earth, and the power of the living to offer prayers and sacrifices for the dead" (NAB, Introduction to the Second Book of Maccabees, 497). Despite this, these books are usually not classified as apocalyptic, but because they are included with the apocrypha, many Bible editions, including the King James version, exclude them from their collections.

Daniel

Perhaps because of its famous description of the coming of the son of man, all Christian Bibles include the book of Daniel, without even calling it apocryphal. This book of loosely related stories was written by an unknown author between 167 and 164 BCE, during the religious persecution of the Jews by Antiochus IV. Although these stories are located in Babylon during the time of exile, some of them describe a religious persecution similar to that in Judea. The first six chapters of the book describe how, because of their belief in God, Daniel and his three friends are given wisdom and help that allows them to conquer various adversities to which they are subjected. In one story, Daniel is thrown into a pit with a hungry lion, but an angel of the Lord protects him. In another, his three friends are thrown into a superheated oven, but again they are protected by an angel and survive unscathed.

The second half of the book consists of various visions and prophecies, some of them relating to Alexander and the Seleucid kings that followed him. One difference between the two parts of the book is the timing of the reward granted to the pious believers. In the fairy tales of the first half, God mediates, and the believer is immediately given an earthly reward. In the apocalyptic stories of the second half, however, which covertly describe the actions of the Hellenistic kings in Palestine, the just get their rewards only after their rise from the grave: "And many of them who sleep in the dust of the earth shall awake; some to everlasting life, and some to shame and everlasting contempt" (Daniel 12:2).

Here we have the first prediction of resurrection in the Hebrew Bible.[15] For the Christians, however, the most important vision relates to the end of the world, when the "Ancient One takes his throne." Then the author describes his vision in which

> One like a Son of man
> came with the clouds of heaven,
> and came to the Ancient of days,
> and they brought him near before him.

And he was given dominion, and glory, and royal power,
that all peoples, nations, and languages should serve him.

His dominion is an everlasting dominion,
which shall not pass away,
and his royal power that which shall not be destroyed.
(Daniel 7:13-14)

Because Jesus often referred to himself as the "son of man," it was easy to identify him with the being in Daniel's vision. As such, he would inherit a special heavenly position and come to earth again as a king to pass final judgment, a basic Christian belief. Of course "son of man" has many meanings and has been used to identify many people. We will discuss these things in chapter 15. At this point, however, two things should be mentioned: first, the angelic figure in Daniel does not do any judging; second, almost all scholars agree that the angelic figure was exactly that, an angel, probably Michael, who would fight for the Hebrews against the angel that would fight for the Greeks.

1 ENOCH

Enoch was Adam's sixth descendant (the seventh generation of humans) and the Bible described that God raised him bodily to heaven when he was 365 years old.[16] He was supposed to have been the only person other than Elijah who never died. When in heaven, he was taken around and shown the workings of the world, all of which were described by unknown writers during the last two centuries BCE. Of importance to the Jews of that time was his insistence that the year consisted of 364 days, not the 360 days counted in the pre-exilic period or the 354 days of the post-exilic period. This of course affected the calculation of the various feast days, and may have been one of the reason for the separation of the Essenes (see below) from Jerusalem.

The Book of Enoch contains perhaps the first judgment scenes in the Old Testament era, with the good being rewarded and the bad being punished. It also includes an expanded description of the descent of angels to earth to take human wives and generate evil progeny, and of their consequent punishment. Because of the manner it portrayed the angels, both the Church Fathers and the rabbis pronounced the book heretical and accursed. As a result it was burned, along with all other books pronounced heretical in those days, and was lost until 1773 when a copy was located in Ethiopia.[17] Of particular interest to the Christians is the appearance of the son of man as one of the judges. We should point out, however, that in this book Enoch himself is on occasion addressed as "son of man" (71:14).

The Testaments of the Twelve Patriarchs

These were probably written towards the end of the second century BCE. They describe the final words that the twelve sons of Jacob told their sons before they died. As such they consist mostly of confessions of their own misdeeds, moral advice to their offspring, and predictions of future events that affect their tribes. Unfortunately we don't possess the original writings, but only copies made by early Christians. As you would expect, the predictions have been modified to point to the coming of Jesus as a savior. References to "the holy Father," the one "born without sin," the "innocent lamb," and the saving of the Gentiles indicate the extensive rewriting that has taken place.

Although the word "Messiah" is not used, there are references to a resurrection of the dead, and to the "shoot of David," which as we shall see later stands for "Messiah." Interestingly, this writing indicates the existence of two Messiahs, one priestly from the tribe of Levi, and one kingly from the tribe of Judah. The dominant figure is the priestly Messiah, who eventually binds Belial, the angel of Satan.

The Dead Sea Scrolls

The two Messiahs also appear in the Essene writings.[18] In these texts, the priestly Messiah takes precedence over the "Messiah of Israel." The Essenes were a group of very exclusive Hebrews who left Jerusalem to settle in the desert by the Dead Sea, away from the sinful Temple. They objected to the high priest not being a descendant of Zadok, David's original high priest, as well as to the different calendar followed by the Hebrews, which resulted in feasts and sacrifices on what they reckoned to be the wrong days. They expected to take part in the final war between the Sons of Light and the Sons of Dark, who corresponded to the angels of God led by Michael and those of Satan led be Belial. The forces are evenly matched, however, so God has to intervene on the side of Light.[19] The scrolls seem to indicate a belief in the afterlife of the soul in the form of an angel, but not in the physical resurrection of the body.

Some Questions to Consider

1. What criteria should one use to decide which writings are divinely inspired and should be included in the canon?

2. Do you think that Jesus would approve of Jews fighting on the Sabbath in order to save their country or their lives? How about fighting on other days?

3. The Jews and, later, Paul believed that kings reigned at the will of God, and hence should be obeyed. We don't believe that anymore, but don't we apply the same reasoning with regard to high religious authority, whether it is the Pope or Ecclesiastical Councils?

4. The book of Jubilees presents a slightly different version of history than the Bible. Why should we prefer one to the other?

5. Life after death, along with the rewards and punishments it entails, was postulated by the ancient Jews to explain the injustices in life. How does this differ from Karl Marx's statement that "religion is the opiate of the masses"?

6. Once the Torah and the other inspired books were written down, the word of God became frozen; there was no more room for prophets. The same thing happened in Christianity after the gospels were written down: the spirit had finished speaking, there was nothing else to say. Why does human writing down of God's words shut him up forever?

Selected References

Charlesworth James H., ed. *The Old Testament Pseudepigrapha: Apocalyptic Literature and Testaments.*

Collins, John J. *The Apocalyptic Imagination: An Introduction to Jewish Apocalyptic Literature.*

Hayes, John H., and Sara R. Mandell. *The Jewish People in Classical Antiquity: From Alexander to Bar Kochba.*

Prophet, Elizabeth Clare. *Fallen Angels and the Origin of Sin: Why the Church Fathers Suppressed the Book of Enoch and its Startling Revelations.*

VandenKamp, James C. *An Introduction to Early Judaism.*

PART II

GOD THE FATHER—THE NEW TESTAMENT

Introduction to Part II

The end of the Old Testament Era left the pious and thinking Hebrew puzzled and perplexed. His religious beliefs held no answers for many of his important questions. Looking back at the last millennium, he could see that the good, all-powerful God in whom he believed had become increasingly more remote and less comprehensible. If God was indeed just, his justice could not be understood by humans; his ability to control the world he had created had become suspect; his directives were impossible to follow; and his alternating bouts of mercy and wrath were completely disorienting.

The world had become ready for a different God. One who would substitute love for justice. One who would care more about a human's intentions than his accomplishments. One whose kindness and understanding would overlook people's failure to rise to his standards. One able to feel human pain and suffering. And more important, a God who could promise mankind a better future than anything currently possible on this earth. In short, the world was ready for Christianity.

In the next eleven chapters we will discuss Jesus, his teachings, and his life. We will also take a brief look at ourselves: what it means to be human, what motivates us, what guides our minds and our actions; are we just physical beings, or can part of us survive the death of our physical body? Is our universe limited to what we can see and measure, or does it contain nonphysical entities ready to interact with us? Are there any indications one way or the other?

CHAPTER 10

MESSIAH

In the play *The Fiddler on the Roof,* as the Jews are preparing to leave their beloved Anatevka, Tevye turns to the rabbi and says, "We have waited here for the Messiah all our lives; wouldn't this be a good time for him to come?" To which the rabbi stoically replies, "We will just have to wait for him somewhere else." But who is this Messiah for whom the Jews have been waiting for over two millennia? Is he the king who will climb on David's throne? Is he the liberator who will lead them to defeat their oppressors? Or is he the messenger of God who will announce the end of the world as we know it and the start of the final judgment? To justify their own expectations, Christians added one other possibility: Isaiah's suffering servant.

Messiah the King

> "Who do people say I am?" Jesus asked his disciples. They replied, "Some say John the Baptist; others say Elijah; and still others one of the prophets." "But what about you?" he asked. "Who do you say I am?" Peter answered:
>
> "You are the Messiah" (Mark 8:29 CEV).
>
> "You are the Messiah, the Son of the living God" (Matthew 16:16 CEV).[1]

> "You are the Messiah sent by God" (Luke 9:20 CEV).

Here the Jewish word "Messiah" has been substituted for the original Greek word "Christ." In either case, however, the word means "the anointed." And who were the anointed ones in the ancient Hebrew state? The kings (and the high priests) on whose heads a prophet, when available, poured an aromatic oil.[2] So in this passage Peter is the first to pronounce Jesus a king.[3]

A week later, Jesus, with his three favorites disciples, climbed a mountain, where he became transfigured: "His face shone like the sun and his clothes were as white as light" (Matt. 17:2). And a heavenly voice said: "This is my beloved Son in whom I am well pleased; listen to him" (Matt. 17:5).[4] Compare these words with those in Psalm 2 that was used in Hebrew royal coronations.

> I will proclaim the decree: the Lord has said to me,
> "You are my son; today I have begotten you."
> (Psalm 2:7)

Hebrew kings were considered to be adopted sons of God, so the longer answer given in Matthew's gospel was just acknowledging this fact. Soon thereafter, Jesus starts his travel to Jerusalem, apparently followed by a large crowd. On the way, a blind beggar calls to him: "Son of David, have pity on me," another obvious acknowledgement of his kingship. Eventually he reaches Jerusalem, where he is greeted by a huge crowd of people who cover the ground with branches and their cloaks shouting:

> Hosanna![5] Blessed [is] he who comes in the name of the Lord! Blessed [be] the reign of our father David that comes in the name of the Lord! Hosanna in the highest. (Mark 11:9-10)

> Or

> Hosanna to the son of David; blessed [is] he who comes in the name of the Lord; hosanna in the highest. (Matthew 21:9)

Or

Blessed [be] the king who comes in the name of the Lord. Peace in heaven and glory in [the] highest. (Luke 19:38)

Or

Hosanna! Blessed is he who comes in the name of the Lord, the king of Israel. (John 12:13)

However, Jesus does not enter the city riding proudly on a horse as a king would have done, but sitting meekly on a donkey, to fulfill the words of Zechariah 9:9:

Rejoice greatly O daughter of Zion;
shout [for joy] O daughter of Jerusalem.

Behold, your king comes to you:
he is just and bearing salvation;

Lowly and riding upon an ass,
and upon a colt the foal of an ass.[6]

And the first thing that Jesus does when he gets in the city is to go to the Temple and chase out the merchants and the money changers, saying: "It is written, 'My house shall be called a house of prayer'[7] but you have made it a den of thieves" (Matt. 21:13). The Temple guards do nothing. Who but a king would dare act like this?

So it is fairly clear that, according to all four gospels, both Jesus himself and the people around him believed that he was the new king of the Hebrews. To deny his kingship is to deny the truth of every word included in these parts of the gospels. It is thus hardly surprising that a little later he was accused of it in front of Pilate.

But when was Jesus anointed? The assumption is that it happened during his baptism when *he* saw the Spirit descend on him in the form of a dove, and a heavenly voice announced, "This is my beloved Son, in whom I am well pleased" (Matt. 3:17). The first two gospels do describe that towards the end of his ministry he was anointed in Bethany, in the home of a man known as Simon the Leper, by an unknown woman with an alabaster jar filled with a very expensive perfume made of pure nard. In John's gospel the event takes place in Lazarus' house, and it is Mary who pours the perfume over his feet and wipes it with her hair, a variation from Luke's gospel where a prostitute does the same thing in the home of a Pharisee (Luke 7:38). In all three cases, however, Jesus attributes this anointment to preparation for his burial, not to his kingship.[8]

MESSIAH THE LIBERATOR

Just as in the days of the Judges the Hebrews called for a king to help defend them against the Philistines, so after their enslavement to the Babylonians the Judeans started dreaming of a king who would save them again. But it could not be just any kind of a king. The Judeans expected the coming Messiah to be a special kind of king.

- He would belong to the house of David (Moschiach ben David).
- He would be a great military leader and would destroy Judah's enemies.
- He would be well versed in Jewish law and obey its commandments.
- He would rule with justice and wisdom.
- He would rebuild the Temple.
- He would bring all the exiled back to Jerusalem.
- He would bring peace in the world. Hatred and war would cease.

In the words of Isaiah 11 (CEV):

Like a branch that sprouts from a stump,
someone from David's family will someday be king.

The Spirit of the Lord will be with him
to give him understanding, wisdom, and insight.

> He will be powerful, and he will know and honor the Lord.
> His greatest joy will be to obey the Lord.
>
> This king won't judge by appearances or listen to rumors.
>
> The poor and the needy will be treated with fairness
> and with justice.
>
> His word will be law everywhere in the land,
> and criminals will be put to death.
>
> Honesty and fairness will be his royal robes.
>
> Leopards will lie down with young goats,
> and wolves will rest with lambs.
>
> Calves and lions will eat together
> and be cared by little children....
>
> Lions and oxen will both eat straw....
>
> The time is coming when one of David's descendants will be the signal for the people of all nations to come together. They will follow his advice, and his own nation will become famous.
>
> When that day comes, the Lord will again reach out his mighty arm and bring home his people who have survived in Assyria, Pathros, Ethiopia, Elam, Shinar, Hamath, and the land along the coast.

At first, the enemies for the Hebrews to defeat were the Babylonians; next were the Greeks, then the Hasmoneans and by the time of Jesus, the Romans.[9] Strangely enough, the Maccabees seemed at first to fulfill most of the Messianic requirements: they defeated the Greeks, rededicated the Temple, and although they did not bring back the exiled, they did reclaim the land of old Israel. But as the Maccabees gained power, they became increasingly transformed from saviors to oppressors. So even if Jesus' followers had expected him to start a military revolution to free Judea, he knew that this was not the right way. In the end it would accomplish nothing. It had been done before and had failed. He knew that only God could bring forth the real, desired change. "My kingdom is not of this world," he said.

THE APOCALYPTIC MESSIAH

Christian scholars have written extensively about Jesus' kingdom of God, and in a later chapter I will add my own thoughts. But for the Judeans, the final kingdom was one that would be ushered in through the intervention of God alone; only in the Book of Zechariah does man play a small part, and Jesus knew that book well.

Just as all prophets complained about the injustices they saw around them—how the rich and powerful prospered while the weak and poor remained downtrodden—so they almost all called for a Day of Judgment and punishment. On that day God would condemn both the arrogant individuals who had abused the poor, and the wicked countries that had oppressed Judah. This event was often preceded or accompanied by the arrival of the Messiah,[10] and nowhere is it more evident than in Zechariah. It is as if this book provided Jesus with a blueprint for his days in Jerusalem. We already saw how he entered the city riding a donkey, "so that what had been spoken through the prophets might be fulfilled" (Mat. 21:4 referring to Zech. 9:9). And we have also discussed that he then proceeded to the Temple to chase away the merchants, fulfilling the final verses in Zechariah: "On that day there shall no longer be the Canaanite [merchant] in the house of the Lord of Hosts" (Zech. 14:21).

Is it possible that Jesus could have missed the most important section between these verses, the actual participation of God in the coming of the final kingdom? We will discuss this in another chapter. But here is a hint for those who can't wait: look up Zechariah 14:4.

MESSIAH, THE SUFFERING SERVANT

The early Christians were at a loss to explain the ignominious ending of Jesus' mission. A Messiah who had failed so miserably could not have been the real Messiah. His humiliating death on the cross rendered him a criminal to the Gentiles and an accursed person to the Jews (Deut. 21:23). If he had indeed been a messenger of God, then God must not have thought much about the importance of the message. So as they always did when they needed explanations, the early Christians examined the old writings until they found a perfect answer in Isaiah, written during the days of the Babylonian exile.[11]

> He bears our sins, and is pained for us: yet we accounted him to be in trouble, and in suffering, and in affliction. But he was wounded on account of our sins, and was bruised because of our iniquities: the chastisement of our peace was upon him; and by his bruises we are healed. All we as sheep have gone astray; every one has gone astray in his way; and the Lord gave him up for our sins.
>
> And he, because of his affliction, opens not his mouth: he was led as a sheep to the slaughter, and as a lamb before the shearer is dumb, so he opens not his mouth.... The Lord also is pleased to take away from the travail of his soul, to show him light, and to form him with understanding; to justify the just one who serves many well; and he shall bear their sins. Therefore he shall inherit many, and he shall divide the spoils of the mighty; because his soul was delivered to death: and he was numbered among the transgressors; and he bore the sins of many, and was delivered because of their iniquities. (Isaiah 53:4-7,11,12)[12]

Notice, however, that the psalm has been written in the present and past tenses. Whoever this suffering servant was, he had lived in days gone by. Just as with many other predictions that the Christians claimed they had discovered in the Old Testament, this passage was not describing the future. "The Jews themselves did not anticipate a suffering Messiah; they usually understood the Servant Song in Isaiah 52:13-53:12 to signify their own suffering as a people".[13]

Nevertheless, this gave rise to the Christian picture of Jesus: a meek lamb, a willing, uncomplaining sacrifice for the sins of humanity. Just as the Hebrews transferred all their transgressions to the scapegoat before leading it away to its death into the desert,[14] in the same manner the Christians transferred all their sins to the sacrificed Jesus. But of course Jesus was destined to rise from the dead and to be elevated and justified in his Father's kingdom. With Paul's help, this became the early Christianity's attraction; we could all emulate him: "The Spirit himself testifies together with our spirit, that we are children of God. And if we are [His] children, then we are [His] heirs also: heirs of God and fellow heirs with Christ; *only we must share His suffering if we are to share His glory*" (Rom. 8:16-17 AMP, emphasis added). And so tens of thousands of Christians marched happily into the Roman arenas to be torn apart by wild beasts in the serene knowledge that if they suffered like Christ, they would be glorified like Christ.

Needless to say, this representation of the Messiah does not correspond to any Jewish expectations.[15] The Messiah would be successful, but Jesus was not: the Romans remained more oppressive than ever; the Temple was not rebuilt, but rather completely destroyed; instead of the Jews returning from the Diaspora to Jerusalem, the Judean inhabitants of the city were expelled and forbidden to ever enter it again. The elusive "Kingdom of God" remained exactly that: elusive.

CONCLUSION

Was Jesus the expected Messiah? Although he may have tried to become a king, he had not been successful. If he did try to liberate the Judeans from the Romans, he was unsuccessful. He was obviously not the apocalyptic Messiah of Daniel, since the world did not come to an end. This leaves us with the suffering servant Messiah, an impotent image that does not appear to accomplish anything except, according to Isaiah, absolve us, somehow, from our sins.

The early Christians tried to solve the problem of the non-Messiah-like Jesus by proclaiming that he would return again in glory to judge the living and the dead. This is a brand-new definition for the word "Messiah." Up to that time the Messiah was expected to come once, at the end of the world, as we know it. Jesus came, but the world did not change noticeably. He may have been the Son of God, and we will discuss this in another chapter, but unless one completely redefines the meaning of "Messiah," one must admit that Jesus was not he.

SOME QUESTIONS TO CONSIDER

1. There is a great disparity in Jesus' actions during his last week of life. From the time of his triumphal entry in Jerusalem on Palm Sunday to the Last Supper on Thursday, he was assertive and self-assured. But from the moment of his arrest he became passive and quiet. Can you offer an explanation? (I will present one unique opinion in a later chapter.)

2. We are told that Jesus will be mankind's judge at the end of the world (Matthew 25:31-46). Yet we saw (in endnote #1) that he assigned part of the job to Peter (Matthew 16:19). In John's gospel he assigns it to the ten disciples—the original twelve, missing Thomas and Judas (John 20:23). In later years, the Church assumed it had the required authority. Why so much confusion?

3. Christians maintain that Old Testament writings predicted the coming of Jesus (like Nostradamus had predicted the coming of Hitler). Do you agree? Explain.

4. Can you oppose the author, and argue that Jesus was indeed the Messiah, and if so which one?

SELECTED REFERENCES

Collins, John J. *The Scepter and the Star: The Messiahs of the Dead Sea Scrolls and Other Ancient Literature.*

Horsley, Richard A., with John S. Hanson. *Bandits, Prophets, and Messiahs.*

CHAPTER 11

THE TIME OF JESUS

Jesus' world was probably not that different from what we find today in the third-world Middle Eastern countries. The gulf between the large number of powerless poor and the very few powerful rich remains as extreme as ever, except that instead of the rich exploiting the poor as in the old days, now they merely withhold from them the national oil wealth. Women's oppression has remained essentially unchanged. Replace the Hebrew priests with the Muslim Imams, and the religious situation continues unaltered. Regarding their emotional perspective, simply substitute the hatred of the United States for that of the Roman Empire. Perhaps the only important difference is population growth. Today's improved health conditions and less devastating wars, coupled with the area's explosive birth rate, have resulted in populations that double every two or three decades making it difficult to visualize any favorable long-term social improvement in the area.

An Agrarian Economy

Nowadays, seven percent of the U.S. population can grow enough food to support itself and all its fellow-citizens, and also generate large surpluses for export. It was not always so. Before the Industrial Revolution, ninety percent of the population was needed for this task. And this was also true during Jesus' days in Palestine, and in the entire Roman Empire. Not only could the farmers barely produce what they needed, but they also had to pay huge taxes. The tax burden imposed on the agricultural community, including religious offerings, probably amounted to at least forty percent of their production.[1] This means that the farmers consumed sixty percent of their produce and the non-productive ten percent of the community consumed the other forty percent. And remember that since every seventh year the farms in Palestine had to remain fallow, people needed to also save resources for those times.

The non-productive group consisted of soldiers, administrators, priests, and the elite aristocratic landowners who numbered about five percent of the entire population. These last kept the agrarian population under control by siphoning off all excess production, preventing the people from saving and thus keeping them continually at the poverty level. And since government taxes had to be paid even in years of drought, farmers often were forced to borrow money to pay their taxes, putting up their land as security. Eventually many forfeited their land to the rich, and ended as hired hands in their own fields.[2] During the sixty-year period before Christ, the number of small landowners decreased by half. This trend accelerated even more in the years after the Jewish revolts, when the land captured by the Roman legions became property of the conqueror (conquered by the sword).

From a governance point of view, people in the ancient world were divided into three classes. At the top was the governing class which made all the decisions: in the Roman world it was first the Senate, and later the emperor; in Judea it was the high priest and, when existing, the king or ethnarch. Next came the retainer class: the managers, the administrators, and the army. All others, including farmers, artisans, and traders, were in the lower class.

It is obvious that a large disparity existed between the lifestyles of the rich and of the poor. The wealthy lived in large villas with many slaves and servants, whereas the poor lived in one- or at most two-room houses, or apartments if in the city. It is interesting to note that although Jews valued manual work, the truly elite despised it.[3] Roman senators, for instance, were not allowed to work. Even trading, despite the occasional high profits that it provided, was thought to be demeaning. According to Roman philosophers, the optimum state was a completely idle life that allowed freedom to think.

THE PATRONAGE SYSTEM

Once when I was starting a new restaurant in downtown Chicago, I called the city administration to request a "No Parking—Loading Zone" space in front of the entrance. "This is a complicated process," I was told. "It requires, among other things, a count of the number of cars that unload passengers in front of your place over a couple of weeks. Why don't you call your alderman instead?" I did so, and in a week or so I had my "no parking" sign. This is an example of the patron-client system. I was the client who needed something, and City Hall was the patron who could fulfill my need. The alderman was the broker, who spoke to the patron on my account. I, the client, received what I needed, and the broker and patron received my thanks, and hopefully my votes in the next election. This is the way things get done in Chicago, and this is how they got done in the Roman Empire.

A patron-client relationship exists when one person, the patron, has the power to do things for other people, the clients. The top patron in Jesus' days was the Roman emperor. He ran the country through his extended family of relations and freed slaves, he was the paterfamilias. The kings and appointed governors of the assorted provinces were his clients. Each king in turn was a patron to his subordinates, and so on. The way to get something done was to ask one's patron for help. And if the patron were too far above his own level, then one would work through an intermediary, the broker. What the patron got out of this relation was loyalty, honor, and praise. It has been suggested that exactly such a patron-broker-client relationship lies behind Jesus' words, "Whatever you shall ask the Father in my name, he will give you" (John 16:23).[4] We are the clients, appealing to the Father, our patron, through Jesus, the broker.

A MAN'S WORLD

Hand-in-hand with the patron-client system went the system of peoples' values. Tell me what you fear losing most, and I will tell you who your God is, is an old saying, going all the way back to Paul. In today's western civilization, what we fear losing most is control—control over our life, of the freedom to do whatever we want to do, whenever we want to do it. Generally this translates into money, but also to lack of infirmities, and for the old, staying away from the nursing home. It also results in later-age marriages and fewer children, something that is becoming increasingly evident now in the newly westernized countries, like Russia and Brazil.

This, however, is not the case in small agrarian societies where people deal with each other face to face. In the Middle East, for instance, today as in the days of Jesus, the most valued thing is honor and the thing to avoid at all cost is its opposite, shame. This is why in Jesus' days, sealing an agreement with a handshake was preferable to signing legal contracts. Rich people competed to build and donate the most ornate temples, monuments, etc., whose cost often exceeded their available funds. King Herod the Great, for instance, was known and honored for the many structures he built in Judea and in foreign countries, while cruelly taxing his own subjects. He even built an entire city, Caesarea, in honor of Caesar because it brought him honor.

The weakest link regarding family honor was the chastity of the women in the household, particularly of the young daughters. Even today in this country, we occasionally read about a Middle Eastern immigrant killing his daughter to cleanse the imagined shame resulting from a sexual indiscretion or even an unapproved marriage.[5] The solution in the old days was to keep one's daughter hidden from the opposite sex and to give her away in betrothal before she became twelve and a half years old, while she still had no right to object.[6] When she was actually married a year later, she was packed off to live with her husband's family.

Religion

The first amendment to the U.S. Constitution, which separates church and state, did not exist in those days. In fact the reverse was true, church and state were inseparable. The Roman emperor was also *pontifex maximus*, chief priest, and offered sacrifices to the gods. Before starting a foreign war, the Romans would sacrifice to their gods for permission to proceed (something that was not necessary, however, for fighting against local insurgencies and revolts). Although Rome had its own local gods, Romans respected all gods, even those of the nations they conquered. Since it was believed that insulting a god would bring down his wrath on the state, everybody was required to show proper respect and offer sacrifices to them. Only the Jews, being an old people who worshipped an old strange god, were excluded from performing obeisance to Rome's gods, something that at first also protected the early Christians. It was only after the first Jewish revolt in 67 CE that Christians were differentiated from Jews, and since they were then considered to be a new, young sect, Christians lost their old Jewish privileges.

The situation with the Hebrew nation was somewhat different. At first the state was a pure theocracy, with all the power in the hands of the high priest. Then from King Saul on, power was split between the king and the high priest. After the Babylonian conquest, except for the few relatively independent Hasmonean years when the king was also the high priest, the Hebrews remained under the political control of foreigners. In general they were allowed to live as they saw fit, provided that the levied taxes were collected, and that peace was kept, for which the high priest, aided by the elders, was held responsible. So Judea again became a theocracy, although one with limited power. Only Herod the Great (37-4 BCE) gained the complete trust of the Romans and reigned in Judea as he pleased. Unfortunately, he was even more repressive than the later Roman prefects and procurators.

Judea's high priests governed with the help of a senate-like group called the Sanhedrin. Although the three synoptic gospels tell of Jesus being brought before such a group for judgment, we don't know much about these organizations. The great Sanhedrin in Jerusalem consisted of seventy-one people headed by the high priest, for a total of seventy-two.

PHARISEES, SADDUCEES, AND SCRIBES

Throughout the gospels Jesus constantly criticizes the actions of Pharisees, who, in turn, together with scribes and Sadducees, challenge him with questions designed to embarrass him. Who were these persons, and why was there so much antagonism between them and Jesus?

The Pharisees and the Sadducees were members of political parties (or associations, as they were sometimes called in those days). They came into existence during the period of the Maccabees, and remained in constant opposition to each other through the next two and a half centuries. The Sadducees belonged to the governing class, and recruited members from among the families of the high priests and the aristocratic landowners of Judea. Not all aristocrats were Sadducees, but all the Sadducees were aristocrats. They believed in the five books of the written Torah, and in nothing else. They did not accept any of the later writings, or the validity of oral tradition and interpretations by others. Specifically, they did not believe in the survival of the soul after death. We can assume that since the Sadducees were completely satisfied with their own interpretations of the Scriptures, they would not have been interested in Jesus' teachings; they would have regarded him, however, as a rabble-rouser who threatened to disturb the peace. And because the Romans apparently held the Sadducees, together with the high priest, responsible for keeping peace in Jerusalem, they reacted strongly to Jesus' almost adversarial entrance into the city, and to the disturbances he caused afterwards.[7] The power of the Sadducees was centered in the Temple, and when it was destroyed in 70 CE, they disappeared from the Judean scene.

Most Pharisees were probably members of the retainer class, who specialized in studying, practicing, and sometimes teaching the Law. The people had great esteem for the Pharisees. So much so, that although the Sadducees had greater power, they had to perform the Temple rituals according to Pharisean instructions, and not as they themselves wished. So when Jesus started preaching in the villages, he encroached upon the jurisdiction of Pharisees by advocating a conflicting message. Only by winning the resulting verbal confrontations could Jesus gain (at their expense) the necessary honor required to win the people over to his beliefs. Although the gospels present them as constant opponents of Jesus, in reality it was Jesus who ignited all the confrontations.

Despite the gospels' presentation of Pharisees in an unfavorable light, they were not really bad. They believed in the

Law and tried to live their personal lives in accordance with it, but (just like today) they felt that the law was insufficiently clear and had to be interpreted. For instance, it was forbidden to work on the Sabbath. But what constitutes work? Hence the famous arguments in the Bible on whether it was permitted on the Sabbath to save animals that had fallen in the well and for Jesus to cure the blind on that day. In contrast to the Sadducees, the Pharisees survived the Temple destruction and evolved into the rabbis of the new Judaism, a religion that does not involve sacrifices or the Temple.

The scribes were not members of a particular political party, as were the Sadducees and Pharisees, but people in a specific job category. They were learned people whose job was to teach and explain the Law. It is understandable therefore, that they would object to an unknown, untrained upstart talking about things in their area of expertise. Although they occasionally tried to ask Jesus embarrassing questions, they generally neither entered into discussions with him, nor met him in social situations the way Pharisees did. Scribes often worked for the governing group, and were thus members of the retainer class. In all four gospels they were members of the Sanhedrin, and followers of the high priest.[8]

SOME QUESTIONS FOR REFLECTION

1. From what you know, has there been much change in the attitude of men towards women in the Middle Eastern countries?

2. Considering the way women were treated, how can you explain biblical stories of strong women like Judith and Esther?

3. Although the Sadducees had more power than the Pharisees, they followed the Pharisees' Temple ceremony instructions because they were popular with the people. Doesn't this justify Jesus' mixing with the common people rather than the powerful?

4. If Jesus was just a broker to God, his father the patron, then he was at a lower level, not one with his father as he keeps maintaining in John's gospel.

Selected References

Ferguson, Everett. *Backgrounds of Early Christianity.*

Hanson, K. C., and Douglas E. Oakman. *Palestine in the Time of Jesus.*

Ilan, Tal. *Jewish Women in Greco-Roman Palestine.*

Jeremias, Joachim. *Jerusalem in the Time of Jesus.*

Saldarini, Anthony J. *Pharisees, Scribes, and Sadducees in Palestinian Society.*

Sanders. E.P. *Judaism: Practice and belief 63BCE-66CE.*

CHAPTER 12

THE TEACHINGS OF JESUS

If we ignore John's gospel where Jesus seems to ramble on interminably, mostly about himself, Jesus' ideas can be summarized in a few sentences: God loves us as a father; we should love everybody as much as we love ourselves; thinking is the same as doing; and with faith and prayer we can accomplish anything. Almost everything else in the first three gospels follows from these few thoughts.

Our Heavenly Father

> "Pray, therefore, like this," said Jesus to his disciples. "Our Father who is in heaven, hallowed be your name...." (Mat. 6:9 AMP).

Perhaps the unique idea in Christianity is the relationship between God and us.[1] In no other religion in the world is God depicted as humankind's heavenly parent. I proposed in an earlier chapter that there are only three things that we could possibly prove about God: that he is interested in us, that he is aware of our entreaties, and that he can help us. Jesus adds that not only is God interested in us, but that he loves us as a parent would. Now this is a concept that we can all understand. True, there are parents who mistreat or even kill their children, but we all agree there is something psychologically wrong with them. Looking at the world around us, that of animals as well as of humans, we recognize that parents love and care for their offspring, and are willing to endanger their lives to protect them. Jesus tells us that this is the way God also feels towards us.

If he is right, it is an incredible empowering experience. We know we have someone to turn to, someone we can trust. Of course parents can't do everything. "Please make the pain go away," pleads the child, but the best the parent can do sometimes is to give an aspirin. I remember once when I was eight, I came down with what turned out to be a lengthy illness. We were out of town, so an unknown doctor came to take some blood out of my arm. He couldn't find the vein and the more he tried the more I screamed, while my father was fruitlessly trying to divert my attention with some funny stories about his father. Finally, although usually a mild person, he lost patience and threw the doctor bodily out of the house. Strangely, I can still remember the stories. And in all fairness to this unknown, and long-dead doctor, nobody has ever successfully tapped the vein in that arm.

If God is like a parent to us, we can expect that he would treat us at least as well as we treat our children.

> Which one of you fathers would give your hungry child a snake if the child asked for a fish? Which one of you would give your child a scorpion if the child asked for an egg? As bad as you are, you still know how to give good gifts to your children. But your heavenly Father is even more ready to give the Holy Spirit to anyone who asks. (Luke 11:11-13 CEV)

What is even more important, we can assume that his punishments will be neither more serious than merited by the offense, nor everlasting. Just as a loving parent will forgive his children any offense, so we can expect that God also will. Assigning us to an everlasting punishment in hell for offenses committed during a finite life on earth does not meet any human love criterion, let alone a divine one. So what do we do with the judgment passages such as the ones we find in Matthew?

> When the Son of man will come in his glory, and all the holy angels with him, then he will sit upon his throne of glory. And all the nations will be gathered

> before him; and he will separate them one from another, as a shepherd divides his sheep from the goats. And he will place the sheep on his right hand, but the goats on the left.... Then he will also say to those on the left hand, "Depart from me, you accursed, into everlasting fire, prepared for the devil and his angels. For I was hungry, and you gave me no food. I was thirsty, and you gave me no drink.... And these will go off to everlasting punishment; but the righteous to eternal life."[2] (Mat. 25:31-33,41,42,46)

Or

> For the Son of man shall come in the glory of his Father, with his angels; and then he shall reward everyone according to his work. (Mat. 16:27)

Such statements belong in the Old Testament, to be spoken by "God the Just," not in the New Testament where "God the Merciful" reigns. Remember also that in Daniel, the Son of man received glory but did not judge. We will look at Jesus' apocalyptic judgment pronouncements in a later chapter. For now, however, I will point out that the obvious conflict can be resolved if we separate what we believe that Jesus may have said from what Matthew may have added. Just as the Old Testament was written by people who had agendas, so also was the New Testament; as time passed, the Evangelists' words were edited again and again, changing the original meaning. We will discuss this at a later time. For now, however, I will reiterate that a merciful God would not dispense disproportionately harsh punishments. So we must read the Bible with our minds open, trying to keep Jesus' true message separate from the Evangelists' writings. Unless of course Jesus was wrong and God does have a dark side.

LOVE THY NEIGHBOR

"Love thy neighbor as yourself," said Jesus. And when he was asked, "Who is my neighbor?" he told the parable of the Good Samaritan. It turns out that everybody is my neighbor: my friends, my enemies, those who live close, those who live far. Of all of Jesus' teachings, this is perhaps the hardest to follow. I am very proud that sometimes I can objectively look at my adversaries' point of view, but love them? That is a long way off. There was only one time, to my knowledge, that people tried to implement this idea. It was back in the days of the Vietnam War and the hippies, when the "flower children" would approach National Guard soldiers who were trying to keep order and offer them flowers. But these were our own people, not an enemy. Even so, at the time it attracted a great deal of attention.

Today, however, when the world has shrunk so much that every country, anywhere, can bid for the manufacture of the toaster on your counter, the word has acquired a real meaning. It used to be that Americans, protected by their unions, could earn fifteen or twenty dollars an hour sweeping the floors in their factories. But now most of these products are made in Mexico, Thailand, or China, where workers are happy to be paid one dollar an hour for their work, and skip health benefits. As a result there are increasingly fewer factories here, and increasingly fewer jobs. And as far as sweeping the floor is concerned, an immigrant, legal or otherwise, will be happy to do it for five dollars an hour. Which brings us to the conclusion that until all our neighbors, everywhere in the world, are as well off as we are, our overall style of living will keep dropping down to theirs.[3] So if we want to improve our lives we should try to improve theirs.

THINKING IS DOING

Jesus preached that thinking about doing evil is equally evil as actually doing it. Jesus must have been a great psychologist. He realized that it is not actions that destroy us as much as thoughts.[4] When I am angry with my brother, it doesn't matter so much if I go punch him or not. What is important is the anger that seethes in me, affecting my thoughts, my metabolism, my entire limbic and physical system. My anger is actually destroying me, while the object of my anger lives completely unaware. There are cases where parents are so angry with their grown-up children that they refuse to visit them or their grandchildren. They live their last remaining years in a fog of anger, bitterness, and loneliness. Their children are probably also hurt, but not to the same extent; after all they still have *their* children to love.

There is also the metaphysical possibility that our thoughts can affect others. My mother used to say that you when you walk in a house you can always tell if there is love in it; you feel it in the air, she would say. Do you remember the days when you were madly in love with someone? How you felt? You loved everybody! You were happy, and you made the people around you happy. On the other hand, if we believe that prayers can help others, the converse must also be true.

What Is in the Heart Is What Counts

This is the mirror image of what we said above. If a bad thought is the same as a bad action, then a good thought is just as important as a good action. Jesus always accused the Pharisees of doing the right things but without the right intentions. "Your hearts are hard," he would say. It is certainly easier to give money to the poor than to love them. I remember talking once to a black deaconess who ran a soup kitchen in the ghetto. "The hardest thing to do," she said, "is to have a dirty, smelly drunk come in for lunch, and to put your arms around him and to honestly say that you love him."

Merely observing rules cannot bridge the huge gap between the righteousness of God and the sinfulness of man. If my heart is not open to welcome God, my fulfilling his commands will not make any difference. After all, as Jesus said, the poor will always be with us. How we deal with them, doesn't count as much as how we feel towards them.

Jesus did not want people to just blindly follow the established religious or social rules. He wanted them instead to look into their hearts and decide what was the right thing. He told his disciples:

> When you go into any land and walk about in the districts, if they receive you, eat what they will set before you, and heal the sick among them. For what goes into your mouth will not defile you, but that which issued from your mouth—it is that which will defile you.[5] (The Gospel of Thomas 14)

FAITH CAN MOVE MOUNTAINS

We have already discussed the power of prayer. "Ask and it shall be given to you; seek and you shall find" (Matt 7:7). Jesus went even further, saying: "He who believes in me will do the works that I do, and he shall do greater ones than these" (John 14:12).

Don't worry about food and clothes and all the other things you need, because the Father will provide, said Jesus. But notice that he did not say to goof off and skip work because the Father will send us a care package. What he said was do your share and stop fretting. In any case there is little else to do since all the planning and worry in the world cannot insure our future. He discussed this in the story of the man who gathered in a large harvest and went to bed thinking that he was now all set for the rest of his life. This turned out to be correct because he died the very same night.

And here I feel that I have to add a personal story. My parents divorced when I was young, but they kept in touch throughout the years, visiting during holidays, sharing their children and grandchildren. But they were poles apart. My father, a hard worker believed in saving and planning for the future. My mother, equally hard working, believed in giving away the money to those in need, believing that when she needed it, it would come. "I don't need to plan for my old years, my children will take care of me," she would say. I remember one time when they were both at my house for Thanksgiving dinner, she turned to my father and asked him for two hundred dollars. Without asking why she needed the money, he reached into his wallet and gave it to her, whereupon he was immediately rewarded with a finger-shaking lecture: "See what I tell you John? You don't have to save money! When you need it, it will come!" Having distributed all her belongings, my mother did indeed die money-less, and her children did take care of her. Unfortunately so did my father. Despite all his remarkable planning, as he grew old he started losing his faculties and made a number of bad business decisions, nullifying all his carefully laid plans.

SAYINGS AND PROVERBS

The following are some short sayings that have been attributed to Jesus:

> Blessed are the [physically] poor and hungry.
>
> It is hard for a rich man to go to heaven.
>
> God has revealed things to the ignorant and hidden them from the wise.
>
> Respond meekly towards aggression.
>
> He who wants to lead must serve.
>
> He who promotes himself, demotes himself.
>
> The Sabbath was created for man, not man for the Sabbath.
>
> What goes out of the mouth pollutes, not what goes in.
>
> Give to everybody who begs from you.[6]
>
> To those who have, more will be given, and from those who have not, even what little they seem to have will be taken away.
>
> There is more to life than food and clothing.
>
> Whoever tries to hang onto life will forfeit it, but whoever forfeits it will save it.

SOME QUESTIONS TO CONSIDER

1. Would you look at the Old Testament's God as a loving father of mankind? If not, how did he change?
2. Is it realistic to expect that some of Jesus' teachings (like loving one's enemies) can be implemented in the real world?

3. Does one give money to the poor in order to help them, or in order to make himself feel virtuous?

4. Jesus said, "The poor you will always have with you." Does this mean that he thought it was an endemic problem of society and beyond human help?

5. Do you find Jesus' statements that are listed at the end of this chapter to be self-evident, or do you think that some require clarification?

SELECTED REFERENCES

Johnson, Luke Timothy. *The Writings of the New Testament: An Interpretation*.

Stein, Robert H. *The Method and Message of Jesus' Teachings*.

CHAPTER 13

PARABLES

In the first three gospels, Jesus peppers his direct teachings with parables, stories that make a point and can be easily remembered. They mostly tell how one should live one's life and prepare for the sudden arrival of the coming reign of God. Although the evangelists have occasionally supplied explanations, it has been usually left up to the listener to derive the intended moral. This, of course, makes them excellent material for preachers' sermons.

Teaching with Parables

If one picture is worth a thousand words, then a good story with a message is worth a hundred pages of admonishments.[1] Not only does a story make the meaning clearer, but it is also easier to remember. Much of Jesus' teaching was done through the use of parables, short narratives that used familiar stories to clarify unfamiliar ideas. Characteristically, they contained hyperboles and ended in a completely unanticipated manner designed to reverse a listener's usual expectations. Although at first there were attempts to explain some of the parables in allegorical ways that incorporated Christianity's beliefs, this has recently gone out of favor. In the rest of this section I will try to place these parables in categories, and give the accepted or, occasionally, my own interpretation of their meaning.

A. RULES TO LIVE BY

How should we act in our everyday lives? Be sensitive to the needs of your fellow-person, don't be afraid to take chances, and use the blessings that God has bestowed upon you, says Jesus.

(1) *The Unforgiving Servant* (Matthew 18:23-35)—The king forgives the huge debt of his servant, but later throws him in jail when the servant forcefully tries to collect a much smaller debt owed to him by another servant.

[Be merciful to your fellow person, as God is merciful.]

(2) *The Good Samaritan* (Luke 10:30-37)—A Samaritan takes care of a wounded man whose needs had been ignored earlier by both a priest and a Levite.

[Help your neighbor; he is anyone who needs help.][2]

(3) *The Rich Man and Lazarus* (Luke 16:19-31)—What happens after death to a rich man and the neglected pauper at his front doorstep.

[Be aware of the needy around you.]

(4) *The Talents* (Matthew 25:14-30)—We are exhorted to take risks and use whatever assets God has given us.

[Use your God-given gifts; take chances.]

(5) *The Ten Gold Coins* (Luke 19:12-27)—Essentially similar to the story of the Talents.[3]

[Same meaning as above.]

B. PRAYER

How should we pray? Pray intently, pray insistently, pray expecting to be heard, says Jesus.

(1) *The Friend at Midnight* (Luke 11:5-8)—Someone pounds on the door of a friend at midnight trying to borrow

some flour, and does not quit until his friend opens the door.

[Be insistent in your prayers, so that you will be heard.]

(2) *The Persistent Widow* (Luke 18:2-8)—An unjust judge refuses to grant a poor widow her rights, but she continuous to pester him until he finally gives in.

[Same meaning as above.]

(3) *The Father's Good Gifts* (Matthew 7:9-11; Luke 11:11-13)—Fathers give their children good gifts, and even more so does our Father in heaven.

[Just as we are good parents to our children, so is God to us.]

(4) *The Pharisee and the Tax Collector* (Luke 18:10-14)—In their prayers, the Pharisee brags about his sinless ways while the tax collector is beating his breast in remorse.

[Be humble and penitent when asking forgiveness.]

C. God Loves Us All

Just as we cherish our loved ones and our special possessions, so God also loves us.

(1) *The Lost Sheep* (Matthew 18:12-14; Luke 15:4-7; Thomas 107)—A shepherd abandons the ninety-nine sheep in his flock to search for the one that got lost.

[Every single one of us is loved by God, especially those who are lost.]

(2) *The Lost Coin* (Luke 15:8-10)—A woman becomes filled with joy when, after a long search, she finds one of the ten (dowry) coins she has lost.

[Same meaning as above.]

(3) *The Prodigal Son* (Luke 15:11-32)—A farmer's younger son asks for and receives his inheritance. After frittering it away he returns home to ask forgiveness and to work as a servant; but before he even speaks, his father rushes to meet him and welcomes him with open arms; this angers the older son.

[Same meaning as above.]

(4) *The Two Debtors* (Luke 7:41-43)—The greater the debt that is forgiven, the more grateful is the debtor.

[Don't be stingy in your good deeds.]

D. Reactions to the Teaching

Some people listened to Jesus' teachings and acted upon them, and some did not.

(1) *The Sower and Its Interpretation* (Mark 4:3-20; Matthew 13:3-9,19-23; Luke 8:5-8,11-15; Thomas 9)—Jesus compares soil conditions to the attitude of his listeners.

[Hear the teaching, accept it, and keep it near your heart.]

(2) *Children in the Marketplace* (Matthew 11:16-19; Luke 7:31-35)—The people were never satisfied; first they said that John the Baptist was possessed by a demon because he was a hermit, and later they complained that Jesus was a glutton and a drunkard because he was sociable.

[People are never satisfied. Both John's and Jesus' teachings were ignored. Personally I don't consider this a parable, but most scholars list it as one.]

(3) *The Two Sons* (Matthew 21:28-32)—One son refuses to do his father's bidding, but changes his mind and does it; the other says that he will do it and does not.[4]

[Those who profess belief but do not follow through with action (like the Pharisees) are worse than the sinners who repent (like the prostitutes).]

E. Time of the Last Judgment

We don't know when we will die or when judgment time will come. Be always prepared, says Jesus.

(1) *The Faithful and the Unfaithful Servant* (Matthew 24:45-51; Luke 12:42-48)—After their master's long absence, the good servant continues to do as he had been directed, but the wicked one starts taking advantage of his fellow servants.

[The time of the last judgment is unknown.]

(2) *The Ten Maidens* (Matthew 25:1-13)—They fall asleep while waiting for the bridegroom and their oil lamps run out of oil; five had spare oil with them; the other five did not and go to town to get some; by the time they return, the hall doors had been locked.[5]

[Be always prepared for the judgment day; keep your affairs and relationships in order.]

(3) *The Vigilant and Faithful Servants* (Mark 13:34-37; Luke 12:35-40)—The master goes away and expects to find his servants awake and working when he returns.

[same meaning as above.]

(4) *The Rich Fool* (Luke 12:16-21; Thomas 63)—A farmer collects a huge harvest and contemplates his retirement in comfort, but dies on the same night.

[Same meaning as above.]

F. Last Judgment

Despite the soft, warm feeling that we associate with Jesus, many of his teachings were full of fire and brimstone. We all have the opportunity to be saved, he said, but if we don't take advantage of it, then we will be lost.

(1) *The Workers in the Vineyard* (Matthew 20:1-16)—Field laborers who worked one hour get paid the same as those who worked the entire day.

[We will be judged not according to our works, but through God's mercy.]

(2) *The Two Foundations* (Matthew 7:24-27; Luke 6:47-49)—A house with a foundation built on rock withstands the elements, but the one built on sand gets carried away by the floods.

[Stay on the righteous path even if at times it seems unnecessary.]

(3) *The Barren Fig Tree* (Luke 13:6-9)—The owner of a fig tree that has not given fruit for three years wants to cut it down, but the gardener convinces him to let it live for one more year.[6]

[It is never too late to mend our ways, but we are running out of time.]

(4) *Attitude of a Servant* (Luke 17:7-10)—The master does not have to be grateful towards his servants just because they did what they were told to do.

[Doing justice is our duty not our merit.]

(5) *The Weeds among the Wheat* (Matthew 13:24-30; Thomas 57)—A farmer tells his servants to wait until the wheat is grown before they pull out the weeds mixed in it.

[The bad and the good live side by side until the last judgment. Also do not attempt to judge your fellow-person.]

(6) *The Judgment of the Nations* (Matthew 25:31-46)—The Son of man rewards those who helped their fellow men and sends to perdition those who did not.

[Helping your neighbor is the salvation criterion. Note that this is in direct opposition to F1.]

(7) *The Tenants* (Mark 12:1-11; Matthew 21:33-44; Luke 20:9-19; Thomas 65-66)—A landowner sends his servants to collect some of the produce from his leased vineyard, but they are beaten up by the tenants; he then sends "his beloved son" hoping that they will respect him, but instead they kill him, whereupon he will come in person, kill the tenants, and lease the vineyard to others.

[Meaning to be discussed in the next section.]

G. WHAT THE KINGDOM IS LIKE

Even if we can't see it coming, God's Kingdom is taking shape right now beyond our horizon, says Jesus, and it is more important than any other thing in this world. Unfortunately, he does not tell us anything else about this kingdom, so Hultgren (2000, 384) believes that people in that era must have been acquainted with the term and concludes that "God's kingdom" must have been a well-known Jewish concept.

(1) *The Seed Grows of Itself* (Mark 4:26-29; Thomas 57)—After a seed is placed in the ground, it grows by itself until it is ready for harvest.

[The Kingdom of God will grow by itself without our efforts.

(2) *The Mustard Seed* (Mark 4:30-32; Matthew 13:31-32; Luke 13:18-19; Thomas 20)—Although the mustard seed is small, it grows into a large plant.

[The small, early actions will eventually yield large results.]

(3) *The Parable of the Yeast* (Matthew 13:33; Luke 13:20-21; Thomas 96)—A small amount of yeast leavens much flour.

[Same meaning as above.]

(4) *The Treasure in the Field* (Matthew 13:44; Thomas 109)—Someone sells all he has to buy a field with a hidden treasure.

[The Kingdom of God is worth everything we have in this life.]

(5) *The Pearl of Great Price* (Matthew 13:45-46; Thomas 76)—A merchant sells everything he has to buy a very valuable pearl.

[Same meaning as above.]

(6) *The Drag-net* (Matthew 13:47-50)—The fishermen pull in the drag-net, and separate the good from the bad fish, which they throw away.

[The kingdom of God is like the last judgment.][7]

(7) *The Great Feast* (Luke 14:16-24; Thomas 64)—A man invited people to his feast, but when it was ready the guests sent their regrets; so he had people brought in from the streets and the highways to be his guests.

[After the elect refused the Kingdom of God, everybody was invited.]

(8) *The Wedding Feast* (Matthew 22:1-14)—Similar to the previous story, with a king throwing the feast. However, here the king burns down the city of the guests who did not come and throws out a guest brought in from the street who was inappropriately dressed.

[Everybody is invited in the Kingdom of God, but the unworthy will be excluded.]

H. Socially Revolutionary Ideas

In chapter 10 we discussed that in Jesus' time there were many differing expectations of what a Messiah was. Christians believe that Jesus was the "suffering servant" Messiah, but what if Jesus considered himself to be the "liberator Messiah?" Such an idea may perhaps account for the following, otherwise inexplicable, parables:

(1) *Building a Tower* (Luke 14:28-30)—One does not lay a tower's foundation, unless one has the means to complete it.

[Do not start what you don't have the means to finish. But read more below.]

(2) *The King Goes to War* (Luke 14:31-32)—A king with only ten thousand soldiers facing an enemy of twenty thousand does not march in battle, but sues for peace.

[Numerical might wins. Do not underestimate the enemy. But read more below.]

(3) *The Dishonest Steward* (Luke 16:1-8)—A rich man tells his steward, who was accused of squandering his property, that he will fire him; to prepare for his employment loss, the steward builds goodwill among the people by arbitrarily reducing the amounts they owe his master, whereupon the master commends him on his prudence.

[Our obligations to our fellow men exceed those to our employer. Read more below.]

Some Hard-to-Explain Parables

There is something wrong with the explanation given to the last three parables. The first one (H1: Building a Tower) certainly sounds reasonable to us, but would Jesus have said it? He is the one who also said, "Do not worry about tomorrow, for tomorrow will have worries of its own" (Mat. 6:34 AMP). The second parable (H2: The King goes to War) sounds even more logical, but would it make sense to Hebrews? The Torah is clear that "When you go out to battle against your enemies and you see horses and chariots and an army greater than yours, do not be afraid of them; For the Lord your God...is with you" (Deut 20:1). In fact, the Hebrews were once punished for King David's sin of counting the men of military age in his kingdom (2 Samuel 24). I know a priest who believes that in this parable Jesus was just pulling the leg of his listeners. These parables are often explained as admonitions to consider the high cost of discipleship, but I don't think this explanation fits the stories, and in any case what moral would we learn?

The last parable in this list, that of the dishonest steward (H3), has foiled any acceptable explanation for two thousand years. Perhaps Calvin came the closest with his interpretation that "We must treat our neighbors humanely and kindly, so that when we come before God's judgment seat, we may receive the fruit of our liberality."[8] Our problem is that we are looking at the parable through our own set of prejudices. Even the name we have given the parable, the dishonest steward, reveals this: the steward had not been accused of dishonesty, but of "squandering" the property. Although he had been using money in ways that did not maximize the owner's property, he had not been stealing it for himself; if he had, he doubtlessly would have become rich, and would not have been so afraid of losing his job. The end of the parable, and the remarks that have been added to it by the redactors, may indicate the way in which the steward had been "squandering" the property: instead of concentrating in maximizing his master's money, he was humanely bending backwards to accommodate the plight of the debtors. Acting on behalf of his master he was essentially being charitable to his fellow-men. Hence the parable's conclusion, "You cannot serve both God and mammon." Ask yourself how often the financial interests of your company oppose the well-being of your fellow-citizens.

But what about the strange final commendation of his master? To explain it we must place ourselves in the society of that day. Seemingly acting on his master's behalf, the steward unilaterally reduced the people's indebtedness. This made his master appear great and magnanimous in the eyes of his debtors, a situation that in the "honor vs. shame" society of that day the master would be loath to reverse by disowning the completed arrangements.

We can understand these three parables better if we try to look at them through the eyes of Jesus. He was walking through a country whose inhabitants were being oppressed by the rich and powerful, and as an acclaimed Messiah he was expected to lead a revolution and free the country. So he told the people: Don't start something you can't finish; the numbers are not in your favor; for the time being, just try to help each other in whatever way your position permits.[9]

The parable of the Tenants (F7) also requires some comment.[10] Here we have a very clear allegory of God's relationship to Israel. He sends his prophets to remind them of what he is due, and they get mistreated. He sends his son, and he is killed. He will now take the promised kingdom away from the Jews and give it to the Gentiles. This is a concise summary of the Church's point of view. But did Jesus say this? The parable appears in all the gospels, attesting to the validity of the story, but notice the change in the tenses of the verbs. The servants *have been* mistreated, the son *has been* killed, but the tenants *will be* killed. The parable itself ends with the killing of the son. The retaliation against the workers is Jesus' prediction of what will follow; it is not a part of the story. What we have here is a narrative that came in existence *after* the death of Jesus, but *before* the destruction of Jerusalem, so it probably originated with the very early Christians.[11]

THE MYSTERY OF THE PARABLES

We usually say that Jesus used parables to make his points clearer and easier to remember, but this is not what he himself said. In the first three gospels the disciples ask Jesus why he speaks in parables.[12] He does it, he replies,

> So that seeing they may see and not perceive; and hearing they may hear and not understand; lest at any time they should be converted and their sins should be forgiven them. (Mark 4:12)

His answer is based on the Hebrew text of Isaiah where God sends Isaiah to talk to the people:

> Go and say to this people: "Hear ye indeed, but understand not; and see ye indeed, but perceive not. Make the heart of this people fat, and make their ears heavy, and shut their eyes; lest they see with their eyes, and hear with their ears, and understand with their heart, and be healed." (Isa. 6:9-10)

In Isaiah, the God of the Old Testament turns people's hearts sluggish, so that they will not be healed—just as earlier he had hardened Pharaoh's heart to stop him from letting the Jews go, so that by punishing him God would reveal his own great glory. This explanation, however, is not tenable when dealing with the God of the New Testament. We don't want to think of a God who deafens our ears so that we cannot hear the message of our salvation. Much has been written to explain away the apparent meaning of these verses. Unfortunately nothing explains away the words "lest" (in case) in the Markan text.[13]

Matthew softens the explanation. His Jesus says,

> Therefore I speak to them in parables, because seeing they see not. and hearing they hear not, neither do they understand. And in them is fulfilled the prophecy of Isaiah, who said, "By hearing you shall hear, and shall not understand; and seeing you shall see, and shall not perceive. For this people's heart is waxed gross, and their ears are dull of hearing, and their eyes they have closed; lest at any time they should see with their eyes, and hear with their ears." (Mat. 13:13-15)

As a result, we are prompted here to conclude that it is the people's own closed minds that stop them from listening and believing, and not a divine action. This is an acceptable argument since we all know that one can never win a theological or political argument through reason alone. Faith comes first, then logic. After all, Billy Graham and other great evangelist preachers do not attempt to address people's logic, but their emotion. "I believe! I believe!" is what the crowd shouts, not "I am convinced!"

The reason for the difference in the two gospels is interesting and tells us something about Matthew to which we had alluded in chapter 10. Matthew was more conversant in Greek than in Hebrew, and so he quoted Isaiah from the Greek Septuagint translation. Perhaps because of an error in translation, the Greek version differs from the original Hebrew text.[14]

In the gospels, Jesus explains the parables to his disciples in private, but ignores the others. It is generally agreed that the parables' explanations given in the gospels have been inserted by the evangelists, and were not part of Jesus' teachings.[15]

SOME QUESTIONS TO CONSIDER

1. Why do you think Jesus spoke in parables?
2. The CEV bible (see note 13) translated Jesus' explanation of why he spoke in parables in a manner that makes sense to us, although it does not correspond to the original text. Do you think this is an acceptable practice?
3. In the talents and coins parables, what do you think the master would have done if one of his servants had invested the money and lost it? Why was this situation not included in the parable?
4. The teachings of parables F1 and F6 contradict each other? Can you offer any explanations? Does parable F4 provide a way out?
5. What do you think of the author's argument that the last three parables discussed in the text might have been prompting the population to revolt against the injustices?
6. Which is your favorite parable, and why?
7. If Jesus' idea of God's kingdom was a Jewish concept, it must have referred to an earthly kingdom of normal people governed according to God's rules. This is far different from the Christian concept of a metaphysical life after death. What do you think?

Selected References

Hultgren, Arland J. *The Parables of Jesus: A Commentary.*

Jeremias, Joachim. *The Parables of Jesus.*

Young, Brad H. *The Parables: Jewish Tradition and Christian Interpretation.*

CHAPTER 14

THE REIGN OF GOD

Although Jesus taught many things, his favorite subject was the rapidly approaching reign of God. But since God had single-handedly created the whole world from scratch, why was he not reigning already? And if he was not, who was? It seems that in the last three centuries BCE, the Judeans had accepted the Persians' concept of God's duality, and Satan had been promoted from being God's prosecutor to being the ruler of the world. It was not clear why God had allowed it, but in any case, it was expected that God would soon be taking over again. And on the day this happened all troubles would disappear and the last would become first. Yet two thousand years later that day still has not come. But just as the Jews did not recognize the arrival of the Messiah, perhaps the Christians do not see God's reign for what it is. Instead of saying that the kingdom of God is coming, perhaps we should be saying that we are all going to it.

The Good News

Jesus sent out his disciples to preach the good news (gospel) about the kingdom of God. But what exactly is the kingdom of God? First of all, it is a misnomer: it should be called the reign of God, as the Greek text makes quite clear. There is a difference between the two: a kingdom is a place; you ask, "Where is the kingdom?" A reign is an event; you ask, "When is the reign?" The change in wording, however, does not explain the meaning of the word. To do this we must look back into history.

In the beginning God created the earth and everything on it. According to the Bible he was in complete charge. Once, when he became unhappy at the behavior of humans, he caused a great flood and drowned all but eight people on Earth. No one will question that in those days he was king of all. His reign was unchallenged.

Later, we see him as being mostly in charge of the Hebrew people. He enters into covenants with them, takes them out of their Egyptian captivity, leads them in their fights against their enemies. Now God is depicted as the king of the Hebrews. There are other gods around, but he is the greatest one; or perhaps he is the only one and the other nations believe in false, non-existent gods. Choose your own interpretation. In any case, he appears to be a king, personally reigning over his select group.

Still later, the Hebrews select their own king from among themselves. God is still their god, but not their leader any more. His reign over the Hebrews, if it still exists, is spotty; sometimes they listen to him, and sometimes they don't. To punish them, he occasionally allows, or even incites, neighboring nations to invade and conquer them. These other nations have their own gods, who to them appear stronger than the Hebrew god. Remember that in those days it was assumed that a nation's god fought its battles, so the winning nation had the stronger god on its side. When the king of Assyria laid siege to Hezekiah's Jerusalem, he told the Judeans:

> "Do not let Hezekiah persuade you to rely on Yahweh by saying: 'Yahweh is sure to save us; this city will not fall into the power of the king of Assyria.... Yahweh will save us.' Has any god of any nation saved his country from the power of the king of Assyria? Where are the gods of Hamath and Arpad?... Where are the gods of the land of Samaria?" (2 Kings 18:30,33,34 JER)

Jerusalem was indeed conquered, but not by the Assyrians. The Babylonians, a new power from the east, in today's Iraq, conquered first the Assyrians and then the Judeans and carted off the Judean upper class to Babylon.

ZOROASTRIAN RELIGION[1]

In Babylon, the Hebrew exiles came in contact with the Zoroastrian religion, and when the Persian king Darius released them to return home, they carried with them many Zoroastrian ideas.[2] The most important ones were the concept of duality of good and evil, the eventual battle between them, and the final judgment that presupposes some kind of life after death.

The supreme god was Ahura Mazda—self-created, omniscient, holy, invisible and beyond human conceptualization. He created various spirits personifying all the good properties such as light, truth, health, and purity. Opposing him was Ahriman, who originally lived in the pit of endless darkness.[3] As soon as Ahura Mazda started creating the good spirits, Ahriman rushed to the surface and created their evil opposites: darkness, falsehood, sickness, impurity, etc. Ahriman was stupid, ignorant, and blind, and he prompted human beings to perform evil deeds. He introduced to Earth noxious creatures such as snakes, scorpions, and lizards, and harmful vegetation such as weeds, thorns, and poison ivy. (Since it was unthinkable that the all-good Ahura Mazda would have created evil things, their existence had to be attributed to the evil Ahriman.)

The struggle between Ahura Mazda and Ahriman extends over the entire seen and unseen world. Light fights darkness; goodness fights evil. But the two opposing forces are evenly matched, and it is humans who will make the difference, so mankind is not a passive watcher of this struggle. Every human action, resulting from mankind's free will, helps tip the balance one way or another. The good actions advance the case of good; the bad actions do the reverse. But note that activities such as asceticism, celibacy and fasting, play no role in Zoroastrian teachings.

When a person dies, his soul stays with the body for three days.[4] On the fourth day, the soul travels to the place of judgment and prepares to cross the "Chinvat bridge," which spans the abyss of hell and leads to paradise. There, at the bridge, all the good deeds of the person are weighed against all the bad deeds. If the good deeds predominate, the soul crosses to heaven in safety. If the bad deeds are more numerous, the bridge becomes the edge of a knife and the soul plunges into the abyss of hell. Paradise is a place of beauty, light, and pleasant scents, whereas hell is one of misery, darkness, and evil smells. If good and evil balance exactly, the soul goes to a place of no torment, but of considerable discomfort.

Ahura Mazda and Ahriman made a pact whereby Ahura Mazda would be lord of the world for three thousand years, then Ahriman for the next three thousand years, and then they would battle during the following three thousand years. During his period, Ahura Mazda created man and the animal kingdom. When his turn came, Ahriman corrupted the earth by creating noxious animals and plants. He then dug a great hole in the ground to serve as an infernal region, and filled mankind with wicked thoughts and evil desires. Fortunately, his period of control ended before mankind was completely doomed.

The start of the third tri-millennium began with Zoroaster entering the world to teach humanity the good religion. After every millennium another savior will appear to re-enforce the religion. Righteousness will prevail more and more until finally, at the time of the resurrection of the dead, all evil and wickedness will have disappeared from the face of the earth.

On Judgment Day all people, righteous and wicked alike, will rise on the spot they perished, and their souls will be reunited with the bodies that will be reconstituted from their original elements.[5] Then the good will be separated from the wicked. The good will spend the next three days in heaven enjoying every imaginable bliss, whereas the wicked will be punished in hell with all sorts of torture. (Note that we have two periods of punishment: one for the souls alone, and one for the combined body and souls). At the end of the three days, both righteous and wicked will be purified by walking through molten metal. Life will be restored as it had been, with every man having his own wife (both returned to forty years of age) and children (returned to fourteen years), without any further procreation.

In the meantime, Ahriman will mobilize a vast army for the great and final Armageddon[6] battle. Much slaughter will ensue, and the earth will burn in a great conflagration. Everything will be destroyed, including Ahriman and his minions, as well as everyone in the pit of hell. Then a new universe will come into being where eternal bliss and the will of Ahura Mazda will reign supreme.

Satan in the Bible

From the day of creation to the day the Judeans were taken into the Babylonian captivity, the Hebrews sinned many times against their God by disobeying his commands. Yet the Old Testament never mentions the existence of an evil one, someone who tempts people away from the path of righteousness. Whenever the people sinned, they sinned on their own; nobody tempted them.[7] When God punished them, he punished the entire nation. The only temptation of which we are told is that of the snake convincing Eve to eat the forbidden fruit. But even there, the Bible presents an Aesop-like tale, complete with talking animals; it was not until the Wisdom writings in the last century BCE, that the snake was arbitrarily identified with the devil.[8]

But during their Babylonian exile, the Hebrews became acquainted with the Zoroastrian concept of an evil god. Soon thereafter we encounter in the Hebrew writings a couple of references to the devil—first in Job, and then in a dream sequence in Zechariah (3:1). In both cases, however, the devil acts as a member of God's retinue charged with determining the true character of humans, a kind of heavenly prosecutor. It is not until the apocalyptic writings of the second century BCE when the two new concepts make their appearance: Satan as an independent evil power, and life after death, during which individuals are judged and rewarded or punished according to their deeds.

By the time the new millennium rolled in, the power of Satan was well established in people's thinking. In the first three gospels, after Jesus' baptism and forty-day fast, Satan "took him up into a high mountain, and showed him all the kingdoms of the world in a single moment. And the devil said to him, 'All this power I will give you, and their glory; for this has been delivered to me and, and I give it to whomever I will. If, therefore, you worship me all shall be yours'" (Luke 4:5-7). In John's gospel there is no actual scene of baptism or temptation, but after the Last Supper discourse, Jesus tells his disciples: "Hereafter I will not talk much with you; for the ruler of this world is coming, [but] he has no power over me" (John 14:30).

These must be very scary passages for those who believe in the literal truth of the Bible. The gospels' words are clear: our world is under the complete dominion of Satan. With a wall placed between us and God, we can understand how the Hebrews must have felt when they realized that their pleas and prayers went unrecognized by God. Think about this for a minute—Satan being in charge of this world. Let the feeling of complete aloneness and helplessness envelop you, and then you will appreciate the importance of Jesus' message: the reign of God is coming, to replace that of Satan. Unfortunately, it is still in the future.

The Kingdom of God

As we discussed earlier, in the days of Jesus the Judeans lived a very precarious existence in a world full of obvious injustice: the rich and powerful were becoming increasingly oppressive, while the poor and downtrodden were losing whatever little security and dignity they had left. The idea that their God had deserted them and that Satan had taken control certainly must have made sense to them. And in this setting Jesus made his pronouncement: "The reign of God is coming." But what did he mean by that?

As with many other matters, the reader of the first three gospels would give a different answer from the reader of John's gospel. For Mark, Matthew, and Luke, the kingdom was here on earth. It has been suggested that the reason Jesus did not describe the kingdom is that all his listeners knew about it. It had been described in one way or another by almost all prophets.[9] It invariably came after a victorious battle between God and Israel's enemies, whereupon Jerusalem and the Temple were rebuilt and everybody lived peacefully thereafter, honoring God and following his commandments.

In the days of the prophets, of course, the Hebrews' enemies were their neighbors. For Jesus, and perhaps for some of his listeners, the enemy was Satan. Jesus apparently did believe that the days of Satan's power were numbered. When the seventy-two disciples returned from preaching around the countryside and told him how they had cast out demons, he replied: "I watched Satan fall like lightning from the sky" (Luke 10:18). And in another instance when he had been accused of driving out demons through the help of Beelzebul, he observed: "No one can enter a strong man's house to plunder his property unless he first binds the strong man" (Mark 3:27). The first three gospels, but not John's, tell us that Jesus believed the reason he and his disciples could prevail (at least temporarily) over demons was that Satan had been beaten already, a sure sign that the reign of God was here, or approaching.

The situation appears less promising in John's gospel. Here there are no exorcisms, no indication that Satan is weakening. In fact, at the very end, Jesus prays to his Father to protect his disciples and those who will believe in him:

> I am on my way to you.... Father, I don't ask you to take my followers out of the world, but keep them safe from the evil one. They don't belong to this world, and neither do I.... I am also praying for everyone else who will have faith because of what my followers will say about me (John 17:13,15,16,20).

What seems apparent here, is that the Earth is Satan's world; God's world is somewhere else, where Jesus and his followers will go. God's protection, therefore, does not extend to our physical comforts on this world, but to the steadfastness of our beliefs. After all, his own son underwent an excruciating death before he went to his Father.

Luke reports one description of the kingdom by Jesus, which differs from everything we have said above:

> Asked by the Pharisees when the kingdom of God would come, He replied to them saying, "The kingdom of God does not come with signs to be observed or with visible display. Nor will people say, 'Look! Here [it is]!' or, 'See, [it is] there!' For behold, the kingdom of God is within you [in your hearts] and among you [surrounding you]." (Luke 17:20-21 AMP Quotation marks added)

Notice that there are two possible interpretations of the Greek in the last sentence. Many liberal theologians accept the first interpretation and propose that the kingdom is really a frame of mind. To the extent that God lives in our hearts, his kingdom has come for us. The problem I have with this explanation is that it is inconsistent with Jesus' and the Baptist's preaching that we should repent because the kingdom is coming. If the argument of the liberals were true, then the kingdom *would not come unless* we repented. I prefer the second part of the explanation, which says the kingdom of God is among us. It fits my own personal beliefs that I will outline at the end of the book.

Some Inconsistencies

Much of Jesus' teaching is focused on the concept of the reign of God. Interestingly, however, he never referred to God as a king. Yet some things are not completely clear. He seems to say that the arrival of God's reign does not occur at a distinct point of time, in an apocalyptic battle that everybody expected to mark the end of the world.[10] The reign is not all encompassing, but instead it takes place inside individual human beings. So at some particular time, it may have arrived for one person, but not for another.[11] But such a partial arrival of the reign would not solve the social problems in the world. Besides, Jesus directed his preaching to the poor, the weak, and the sinners. Even if he had been able to convert all his listeners, the oppression in the world would still not have decreased, and this is what his audience really longed for.

In any case, Jesus was not very successful in persuading people. We can deduce this from his tirade against the cities of Chorazin, Bethsaida, and Capernaum, which had witnessed many miracles and yet still did not believe (Luke 10:13). This is clear also from the parables of the mustard seed and the seeds that grow on their own, which seem to promise that with God's help, big results will eventually come from small beginnings, all by themselves. Yet today, two thousand years later, the reign of God still seems as blurry as it did back then.

Perhaps Jesus was not a good teacher. After all, what is the reign of God? Apparently it is something more valuable than anything else. But what exactly? Certainly not a pearl, or a treasure hidden in the ground. Twenty years later, Paul was a far better salesman:

> This I say, brothers, that flesh and blood cannot inherit the kingdom of God; neither does corruption inherit incorruption. Behold I show you a mystery; we shall not all fall asleep, but we shall all be changed, in a moment, in the twinkling of an eye, at the last trumpet; for the trumpet shall sound, and the dead shall be raised incorruptible, and we shall be changed.[12] For this corruptible must put on incorruption, and this mortal must put on immortality.... Therefore, my beloved brothers, be steadfast, unmovable, always abounding in the work of the Lord, knowing that your labor is not in vain in the Lord. (1 Cor. 15:50-53, 58)

I am not sure, however, that Jesus would have agreed with Paul's anticipations. According to the gospels, Jesus probably expected a continuation of the same kind of earthly life, only better.

But I have left out something important from Jesus' message. Again and again he called for repentance. However he did not exhort the people to repent and follow the laws of the Torah, as John the Baptist had. Instead, he kept telling them to repent and to believe in him and his message. Repentance was equated with belief in Jesus, which means that everybody had to repent, sinners or not. And they had to repent because the reign of God was coming. But this is strange because we said above that the coming of the reign was a personal event, specific to each person. Jesus was saying that someone could repent and open his heart to welcome the reign of God, but that if he chose not to, the reign might still come and catch him unprepared! There is an inconsistency here: the reign cannot be simultaneously something personal inside an individual person, and also an external event.

Is it possible that Jesus did expect an Armageddon type of occurrence? It is true that both he and the Baptist went around proclaiming, "Repent! The end of time is near!" But two thousand years later, such exhortations sound a little pathetic. Somewhat like the cartoon of a bearded man on the street carrying a sign saying "The End Is Near!" Besides, scientists would frown at such a discontinuity in nature's laws. Anything that occurs is a result of existing physical laws and thus, presumably, predictable. The sudden end of civilization caused by a planetoid impact is understandable; but a clash between heretofore unobserved and unobservable deities does not fit in a scientist's neat rulebook.[13] Since Jesus' predictions did not come true, we can conclude that he was wrong; the end was not near then, and it is probably not near now, unless...

Unless we are looking for the wrong thing in the wrong place. Maybe instead of looking at the entire world, we should look at each individual's life. Here we do find the occurrence of a discrete event that changes everything, and yet one that is in perfect harmony with nature's laws: death. Was Jesus perhaps saying that if we die before we repent, then we lose the opportunity to accept God's reign? It certainly makes sense. And if we look at the apocryphal writings of Enoch and the others, people's real judgment was said to take place at the time of death. The later judgment scenes before God's throne were not an evaluation, but just the rendering of the sentence.

We can then interpret Jesus' message as telling us that we should each open our heart and let God reign in it, for it may be later than we think; none of us knows when his end will come. Perhaps it was not the bridegroom for whom the ten maidens were waiting in the parable, but St. Peter.

JUDGMENT

And what happens if we die before we have time to repent and open our hearts to God? The prophetic writings and Hebrew apocrypha are not very reassuring. They present God handling the lost souls with chilling vindictiveness. The last verse of the book of Isaiah describes saved mankind going to view the remains of the punished ones.

> And they shall go forth, and look upon the carcasses of the men who have transgressed against me; for their worm shall not die, neither shall their fire be extinguished; and they shall be abhorrent to all mankind. (Isa. 66:24)

Christian writings foresee even worse. The *Apocalypse of Peter* contains such gory descriptions of hell that if librarians ever read it, they would surely take it off the shelves. In the gospels themselves, hell is consistently described as the place "where there will be wailing and grinding of teeth."

Most, though not all of today's Christians, downplay hell and the gospel references to it. The argument is that a merciful God would not subject even the worst of his children to such unending torments. After all, as we saw above, even the Zoroastrian god was more merciful, limiting the time of punishment. Scholars maintain that hell references have been added into the gospels by the evangelists and represent their individual anger at their opponents against whom they were in constant fight for converts. I said it before and I cannot say it enough times: as one reads the gospels, one must keep in mind that they were written by people with specific goals and axes to grind. The words they place in Jesus' mouth mirror their own opinions and prejudices, not Jesus' ideas.

There is very little to be said about heaven, because very little has been written about it. An occasional description of the saved souls standing around God's throne for an eternity singing "Hosanna to the highest" is even more chilling than the descriptions of hell. My mother thought that everyone would go to heaven and do whatever they did not have the opportunity to do on earth. She herself planned to take up ice-skating. I doubt that heaven is like that, but it certainly sounds like a better idea than singing unending choruses, unless of course you had always wanted to be a choir singer.

It would be logical to continue the discussion by examining this "soul" and the idea of life after death. But instead, we will delay it by a chapter to first discuss who Jesus was.

SOME QUESTIONS TO CONSIDER

1. The Zoroastrians answered the question of why a good god created bad things by postulating an evil creator. How does this idea fit with our Christian religion?

2. The idea of judgment after death has been used for millennia to soothe mankind's realization that life is un-

just. Yet this proposed solution is devoid of mercy and perhaps equally unjust in its severity. Can you suggest a better way of looking at it?

3. How do you explain the similarities between so many Zoroastrian and Judeo-Christian beliefs?

4. If Satan acts as a prosecutor reporting to God, can we still consider him as an independent evil entity that opposes God?

5. How could Satan say that all the power and glory in the world were given to him? Did God really give them to him, and if so, why? If not, who did, and why did God allow it?

6. If the Jews believed that the soul did not leave the body until the person had been dead for three days, how could the Christians claim that Jesus was resurrected from the dead when scarcely thirty-six hours had elapsed between his death and his resurrection?

7. Some authorities maintain that when Jesus said, "the ruler of the world is coming" he was referring to the Romans. What do you think?

8. The Good News (gospel) is that the kingdom of God is at hand. Yet two thousand years have brought little discernible change in the world and its inhabitants. What would *you* say is the kingdom of God, and is it two thousand years closer now than it was in Jesus' days?

9. If the kingdom of God is a personal event in everybody's life, then in what way is Jesus part of the Good News?

SELECTED REFERENCES

Ehrman, Bart D. *The New Testament: A Historic Introduction to the Early Christian Writings.*

Nigosian, S. A. *The Zoroastrian Faith, Tradition and Modern Research.*

Page, Sydney U. T. *Powers of Evil: A Biblical Study of Satan and Demons.*

Reiser, Marius. *Jesus and Judgment: The Eschatological Proclamation in Its Jewish Context.*

CHAPTER 15

SON OF MAN OR SON OF GOD?

Maybe Jesus was the Messiah, and maybe not. But who was he otherwise? He often referred to himself as the "Son of man,"[1] but it isn't clear what that means. It could be a reference to an apocalyptic figure expected at the end of the world or it could just mean plain "man." Christians, on the other hand, believe he was the "Son of God," and although he himself never said that, he did say that God was his father. But he also said that God was everybody's father, which makes us all sons of God. And in John's gospel, he made the most unusual pronouncements: He was in God, and God was in him, and whoever saw him saw God. As we will discuss, according to human logic this means that he and God are one. And since he also said that he is in us and we are in him, the same logic would support the conclusion that according to John's gospel, God, Jesus, and we are all one.

Son of Man

The Bible shows people addressing Jesus by a number of titles: teacher, rabbi, lord, master, and sir, the last almost exclusively in John's gospel. But the variety is not as large as it appears. A rabbi is an outstanding teacher of law, although at the time of Jesus the term might have been just another form of address.[2] The other three, lord, master, and sir, represent different translations of the same Greek word *kyrie*, and their use varies among the different Bible versions. Nobody, however, refers to Jesus as "the Son of man" except for himself. And although he does it often through the first three gospels—though never in John's—he always uses the personal pronoun "I" when he wants to stress the contrast of his new theology to that of Moses.[3] So what does this expression mean?

In Aramaic, which Jesus presumably spoke, the term *bar-nash* (or perhaps *bar-enasha*) son of man, means simply "man," although there is no way to tell whether this was true as far back as Jesus' days. Logically, however, the meaning of the expression "son of man" is "man"; because after all, that is what man is. The only exception is Adam, whom the Bible calls "Son of God." In the book of Ezekiel, God consistently addresses the prophet as "son of man," a practice that the New American Bible explains as follows: "A formal way of simply saying 'man'; God's habitual way of addressing the prophet throughout the book. Probably the term is used to emphasize the separation of the divine and the human" (NAB, note 2:1, 875). Even more telling is that in identical passages, the different gospels identify Jesus by either the personal pronoun "I" (or "me") or "the Son of man" expression in an apparently interchangeable manner.[4] This would indicate that the evangelists themselves did not attach any profound meaning to the expression. Semites used the "son" prefix to highlight the relationship to the rest of the expression. "Son of strength," for instance, identifies someone very strong. So a "son of man" would be a perfect representation of man, an archetypical man.

In the Book of Daniel, there is a vision where at the end of time, "One like a son of man" comes on the clouds of heaven, is presented to God, and receives dominion, glory, and kingship over all nations. The New American Bible explains the term as follows: "In contrast to the worldly kingdoms opposed to God, which appear as beasts, the glorified people of God that will form the kingdom on earth is represented in human form" (NAB, note 7:13f, 926). Christianity has historically identified this figure with Jesus, anchoring in this description its contention that Jesus will return and become lord of all.

In the Similitudes of 1 Enoch (46-62), written at the beginning of the first century BCE,[5] "The 'Son of Man' was given a name even before the creation of the stars. He will become a staff for the righteous ones in order that they may lean on him and not fall....He will remove the mighty ones from their comfortable seats, and the strong ones from their thrones...for they do not extol and glorify him, and neither they obey him, the source of their kingship. He is the light of the gentiles and he will become the hope of those who are sick in their hearts. All those who dwell upon the earth shall fall and worship upon him.... The Lord of Spirits placed the elect one on the throne of glory; and he shall judge all the works of the holy ones in heaven above, weighing in the balance their deeds....The Son of Man was concealed from the beginning, and the Most High One preserved him in the presence of his power; then he revealed him to the holy and the elect ones" (Charlesworth, 1983, 34-43). Notice how close this description agrees with the *Logos* in John's gospel (the power that pre-existed creation and through which the act of creation took place, the final judge of mankind). Unfortunately, to complicate matters, Enoch himself was called "son of man" by an angel (1 Enoch 17:14), something that confuses the meaning of the term even within this one short writing.

What did Jesus have in mind when he used this expression? Better yet, let's ask the question the other way around: what did his listeners think when they heard him use the expression? When Jesus said, "Foxes have holes and birds of the air have nests; but the Son of man has nowhere to rest his head" (Luke 9:58), did they identify him with an eternal, divine creature who for some bizarre reason was walking destitute on earth? Nobody ever reacted to him as if he was anything other than what he appeared to be, at best an inspired prophet. Even Peter, in his moment of profound insight, proclaimed him to be only the Messiah, a human being with a normal human ancestry, not God.

Is it believable that Jesus would have referred to himself throughout his teaching period by using a *double entendre?* If people could not understand his name, how could he expect that they would get his teachings right? And this, of course, is exactly what happens in John's gospel where Jesus keeps trying to convince his listeners that he is God's son, that he and God are one, and that people must believe this in order to be saved.

Not unexpectedly, he encounters minimal success despite his miracles. Is this surprising? Suppose the magician David Copperfield started proclaiming divine powers? How many people would believe him?

It has been often proposed that Jesus deliberately used a word of many meanings in order to protect himself. If he proclaimed that he was the Messiah, he would immediately attract the wrath of the Romans, but no authority would be concerned about someone who simply said that he was a man. If true, there is no way to tell what Jesus really meant by this term. But attributing a metaphysical meaning to the expression makes it difficult to describe either Jesus' presence or his actions on Earth. The apocryphal writings clearly present the "son of man" as a heavenly king and judge, not as an earthbound evangelist.

Son of God

Just as only those in need of healing called Jesus "son of David," only the demons he expelled called him "Son of God." Jesus never called himself that, although he consistently referred to God as his father. He also referred to him as "the father," and "your father," apparently considering him to be the father of all. In the Gospel of John he clarifies the various relationships with a series of short statements that would fill with joy any student of algebraic set theory. On a piece of paper draw three circles, as in the figure on page 163, and label them 'Father,' 'Jesus,' and 'We.'

- Jesus' statement, "I am in the Father" (John 14:10) is represented by drawing a small 'Jesus' circle inside a larger 'Father' circle.

- Jesus' statement, "The Father is in me" (John 14:10) is represented by drawing a small 'Father' circle inside a larger 'Jesus' circle.

- The combined sentence "I am in the Father and the Father is in me," has to fulfill both previous requirements: the 'Jesus' circle must lie inside the 'Father' circle, and the 'Father' circle must lie inside the 'Jesus' circle. The only way to accomplish this is by drawing equal size 'Father' and 'Jesus' circles, one on top of the other. When we now look at these circles, we only see one. This corresponds to the statement, "The Father and I are one" (John 10:30), and, "Whoever sees me sees the one who sent me" (John 12:45).

- To illustrate the statement "I am in my Father, and you are in me, and I in you" (John 14:20), we must start by drawing the 'We' circle inside the 'Jesus' circle, and the 'Jesus' circle inside the 'We' circle. Just as in the case above, we can only accomplish this by drawing two identical circles, one on top of the other. We then draw the 'Father' circle in the position we determined earlier, equal to the 'Jesus' circle and superimposed on it. The final result is one circle, consisting of three superimposed circles labeled 'Father,' 'Jesus,' and 'We.' WE ARE ALL ONE!

Sacrilegious? Perhaps, but these were Jesus' words. However if someone absolutely cannot agree with this conclusion, he may wish to interpret the statements a little differently: Jesus might have meant that he was partly in the Father, and that the Father was partly in him; similarly that he was partly in us, and that we are partly in him. He can then draw three circles in line next to each other, with the 'Jesus' circle in the center, partially overlapping the 'Father' circle on the left, and partially overlapping the 'We' circle on the right. Now the only way in which we can relate to the Father is through Jesus, just as the Christian Church likes to teach! The problem with this interpretation is that it renders false two of Jesus' statements: "The Father and I are one," and, "Whoever has seen me has seen the Father." Partial overlap of the circles does not validate these statements.

This brings us to what, for me, are the most powerful words in John's gospel: when Mary Magdalene met the just resurrected Jesus,

> Jesus said to her, "Mary!" Turning around she said to him in Hebrew, "Rabboni!"—which means Teacher or Master. Jesus said to her, "Do not cling to me [do

> not hold me], for I have not yet ascended to the Father![6] But go to My brethren and tell them, 'I am ascending to My Father and your Father, and to My God and your God.'"[7] (John 20:16-17 AMP Quotation marks added)

Jesus is clearly saying here that we are his brothers and sisters, and that we have the same father and God as he has.

Some Questions to Consider

1. What do you think that Jesus meant when he referred to himself as "son of man"?

2. Since different gospels use the personal pronoun and the term "son of man" in identical passages, why should we believe that the "son of man" term is anything other than a stand-in for "I" or "me" used by the evangelists? How can we tell what were Jesus' actual words?

3. How can someone and his father be identical, one and the same? (Remember this answer when we discuss the Trinity in chapter 24.)

4. If the evil spirits recognized Jesus and called him "Son of God," what does this tell us about them?

5. Can Jesus be simultaneously both the Messiah and the Son of God?

6. Jesus said: "Whoever has seen me has seen the Father." What do we know about Jesus, and what does this tell us about the Father?

Selected References

Ehrman, Bart. *The New Testament: A Historic Introduction to the Early Christian Writings.*

Stein, Robert H. *The Method and Message of Jesus' Teachings.*

Vermes, Geza. *The Changing Faces of Jesus.*

GRAPHICAL REPRESENTATION OF JESUS' RELATIONSHIP STATEMENTS

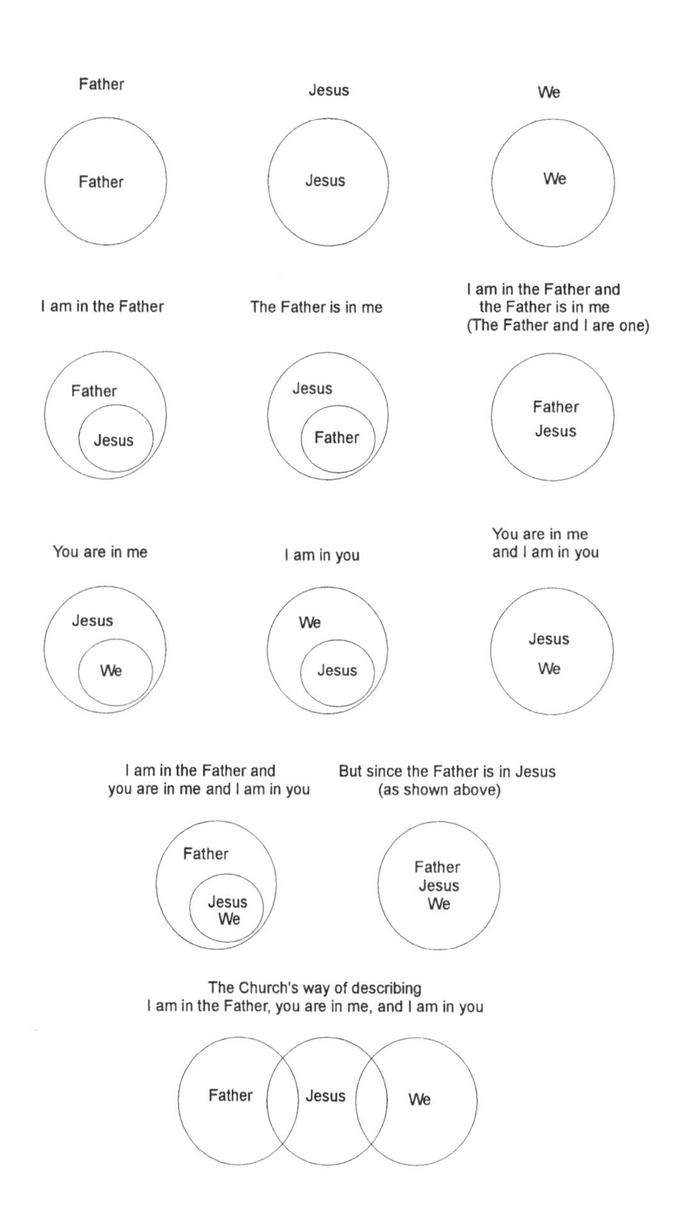

CHAPTER 16

OUR SELVES AND THE WORLD

Jesus said that he was the Son of man. But what is man? A scientist might say that man is an animal capable of abstract thought made possible through the greatly developed cortex area of his brain. But a Christian believer will rush to respond that man is not an animal but a being created in the image of God. The problem with this answer, of course, is that we don't know what God is. In fact, in our mind we often reverse the believer's syllogism: if we have been created in God's image, then he must be like us, only better, which leads us to construct a God in *our* image.

In this chapter we will look at some of the things that scientists say about mankind, our minds, our senses, our emotions, our psychology. We will see that the human brain is a biological supercomputer that is continuously interpreting the information that flows in from the body receptors to keep its owner informed of the surrounding dangers and enticements. Unfortunately, being a biological construct, even minor chemical imbalances affect and disrupt the brain's processes.

The *Mind* is the name we give to the operation of the brain. It is like a computer program, created throughout the course of human evolution and whose database, compiled during its owner's lifetime, contains numerous errors and misconceptions. Our minds' purpose is to provide us with information that will help our survival, rather than with objective analyses of our surroundings. In the process, however, the mind also attempts to screen us from any information that might challenge our preconceived knowledge of ourselves. It does it by filtering, manipulating, and thus tainting both inputs and outputs, so that they will correspond to ideas and desires hidden in our unconscious, which is completely inaccessible to us.

We have less understanding of our emotional side, much of which is tied to our evolved inherited physical makeup and to hidden memories of past experiences. Emotions are not only inputs into our brains' processes, but quite often directly affect our bodies' actions, completely bypassing our conscious minds. So although we may believe that we are in complete control of ourselves, we are often just puppets of things that we can barely name, let alone understand. As a result, we remain oblivious of nine-tenths of our minds' and bodies' actions and processes. This may relegate our postulated free will to the province of philosophy.

WHAT WE ARE: THE MIND

Aristotle believed that the brain is a heat exchange chamber used to cool the blood. His idea was perhaps reasonable if one considers that Greeks and other Mediterranean people are so hotheaded that they probably have higher blood-cooling needs. In any case, it certainly justifies the opinion that ancient Greeks should have confined themselves to matters of philosophy and art and left the teaching of biology to others.

As we all now know, the brain (and its operation, which we call the mind) is the one thing that distinguishes humans from other animals, and for that matter, animals from plants (ignoring for a minute the not-so-small matter of the plants' lack of locomotion). All thinking and reasoning activities take place in the brain, and most of the directives for bodily actions originate there. As the biological species evolved from amoebas to humans, so also did their brains (amoebas, of course, having no brain at all).

Nature's first brain design is called the *reptilian brain*, because it is found in reptiles, one of the oldest groups of animals in the evolution ladder. This brain's main concern is the self-preservation of its host. It receives inputs from the body's sensory cells and olfactory organs, since at that stage of evolution tactile feel and smell were the most important senses both for the animal's protection and for the location of its prey. For the early mammals in evolution's next step, nature designed a brain that was concerned with such matters as nursing, parental behavior, socialization, play, and procreation. We call this the *paleo-mammalian brain*. Collectively, the components of the paleo-mammalian brain, together with its nerve extensions throughout the body, are called the *limbic system*. In humans we refer to it as the emotional brain since it reacts to emotions such as fear, love, etc. Nature, however, did not dispense with the earlier reptilian brain; nature does not like to waste effort. Instead, it wrapped the newer paleo-mammalian brain around the older reptilian brain. The animals could then use both brains at once, each one for the purpose for which it had been originally designed.

Nature's final brain development, the *neo-mammalian brain*, is represented by the huge expanse of the cortex that physically surrounds the other two brains. As was the case before, everything is being used, nothing is wasted. We consider the cortex as the chief executor that dominates the other components, and it is extraordinary. It consists of billions of neurons joined to each other through trillions of connections. When a sufficiently high electrical charge is applied at one end of a neuron, it fires and transmits this charge to the hundreds or thousands of other neutrons to which it is attached. Sometimes these attachment points attempt to incite the nearby neurons to fire and at other times they try to stop them from firing. This pyrotechnic display can be visually observed through the various PET- and CAT-scan systems that have been developed.[1] Chemistry plays a great role in all this because the neurons stop just short of touching each other and the space in between is filled with a so-called synaptic fluid. It is the chemicals in this fluid that transmit the electrical charges. Change the chemicals slightly, and the operation of the mind will change.

Our three brains respond at different speeds, related to the degree of their evolution; the less evolved the brain, the faster its reaction. Place your hand over a flame, and the reptilian brain will order it retracted before you even feel the heat; the pain sensation will come later. See something threatening and you will react before you even recognize it; the limbic brain makes an immediate decision between good and bad; the cortex takes longer to recognize the object and to send more detailed information to the limbic system for a better-informed decision.

Normally we are not even aware of what takes place in our lower brains. Our emotional reactions, for instance, occur at the limbic level, completely skipping the cortex and thus escaping our consciousness. We often don't realize how excited we are until we hear the loudness of our own voice or feel the flush's warmth spreading over our face. This is why when somebody gets in a particularly stressful situation bystanders can easily recognize that he is "losing his cool," whereas he himself remains completely unaware of it.

Recent experiments have revealed another interesting, and also puzzling example of our conscious mind not being in charge of our body. Suppose we place a volunteer in a brain-scan apparatus that will detect signals at the brain locations that control hand muscles. We then ask the volunteer to move his hand whenever he decides, but to notice the time on a clock placed in front of him. The strange result of this experiment is that three-tenths of a second before the volunteer consciously decides to move his hand, the related neurons in his brain become activated; half a second later the hand moves. The brain apparently received orders to move the hand 0.3 seconds before the person made the conscious decision to do it![2] This experiment has been duplicated many times, with the same results. So if the conscious mind did not make the decision, who did?[3]

Sensing Our Environment

Our brains are continuously responding to external stimuli. Some of these stimuli trigger sensors attuned to the outside environment, and others trigger sensors that monitor the chemical and physical conditions inside the body. The brain's function, however, is not to observe the environment, but to help the body survive, and this is the criterion it uses when interpreting the inputs. The body's skin sensors, for instance, are affected by the kinetic energy of impacting air molecules, but the brain interprets this as an air temperature, warm or cold, prompting a person to dress accordingly.

Similarly, electromagnetic radiation at certain frequencies activates cells on the eye's retina, which then transmit this sensory output to specialized parts of the brain. There it is analyzed through intricate neural circuits, and eventually an image concept is presented to the conscious part of the brain. But everything is a result of learned responses. People born with opaque lenses in their eyes, which prevent the electromagnetic radiation from reaching the retina, are unable to see even after corrective surgery, if that surgery has been delayed too long. After ten years of congenital blindness, electromagnetic radiation cannot be transformed by the brain into patterns and shapes, but is sensed only as localized pain.[4]

It follows from all this that my brain cannot use the input of my senses to tell me what the outside world really is, but only what it thinks it is. Of course since I am color-blind, I have always been somewhat aware of that, just as I am aware of the fact that at night I can't distinguish the red traffic light from the orange. Color-blindness is fairly common; but one out every hundred thousand people also sees colors whenever he hears sounds.[5] Then there are people who, when tasting foods, sense shapes with their hands: sweets have a round shape, sour tastes feel pointy.[6] This simultaneous activation of different senses is called synesthesia. Again we see that the experience of an external event does not depend on the event itself but on our body's interpretation of it.

And then there is the matter of chemicals. I said earlier that electrical signals get transmitted in our brains from one neuron to the next through a fluid-filled gap. But the entire brain is being continuously bathed in chemicals. A slight change in any of these chemicals plays havoc with the brain's normal operation. "A few molecules of LSD can elicit vivid hallucinations in which sounds may be seen and our everyday picture of reality disintegrates."[7] Does that sound a little like the synesthesia we discussed above?

The point I am trying to make is that the (virtual) world we sense has little relation to the actual world that exists out there. We see red, but there is no color or even light out there, just electromagnetic waves. We hear sounds but there is no noise out there, just pressure waves in the atmosphere. We see in three dimensions, yet (as we discussed in chapter 3) some scientists are telling us that the universe has eleven dimensions. True, the scientists also say that these additional dimensions have contracted into single points, but even if they had not, how would we go about designing instruments to look into them? The inescapable conclusion is that the physical world in which we live is completely unrelated to the one that our senses have constructed.[8]

One important property of the brain is that it is a meaning-seeking machine that is constantly analyzing all the information it possesses, trying to sort it into a meaningful pattern.[9] (Interestingly enough, this is a characteristic of even the simplest computer simulations of artificial intelligence.) Perhaps this is why the brain can transform simple electrical impulses into the rich images that fill the mind.

WHO WE ARE: THE PSYCHE

Sigmund Freud first made people aware of the mental constructs that reside in us and govern our lives. Just as our bodies are controlled by unconscious operations of our minds, so also are our thoughts affected by unconscious logical reasoning designed to satisfy our hidden ego states (selves).

Many psychological theories have been proposed to describe the hidden drives of our minds, but my favorite one is Eric Berne's Transactional Analysis, developed in the 1950s and popularized by himself and others during the following two decades.[10] Briefly, Berne postulated that a human persona consists of three independent personalities: 1) the child, representing the mentality of one's early age, where playfulness and gratification of personal desires reign supreme alongside feelings of insecurity and lack of knowledge; 2) the adult, representing the thinking of someone who always makes decisions on strictly logical grounds and who is concerned with everyday matters of life; and 3) the parent, a storage for all the directives, instructions, and prejudices heaped upon the growing child before he had the capacity to evaluate them independently; here is stored the knowledge of good and evil, of sin, of prejudices, all assimilated before the development of the capacity to reason. At any given instant, we can express the thoughts and reasoning of one and only one of these personalities.[11] (Keep in mind the concept of these triune personalities when we discuss the Trinity in chapter 24.)

An important part of the theory regards our evaluation of ourselves compared with our evaluation of others. Positive personas are judged to be OK, and negative ones, NOT OK. Harris (1973), a follower of Berne, described the four possible categories, which I am listing below with my own brief explanation: 1) I'M OK-YOU'RE OK, which represents the only acceptable position; 2) I'M OK-YOU'RE NOT OK, which was the attitude of the praying Pharisee in Jesus' parable; 3) I'M NOT OK-YOU'RE OK, the belief of the praying publican in the same parable; 4) I'M NOT OK-YOU'RE NOT OK, the self-evaluation of killers who go on murder sprees, and at the end kill themselves; it was also Judas's attitude as described in the Bible.

Unless we have had outstandingly astute parents, most of us spend our childhood in the I'M NOT OK-YOU'RE OK position, since during our early formative years we are and feel physically and mentally inferior to our parents. After all, a baby can do nothing by itself. Perhaps here lies the origin of the concept of original sin. Just as we individually started life from the NOT OK state, so did the whole human race. We were inferior to God and we did not follow his rules, so he kicked us out. We are definitely NOT OK, and he is OK.

Jesus went around trying to change the OK and NOT OK status of people. He told the sinners who believed they were NOT OK that repentance made them OK. He told the Pharisees who believed they were OK that pride and lack of repentance made them NOT OK. He even told the rich young man who had followed all of Moses' rules during his lifetime, and hence presumably had nothing to repent for, that he was NOT OK not because he did not repent but because he was rich (Matt. 19:16-22). To Jesus, being rich was definitely NOT OK, something that today turns off the well-to-do parishioners in the suburbs.

The part of Berne's theory that is perhaps the most difficult to accept deals with our destiny. He maintained that during our early childhood years we form a decision regarding our role in life. Somehow our child's unconscious mind picks up hints from our environment, clear or hidden, which help us form an opinion about ourselves. We then identify with a character in a fairy tale or some other situation in our environment and proceed to pattern our lives according to that script.[12] Fate, the Arab *kismet*, the Greek *moira*, or Berne's script, whatever you want to call it, it guides us inexorably toward our individual destiny, whether it is to become the president of the country or the town drunk. According to this theory's proponents, it is only when we identify our script and succeed in consciously refusing to follow it that we can freely decide what to do with our lives. However, this is easier said than done.[13]

FREE WILL

Are we responsible for our actions if we live our lives according to scripts that we wrote at the age of six? The concept of sin and punishment is based on the assumption that we are free to make decisions in our lives. One person sins (or breaks the law) whereas another under the same circumstances does not. Christians expect that the former will be punished and the latter rewarded. But why is there a difference in the actions of these two?

As any college freshman will argue, we make decisions based on how we evaluate a situation on the basis of knowledge and standards that we acquired in the past and which are thus beyond our current control. Our genetically inherited disposition also affects our behavior but is equally beyond our control. In addition, as we discussed earlier, it appears that our subconscious holds knowledge and opinions of which we are completely ignorant but which nevertheless influence every decision we make without us even becoming aware of them.[14] And finally, we just postulated that we all go through life adhering to scripts that we constructed before we reached the age of reason. Nowhere in all this is there room for exercising free will.

The Greeks believed in logic, so we could expect that their philosophy would leave little room for free will. According to Aristotle, a voluntary action is a willing action. An unwilling action involves external compulsion or ignorance. But I have argued above that there are no actions free of compulsion.

The Stoics believed in determinism: a person is carried by the river to perfection; the free will is like the eddy that resists but, like the stream, it will eventually be carried away. The early Hebrews, on the other hand, believed in free will because they maintained that human beings were created free. This, however, was a pure assertion, not a fact that could be proven. The Hebrews in Jesus' days were divided: the Essenes were strict believers in fate; the Sadducees, who did not accept life after death, believed that all of man's actions were free; and the Pharisees straddled both camps.

To the Early Christians, however, the matter was clear and unquestionable. Man had been created with a free will, and whoever sinned did it through his own free choice. Strangely, however, Christians generally agreed that God could look at a man's makeup and determine whether he was going to sin or not. This was not an act of predetermination on God's part, but an act of evaluation and prediction. I find that this statement describes the nonexistence of free will. If someone can look at you and what makes you tick and can then deduce how you will act, it follows that from that point of time on you have no free will; you will be acting as a puppet of your makeup. But since this inspection can be moved to earlier and earlier points of your life, there never was a time when you were capable of free will. There is, however, one reasoning that will allow God to know our future actions and still leave us with free will: God exists beyond time; in the present, in the past, and in the future, and since he has already seen the future he knows what we have done, despite the fact that in our own time we have not done it yet. This is a relatively recent argument, however, and was never made during the early days of the Church.

This whole discussion assumes that man acts logically, guided by his computer-like brain. However humans are not logical, computer-like robots. We have emotions that affect us in ways that nobody can predict. Consider, for instance, the difference between "like" and "love." We *like* somebody *because* of his good qualities. We *love* somebody *despite* his bad points. Obviously love is an illogical emotion, but it is precisely the one trait that Jesus valued above all. As he told his disciples, "This I command you, that you love one another" (John 15:17). Perhaps just as our logic lifts us from the level of animals, so our love elevates us from the level of humans. A strange statement, since love and other such emotions are generated in the lower, older, middle-level brain.

Am I proposing that our free will resides in that part of our brain over which we have no conscious control? Perhaps, but according to Wilson (2002, 211-212), it is possible to change our nonconscious inclinations by changing our conscious behavior. He quotes Aristotle saying that to become courageous we should start by performing courageous acts. (Remember the old song, that advises whistling when you're afraid?) We can extrapolate this and conclude that to become loving we must practice love. In fact I have known clergy who practice among the poor, who say that they make a conscious effort to love the dirty, smelly, drunks they come across.[15]

THE SOUL

And where is the soul in all of this? First of all, what is the soul? Everybody will agree that, if it exists, the soul is whatever survives an individual human being's physical death. But it cannot be just a "living principle," or "the breath of life" that God was supposed to have blown into Adam's nostrils to bring him to life (in Gen. 2:7). Such a concept might have satisfied the early Hebrews who expected that the afterlife consisted of silent shadows scurrying in the dark; but for those religions which expect that individuals will be rewarded or punished in some after-death judgment, a soul must also carry the identity and bear the responsibilities of the being to which it had been attached.

Even if we exclude the matter of rewards and punishments, and hence that of responsibility for the performance of our earthly acts, we must still define survival after death as the survival of our individuality. If the soul is not related to our conscious self, its continuing existence after our death is irrelevant. If it goes on living when we are dead, but cannot remember who it used to be, then for all practical purposes we don't exist anymore. To be meaningful, life after death must involve the continuation of our memories because, after all, this is what we have been.

But in addition to our memories, our individuality consists of the way that we do things, the programs that run in our brains.[16] In a computer, programs and data are stored in the memory, so according to this argument the soul must be some type of memory that survives death. Perhaps it's like ideas in a book: you read a book and you get some ideas; you burn the book, but the ideas remain in your head; you tell the ideas to your friends, and even after you die the ideas survive.[17] Even people who don't believe in personal immortality often say that after their death they will live in the memories of their family and friends. Children, in particular, represent immortality: "Even when the father dies, he might well not be dead, since he leaves his likeness behind him" (Eccl. 30:4 JER). But a soul must be more than that. It must be some independent surviving entity.

Both Descartes and Plato considered the soul to be an immortal component element of a human being, but distinct from the mind. Since we have defined "mind" as that which occurs when the brain is operating, the brain process, it follows that when the brain dies, the mind disappears. So if the soul survives the death of the brain and the cessation of the mind, what can it be? It must be something that contains at least part of the program that ran the dead person's brain as well as the important accumulated memories. But is there any evidence that such entities survive the death process, and if so, in what form? We will consider this in the next chapter.

SOME QUESTIONS TO CONSIDER

1. If the outside world that each of us perceives depends upon his own brain, we all operate in our own personally generated world. How can we then interact with each other if we all live in our own individual worlds?

2. Can you argue that humans have free will? On what do they then base their decisions if not on logical and emotional evaluations? Can you give an example?

3. The author proposed that the soul survives the body's physical death. Many believe that it might have also preceded the body's birth and that it later became incorporated with it. How would that change the author's description of the soul?

4. What do you think of the experiment that purports to show that brain activity occurs a fraction of a second

before we consciously use our mind? Is our conscious mind a slave to the nonconscious which makes all the decisions?[18]

5. Following T. Wilson's theory, if we act with love towards each other, we will end up loving each other. Do you agree? How should we change our lives to implement this idea?

6. What do you think of Jesus' teaching that rich people are NOT OK?

7. The idea that we form a life script at a young age and then adhere to it sounds far-fetched. Is it? What was your favorite childhood tale and do you think that you are living the story of someone in it?

SELECTED REFERENCES

Berne, Eric. *Transactional Analysis in Psychotherapy.*

_________ *What Do You Say after You Say Hello? The Psychology of Human Destiny.*

Carter, Rita. *Exploring Consciousness.*

Cytowic, Richard E. *The Man who Tasted Shapes.*

Gazzaniga, Michael S. *Nature's Mind: The Biological Roots of Thinking, Emotions, Sexuality, Language, and Intelligence.*

Johnston, Victor S. *Why We Feel: The Science of Human Emotion.*

Schacter, Daniel L., and Elaine Scarry, eds. *Memory, Brain and Belief.*

Steiner, Claude M. *Scripts People Play: Transactional Analysis of Life Scripts.*

Wilson, Timothy D. *Strangers to Ourselves: Discovering the Adaptive Unconscious.*

CHAPTER 17

SURVIVAL OF THE SOUL

What happens after we die? Nothing, according to the ancient Hebrews. According to Christians and some of the Jews in Jesus' day, some time after we die we become physically resurrected and live forever. We will discuss this idea in chapter 20. Here, however, we will limit our examination to the possibility that our identity, or our soul as we defined it in chapter 16, survives our death and remains conscious. I should make it clear that this is not the Christian concept of life after death but is closer to that of Plato (419-327 BCE) and the latter-day Platonists (first century BCE to fourth century CE) who believed in the existence of a deathless soul distinct from the body.[1]

Scientists don't believe that anything survives one's death. Yet there is an intriguing mountain of scientifically unaccepted stories and sightings indicating that one's consciousness can survive death in a noncorporeal state. The huge number of stories regarding hauntings, apparitions, communications with the dead, near-death experiences, even stories of reincarnation, have convinced many of the reality of this phenomenon. Most scientists, however, remain unconvinced because nobody has performed proper laboratory experiments in which consciousnesses of dead persons can be produced at will. Even more important, they ask how can a consciousness maintain its identity when the repository of memory, the brain, ceases to function.

There is only one answer to this last predicament: memory information is not stored inside the physical brain, but outside of it, possibly in the form of a memory field that is accessible to the brain similar to the way some electromagnetic fields are accessed by radios. In fact, a new generation of reputable scientists are proposing that there is nothing physical in our universe. Since the elementary particles (electrons, quarks, muons) are energy clusters rather than physical entities, everything constructed from them is also an energy representation. I probably don't need to point out that this chapter deals with matters that rest at the boundaries of science, which will require much more research before gaining wide acceptance.

The Bible's Position

> The Lord told [the man], "You may eat fruit from any tree in the garden, except the one that has the power to let you know the difference between right and wrong. If you eat any fruit from that tree, you will die before the day is over!" (Gen. 2:16-17 CEV)

And the man ate the fruit from that tree and he died. This is how it went: first life, then death; afterward almost nothing. If the dead had life, it was only a shadowy existence in some dark and eerie place. There is nothing in the Old Testament to dispute it.[2] Humans spent their lives on earth, and when they died they were gone forever.[3] The obvious conclusion was that if God were indeed just, then he would reward the good people and punish the bad ones in this world. It was during their Babylonian exile that the Hebrews first encountered the concept of a soul as something separate from the body—the idea that a limited lifespan on earth would be followed by an everlasting life of reward or retribution. Even so, nothing was written about it until the second century BCE, first in Daniel, and then in the apocryphal book of Enoch. As we discussed before, by that time it had become painfully obvious to the Hebrews that if life on earth was all there was, then life was not fair, and neither was God. The answer given to Job, that man was too stupid to understand God's ways, did not seem to satisfy any longer.

By the time Jesus came, the Hebrews were of mixed opinion: the Pharisees believed in the resurrection of the dead; the Sadducees did not. When they asked Jesus about it, he quoted what God had spoken to Moses out of the burning bush: "I am the God of your father, the God of Abraham, the God of Isaac, and the God of Jacob" (Exodus 3:6); and Jesus concluded: "He is not a God of the dead, but of the living, for to him all are alive" (Luke 20:38). This statement, however, does not necessarily imply life after death. If God is beyond time, his present existence includes what to us is past, present, and future. To God we are always alive because he is always there with us. In the first three gospels Jesus mentions resurrection only once: when the Sadducees asked him who would be the husband of a woman who had married successively seven brothers, each one dying after the previous one, he replied, "When they shall rise from the dead, they neither marry nor are given in marriage, but are as the angels who are in heaven" (Mark 12:25). Luke added: "They are the children of God, being the children of the resurrection" (Luke 20:36).[4]

Jesus' teachings about the kingdom of heaven are a little unclear regarding the kind of life that will follow death. Most of the time his teachings can be interpreted as referring to the implementation of God's reign on this world. When the Zebedee brothers asked Jesus to sit by his side after he came into his kingdom (Mark 10:37), they probably thought it would be in this world, not in another life. Jesus did mention occasionally a place "where there is gnashing of teeth," but that could also have existed in the present. At other times, however, he seems to be referring to a heavenly world. In the parable of the afterlife of the rich man and Lazarus, Jesus presented Lazarus as lying on the bosom of Abraham while the rich man was suffering torments in the netherworld. Similarly, at the end of Luke's gospel Jesus tells one of the bandits hanging alongside him: "Today you will be with me in Paradise," although this is probably a fabrication by the evangelist. In the last of the Beatitudes, however, Jesus is clearer: "Rejoice and be exceedingly glad, for great is your reward in heaven" (Mat. 5:12). In John's gospel he is even more specific: "If I go and prepare a place for you, I will come back again, and receive you to [it] myself, so that where I am, there you may also be" (John 14:3). That gospel, however, does not describe Jesus' words, but those of the evangelist as he develops the new religion of Christianity. So although in the gospels Jesus refers to afterlife, he doesn't give it much importance. He certainly doesn't describe it.

We owe to Paul the great importance that resurrection and life after death play in the Christian religion. It was Paul who first claimed that Jesus' death on the cross saved us from our sins, and that by his resurrection Jesus defeated death. According to Paul, in Jesus' second coming, or parousia, all true believers will trade their earthly, corruptible bodies for incorruptible heavenly garments (1 Cor. 14:53).[5] It is probable that at first Paul associated resurrection only with Jesus, not with humankind. He expected Jesus' return to take place momentarily, at which time believers would just progress from an earthly existence to a heavenly one. It was only when the Coming was apparently delayed and some believers started dying that Paul stressed our human resurrection also. As we will see in chapter 22, this confused his theory a little, since he had been preaching that by Jesus' resurrection man had escaped the hold of death. Apparently the death of believers must have caught him by surprise.

In today's America, forty percent of Jews, eighty percent of Protestants, eighty-four percent of Catholics, and even fifty percent of those without religious affiliation believe in some kind of conscious life after death.[6] But is there any proof?

Evidence of Survival

The first problem when discussing evidence of any kind is its reliability. Three people look out of the window: the first says, "I see a bright disc"; the second says, "I see the full moon"; the third says, "I see a UFO." Here the first person's report seems error-free: he reports what he sees. The other two are interpreting what they have seen, introducing a possible error in their statement. Since full moons occur once a month, while UFO sightings are rare and most people don't believe in them anyway, we would assume that the second observer's conclusion was more error-free than that of the third's. But if we had checked the calendar and noticed that the full moon was many days away, then the third person's interpretation might be preferable to that of the other. All three cases, however, are still subject to the fabrication error: maybe there was nothing out there to see. Such fabrications are particularly probable when a person offers evidence that, if true, would prove the point he is trying to make. I am surer of the existence of a ghost that I have personally seen than of one that a friend tells me he has seen. I am even less sure about the existence of a ghost I only read about.

We will now enter an area where imagination runs rampant. People see what they want to see and, often, if they don't see it they fabricate it. Every so-called fact has to be taken with a great deal of salt.

A. *Ghosts*

One of the most common metaphysical events is the apparition of ghosts. I would venture to say that during their lifetime, at least half the population will either see a ghost or will talk to a person they trust who has seen one.[7] A ghost is defined as a disembodied consciousness (or soul, if you wish) that appears in bodily form to one or more people. They can be grouped in various categories:

Crisis Apparitions. An apparition of a person who has just died is seen by a loved one, but not by others; or an apparition of someone who either needs help or wants to help is seen by persons who might be able to provide or use help. The first type is usually explained away as a telepathic event between loved ones, although traditional science does not yet accept the existence of telepathy.

Because of the inherent time delay, the second type cannot be explained as plain telepathy between two living consciousnesses, since the sender is presumably already dead. John Fuller documented an example in *Ghost of Flight 401*. Following the crash of an Eastern Airlines plane in the Everglades in December 1972, apparitions of its captain and flight engineer were reported on other Eastern Airlines planes that had been fitted with replacement parts removed from the doomed plane. Fuller argued that the apparitions were trying to prevent a similar malfunction in other planes of the fleet. Despite the airline's attempt to hush up the situation, it became a well-known secret until eventually the airline was forced to remove all the replacement parts cannibalized from the fallen plane. Of course, the reader must take the author's word for the veracity of the entire story, although independent confirmations of the events do appear to exist.

Many other documented stories describe cases where the ghost of a person killed in an accident attempts to get help for someone else lying in the wreckage, or where the ghost of a murdered person attempts to point out the killer. Similarly, almost sixty percent of widows or widowers report having seen the ghost of their dead spouse at least once (but never more than three times) after the funeral.[8] Such apparitions are usually explained as the residual energy of a consciousness that forces itself to become visible to someone in order to accomplish a specific goal.

Recurring Apparitions. These situations describe ghosts that perform the same repeated actions, at the same location, and which are visible to most nearby observers. These are attributed to the residual energy imprinted on its surroundings by the living personality during a particularly stressful event in its life. Some say a time slip of some kind may also be involved. Castle- and house-haunting cases are usually of this type.

A different explanation that has been proposed is that the consciousness of the dead person has not realized its own death, and so has not moved on along the path normally taken by the spirits of the dead. It thus remains in its old environment, usually in its house, place of work, or favorite place of entertainment. It often objects to new people moving in and to alterations in the physical building. More often than not, these situations do not involve visible apparitions, but noises, lights, and swinging doors and windows. These ghosts often leave if they are forcefully told to do so.

Poltergeist Activity. These involve flying pictures, dishes and other objects in the house. Researchers have almost always associated the events with a specific person in the house, usually someone young who presumably cannot control his hidden paranormal abilities. Thus no ghosts are involved. But although ghost investigators are not interested in the phenomenon, shouldn't someone else be? If these events are indeed real, how would the laws of physics have to be rewritten to explain how our anger can cause a dish to levitate and smash onto a nearby wall? What type of forces are involved? How would we have to change our previous discussion of the mind, or perhaps even of prayer? But perhaps it is not our mind that causes such activities. Maybe demoniac possession of the type we will be discussing in the next chapter causes them. In that case where is the dividing line between ghosts and spirits?

B. Near-Death Experience

In 1975, Dr. Raymond Moody published his astounding book *Life after Life*, relating incredible stories told to him by people who had been very close to dying or who had been actually pronounced clinically dead by attending physicians.[9] According to these stories, people close to death had amazing experiences of dissociation from their bodies, meetings with disembodied beings, and encounters with dead relatives. Soon thereafter Drs. Kübler-Ross, Ring, Morse, and others published books with similar stories. What all these writers claimed is that people who came very close to death had some common experiences during the period they were presumably unconscious:

> They saw a bright light shining upon them. They left their body and floated upward, with or without help from the light. They experienced a strong feeling of peace and security. They looked down and saw their own prone body as well as any medical or other personnel that were attending them.[10] They floated away from the scene, usually at high speed and sometimes through a tunnel. (One author reported that Hindus described making the trip on top of a sacred cow.) They met a creature of light, which adult Christians identified as Jesus, and some children thought was their grandfather. (People from other traditions made culturally appropriate identifications.) They were filled with love.
>
> In these places where they visited, those who had near-death experiences saw their entire life pass before them, and felt the hurt they may have caused others during their life. Significantly, however, they were

> always the only ones to judge their past actions. There were no exceptions to this; there never was a judge. They may have met dead relatives and others in a place where everything was shining, and where there were no shadows. Sometimes they were given explanations regarding the workings of the universe, but these were always forgotten upon their return to their earthly bodies.[11] Sometimes they were offered a choice whether to return or not, but more often they were told that their time had not come yet. Usually, however, they liked being there so much that they wanted to stay forever. The trips always ended with the people suddenly finding themselves back in their body, frequently in considerable pain from whatever had been afflicting them.

Few of these people talked about what had happened, and those who did stopped very quickly, since they realized that nobody believed them.[12] The event, however, had tremendous, lifelong effects on them: they became much more spiritual, but they usually stopped going to church when it became obvious that their new experiences conflicted with the conventional religious teachings; they became unafraid of death; they decided to enjoy life in a more active matter.

Not all episodes, however, were uniformly pleasant. Tormented people whose near-death experiences were the result of suicide attempts, were equally tormented in this other world, being able to see the harm that their actions had caused. On the other hand, there were some cases of young people who later attempted suicide in order to return to this wonderful, unearthly place.

What do these things mean if they are indeed true? At a minimum, that consciousness does not get snuffed out like a light at the end of life, that something peaceful and loving exists beyond the constraints of this world. As you might have expected, most scientists don't agree with such interpretations. They point out that the same experiences can occur from oxygen deficiency or the release of endorphins from the system when the body feels the onset of death; also by the use of drugs in medical interventions, through stimulation of the brain by electrodes in experiments, or when subjecting test pilots to centrifugal accelerations of 6 Gs. Furthermore, people with synesthesia (discussed in chapter 16) have also reported the experience of watching themselves walk as seen from above.

The advocates of the existence of near-death experiences respond that these situations only describe additional techniques that enable the mind to escape its physical constraints and to enter into the nonphysical world. They point out that in the past, aboriginal tribe members often used mescaline and other drugs to help shut down the interference of the body's physical inputs, leaving their minds free to float in this other, real world. Besides, the fact that many roads lead to a certain place does not mean that this place does not exist.

C. After-Death Communications

Near-death experiences are stories told by people who probably were never actually dead, just close to it. In contrast, after-death communications involve people who are dead. In 1995, Bill and Judy Guggenheim published their bittersweet book *Hello from Heaven*, which relates a few hundred of the more than three thousand stories they say they collected from interviews with people who reported encounters with the dead.

The reasons for such messages fall in various groupings: saying good-bye, apologizing for past misdeeds, warning of specific problems ahead, giving specific advice, seeing relatives born after someone's death, or just saying "I love you." Sometimes the body appeared to be completely real, sometimes it was transparent; sometimes only a portion of it was visible, and sometimes it remained completely invisible but could be identified by its voice or distinctive perfume.

By far the most common occurrence reported was the "I am OK, good-bye now" type. Because of this, these experiences are often called grief-induced hallucinations. A loved one dies suddenly before he had a chance to properly say good-bye, so he comes back to correct the omission. Or he comes back to say that everything is well and happy in the world he now finds himself. A young bride is walked down the aisle by her stepfather and she catches a glimpse of her dead father smiling approvingly. A dead grandmother appears and smiles happily at her new grandson. A dead father tells his daughter that everything will turn out all right and to not worry because he will be watching over her. All these situations serve to satisfy our anxiety, so a skeptic would argue that they are products of our subconscious, designed to make us feel better.

The cases in which the farewell messages were delivered before the person who received them even knew of the other person's death can be explained, as I said earlier, by postulating some type of telepathy. A dying person attempts to say good-bye to his loved ones. But science does not accept the existence of telepathy any more than it accepts life after death.

Some apparitions, however, cannot be explained away so easily. There are cases in which the apparition advised someone to take action that saved him from harm. If true, the only possible refutation could be that the advice giver was not a dead person but a disguised angel of some type—another thing in which orthodox science does not believe, and something that we will discuss in chapter 18. The only occurrences that cannot admit of any explanation other than outright fraud are those where the apparitions were dressed in some of their favorite clothes that the viewers had never seen before; or when the apparitions said something that made no sense to the viewer but which others could relate to the dead person.

Two interesting features occasionally appear in the stories related to the interviewers: "Let me go," plead the apparitions; and "I have to go now" or "I can't stay any longer." The first shows that the dead person may be somehow tied up with the living. Just like parents have to let go of their children when they grow up so that the children are free to go their own way, in the same way we also have to stop crying for the dead and release them.[13] The second statement may indicate either that the spirits are busy and have things they must do or that their ability to manifest themselves is limited.

So how should we evaluate these phenomena? We can say, perhaps, that they are all fabrications. This seems highly unlikely when we consider the number of years that went into the documentation of these cases and the number of interviewers involved. We can say that the respondents were fooled by their subconscious desires, and doubtlessly some were. We can say that responders said outright lies, and perhaps many did.

But if even only one out of these thousand cases were true, if only one case corresponded to an actual apparition of a dead person, then the authors' position would be substantiated. Only one dead person needs to appear in order to prove the existence of life after death.

D. Aided Communications with the Dead

At the beginning of the nineteenth century, and particularly in London, much attention was given to mediums and other psychics who maintained they could communicate with spirits of the dead. But because so many of the psychics proved to be frauds, this belief soon fell out of favor. Besides, even those who were not outright swindlers could have been detecting telepathic signals from their customers, helping them guide their answers. Furthermore, how can anyone identify who is really talking from the other end? In any case there are probably sufficient reasons to discount mediums or ouija-board experiences.

The Problem of Memory

The easiest and most successful way to discredit the idea that one's consciousness persists after death is to ask how it can still remember its past. Most scientists believe that they know how memory works: it is embedded in the connections between brain neurons. We have already discussed in the previous chapter that when the synapse (the connected end) of a neuron fires, it activates the neurons connected to it, whose synapses then also fire to activate still other neurons, and so on, until a section of the brain lights up. According to these scientists, specific memories are encoded in these so-called engrams of the activated synapses in our brains. Computer and TV screens are crude analogies. Each one contains a few thousand pixels and, depending on which pixels are activated, an image is formed; an almost infinite number of pictures can be generated by a small, finite number of pixels. But break the screen, and you will destroy its ability to present any more pictures. According to most scientists, therefore, we should expect that when we die and the neurons in our brains stop firing, our memories would disappear.

So how can our memories survive our death? The only possible way for this to happen is if a person's memory is stored outside his brain! Strange as it may seem, this is not a new idea. As far back as the early 1900s, psychiatrist Carl Jung concluded that certain aspects of dreams could only be explained by postulating the existence of a collective unconscious, shared by all people. How far-fetched are such ideas?

A Universe of Fields

The space around us is full of information, some of which we are constantly accessing, encoded in electromagnetic and other type of fields.[14] Electromagnetic waves of certain frequencies are recognized by the retinas in our eyes, rendering visible the objects in our environment that emit or reflect them. Other frequencies are sensed by our bodies' skin receptors as heat or cold. Others carry music or pictures to our radios and TV sets. Gravitational fields cause planets to be attracted to stars and keep us from flying off the earth. The postulated, but never discovered, Higgs field that we discussed in chapter 3 is supposed to endow us and other physical objects with mass.

Fields are associated with energy and thus with mass, since, according to Einstein, energy and mass are interchangeable. As physicists continued developing the theory of quantum mechanics, they found that the basic physical particles could be described better as packets of energy than as little balls of matter. Physicists insist that these fundamental entities act as physical particles with specific properties only when man observes them. In fact, Niels Bohr, the founder of quantum mechanics, went as far as to say that these fundamental particles have no properties until they are examined.[15] In other words, it is our own interaction that gives them physical substance.

A new generation of scientists has carried this thinking even further. Some of them think that perhaps even the entire universe can only be described mathematically. Stephen Hawking, in particular, perhaps the greatest physicist of the late twentieth century, thought that the universe has its own wave function, which would indicate the existence of an infinite number of universes of dream-like substance, universes defined by mathematical equations but invisible to humans.[16]

David Boehm and Karl Pribram went further, proposing that the universe and everything in it can be described by mathematical fields similar to virtual holograms. A hologram is a picture taken with lasers impinging on the object from two different angles. The result, when viewed under laser light, is a lifelike, three-dimensional representation of the object. Holograms have a strange property: the entire picture is captured in even the smallest portion of the picture—a property which they share with fractals. Cut a small part of the picture and view it under laser light, and you will see the entire object. Thus the image of any part of the picture is stored at all other locations in the picture.[17] This is somewhat similar to field equations that describe the field properties everywhere in space, not just at a particular location.[18]

All this affects greatly our concept of memory. If the universe is a holographic construction, reasoned Michael Talbot (1991), then so also are our brains; as a result, our memory is not localized in the gray matter inside our heads but is spread out throughout the entire universe. Looking at it in a simplistic manner, just like our TV selects from space the electromagnetic waves of a specific frequency and then decodes them to present on the screen our favorite shows, so also our brains first broadcast and then retrieve our memories from the universe around us. And since every location in holographic records (and wave fields) contains the entire recorded information, every point is interconnected with every other point. So if one could adjust the tuning frequency of his mind, he would be able to sense everybody else's thoughts—this is telepathy.

A memory field external to our minds can also explain reincarnation stories. There have been many apparently well-documented cases in which young children, two to five years old, remembered having lived before as other people.[19] They were able to describe people, localities, and events in remote towns, assertions that were then confirmed to be true. It is perhaps possible that a young untrained mind may be able to tap by mistake into the wrong external memory database. In all cases, such personal reincarnation beliefs disappeared by the time the child was ten years old, when either the child had presumably attained better control over his own memory-retrieval mechanism or the external field had faded. What this does not explain, however, is that in all cases the foreign memories belonged to recently dead persons. Why were children not able to tap into the memories of living people? Is it because the memories of the dead are more accessible, not tied down to somebody?

Let me diverge for an instance from memory fields and talk about "personal" fields. Some spiritualists maintain that the physical body (or perhaps a duplicate ethereal body to which it is attached) emits a form of radiation whose intensity and color relate to the inner quality of the individual. Hindus call this a person's *chakra*.[20] They insist that it is possible to judge the spirituality or even the physical health of an individual through the strength and color of his chakras. It has been proposed that the reason saints have always been drawn with halos around their heads is that their holiness made their chakras visible to all. It is also said that sometimes the chakras look like tongues of flame, and that explains (did you guess?) the appearance of the Holy Spirit on the head of the disciples during Pentecost. As I said at the outset, some things are harder to believe than others, unless, of course, you see them yourself.

In the Creation chapter, I discussed the scientists' opinions that matter and energy were self-created in accordance with laws of physics, and then asked who had created these laws. Mathew Fox (1996, 9-10) offers an answer:

> The evolutionary cosmology throws the old idea of eternal "laws of nature" into doubt. If nature evolves, why shouldn't the laws of nature evolve as well? How can we possibly know that the "laws" that govern you and me—the crystallization of sugar, the weather and so on—were there at the moment of the Big Bang? In an evolutionary universe it makes more sense to think of the laws of nature evolving too. I think it makes better sense to regard the regularities of nature as habits. And the habits of nature evolve. Instead of the whole universe being governed by an eternal mathematical mind, it may depend on inherent memory.

It has become popular nowadays to describe the universe as an evolving, and hence a living, entity: "The ground of all being," as Paul Tillich was fond of saying. We will discuss his ideas in chapter 28.

SUMMARY

Paul never talked about the kingdom of God, only about the return of Christ Jesus. When it happened, all those baptized in his name would be transformed into new, ethereal bodies and have eternal life. If they had died beforehand, they would be resurrected for the event. The gospels, on the other hand, paid greater attention to the coming kingdom of God when all evil would disappear from the face of the earth; they did not discuss everlasting life in any detail. Two millennia have elapsed since Jesus' death, and there still has been no clearly discernible kingdom of God. So the question arises, is there life after death?

Although scientists say that they have no proof that anything survives one's death, there exists a large number of stories describing the appearances of disembodied dead people and their interactions with the living. If true, these stories would indicate the survival of our consciousness, including memory, after our death. Our present physical understanding considers this to be impossible, but in recent years many acclaimed pioneers in the area of abstract physics have proposed that our world and everything in it are constructs of our imagination rather than fixed reality. These scientists suggest that all things are energy manifestations of electromagnetic and other fields, which are construed by our brains as tangible physical things. Such theories would explain many spiritual and paranormal occurrences whose existence is currently disputed by conventional scientific wisdom. Memory could be just an entity in a memory field, attached to an individual being during his lifetime but able to survive his physical death.

Some Questions to Consider

1. The Hebrews encountered the concepts of soul and life after death during their Babylonian exile in 600 BCE. Why did they not write anything about it for four hundred years?

2. Have you ever seen a ghost? Do you know of anyone who has?

3. Some ghost manifestations are explained away as results of paranormal phenomena whose existence, however, science does not accept. Is science wrong in refusing to accept the existence of paranormal events or in refusing to accept the possibility of life after death?

4. In chapter 3 the author insinuated that scientific experiments couldn't be trusted because their results always corroborate the existing theories but are later modified to justify new developments. Now he presents some half-baked ideas that completely lack any credible documentation. Science seems to work everywhere around us, why should we mistrust the scientists?

5. Can you connect this discussion about memory fields distributed throughout space with the discussion in chapter 7 where we talked about people getting their inspirations seemingly out of thin air?

6. Scientists tell us that subatomic particles don't even have properties until we try to measure them. Does this imply that our very act of discernment creates that which we are trying to discern?

Selected References

Atwater, P. M. H. with David H. Morgan. *The Complete Idiot's Guide to Near Death Experiences.*

Bechtel, William, and Adele Abrahamsen. *Connectionism and the Mind.*

Guggenheim, Bill, and Judy Guggenheim. *Hello from Heaven.*

Kovach, Sue. *Hidden Files: Law Enforcement's True Case Stories of the Unexplained and Paranormal.*

Moody, Raymond A., Jr. *Life after Life.*

Ring, Kenneth, and Sharon Cooper. *Mindsight: Near-Death and Out-of-Body Experiences in the Blind.*

Sheldrake, Ruppert. *The Rebirth of Nature: Science and God.*

Vincent, Ken R. *Visions of God from the Near Death Experience.*

CHAPTER 18

ANGELS, DEVILS, AND MIRACLES

Although the Bible rarely mentions our surviving souls, it often refers to angels and demons. In the Old Testament, angels are presented as God's creatures, acting as his messengers and, sometimes, as members of his council, with Satan cast in the position of prosecuting attorney. Starting, however, with the apocalyptic writings around 200 BCE and throughout the New Testament, Satan is presented as a fallen angel, God's enemy, who with his army of satans and demons tries to bring evil to God's creation. Of course our scientifically minded society does not believe in the existence of angels or satans, or in God for that matter. But many of the so-called "enlightened" Churches also do not believe in a sentient evil power, and consider exorcisms and other such ideas to be Middle Age superstitions.

Miracles are incidents that our modern society and even some of our Churches disbelieve. But defined as events of extremely low probability, there is no scientific reason to dispute their occurrence. It is their fortuitous timing that renders them so extraordinary. And if we look carefully at our lives, most of us will see that they are full of little and big miracles. "It was a miracle" is a fairly common expression in our vocabulary.

ANGELS IN THE BIBLE

> So [God] drove out the man; and he placed at the east of the Garden of Eden the cherubim and a fiery sword which turned every way, to keep and guard the way to the tree of life. (Gen. 3:24 AMP)

This is how the Bible first introduces angels. Unless, of course, one counts the earlier times when God spoke in the plural: "Behold, the man has become as one of *us*, to know good and evil" (Gen. 3:22, my emphasis). Or even earlier: "Let *us* make man in *our* image, after *our* likeness" (Gen. 1:26, my emphasis). Much has been written regarding who God's listeners were. Those who suggested that he was perhaps talking to his fellow gods were sternly censored by the monotheistic Christian church, which asserted instead, that he was talking to angels. But if this is really the case, then the angels' makeup must be pretty close to God's. Another explanation, of course, is that he was talking to the pre-incarnated Jesus, the *Logos* that existed before time.

And when were the angels created? The Bible does not say. By the seventh day, however, "the heavens and the earth were finished, and all the *host* of them" (Gen. 2:1, AMPL). So we can assume that they were created sometime during the first six days.

The word "angel" (*angelos* in Greek, *mal'ak* in Hebrew) means messenger. Angels are God's messengers, bearing messages from him. In the books of Genesis and Exodus, however, they are often indistinguishable from God himself. Time and again, the text describes the appearance of an angel, only to refer to him later as God. When Moses sees the burning bush, for example, we read: "And the angel of the Lord appeared to him in a flame of fire out of the midst of the bush....And when the Lord saw that he turned aside to see, God called to him out of the midst of the bush" (Ex. 3:2-4).[1] Even in the later book of Judges we read that an angel appeared to Manoah to announce the birth of his son Samson, and at the end of the incident "Manoah, realizing that it was the angel of the Lord, said to his wife: 'We shall surely die, because we have seen God'" (Judges 13:22); to him God and the angel were one. Of course since

God is not a physical being, when he does become visible, he must look like something that he is not, and the closest thing might be an angel. Although artists have usually drawn angels as ethereal creatures with wings, in the Bible they usually appear like normal men. Perhaps the most confusing apparition of God was when he appeared to Abraham as three separate men (in Genesis 18:2). At first it was not clear whether God was one of them or all three, since they all spoke in unison. Later, however, the text identifies one as the Lord, and the other two as angels.

Angels are supposed to be lower than God but above man. They are not that different from men, however, because soon after creation they came down to earth and intermarried with human women. "And it came to pass, when men began to multiply on the face of the earth, and daughters were born to them, that the sons of God saw that the daughters of man were fair; and they took [for their] wives all those they chose" (Gen. 6:1-2).[2] Since these marriages resulted in (giant) offspring, it would appear that biologically, angels and humans are pretty close. In any case this contradicts Jesus' later statement that angels are sexless. When asked about resurrection he said: "When [people] rise from the dead, they neither marry nor are given in marriage; but are like angels who are in heaven" (Mark 12:25).

Only two angels are named in the Bible: Gabriel is the messenger who talked to Daniel (Dan. 9-21), to Zechariah (Luke 1:19), and to Mary (Luke 1:26); Michael is the warrior, the protector of the Hebrew people. In the Book of Daniel, Michael is clearly described as the Messiah expected at the end of time. A little after the famous "son of man" passage, whose protagonist remains unidentified, comes the following, clearer text.

> At that time Michael will stand up, the great prince who mounts guard over your people. There is going to be a time of great distress, unparalleled since nations first came into existence. When that time comes, your own people will be spared, all those whose names are found written in the Book. Of those

> who lie in the dust of the earth many will awake, some to everlasting life, some to shame and everlasting disgrace. (Dan. 12:1-3 JER)

Just as the Hebrews had their guardian angel, so apparently did all the other nations in the world:

> [The] God Most High gave land to every nation. He assigned a guardian angel to each of them, but the Lord himself takes Israel. (Deut. 32:8-9 CEV)

Angels served as God's court and on occasion as his advisers; perhaps because they were closer to humans they could think more like them. There is a scene in the Bible where prophet Micah describes how God assembled his host and asked them how they could deceive the king of Israel so that he would attack one of his enemies and in the process be destroyed. One of the angels suggested that he could become a lying spirit in the mouths of all the king's prophets and give him the wrong advice. God agreed, and predicted success for the operation (1 Kings 22:20-23). A similar council is described at the beginning of the book of Job.

Not counting their appearances in connection with events related to the births of Jesus and John the Baptist, angels are not mentioned often in the New Testament. We do read that when Jesus was chased into the desert following his baptism, angels ministered to him (Mark 1:13). Later, Satan tempted him to jump from the temple parapet since God's angels had been instructed to protect him. There are also two other intriguing mentions of angels. In Matthew's gospel, Jesus calls a child and places it in the midst of his disciples and tells them: "See that you never despise any of these little ones, for I tell you that their angels in heaven are continually in the presence of my heavenly Father" (Mat. 18:10 JER). He thus seems to imply that they each have a guardian angel in heaven.[3] In Luke's Acts (12:7-15), Peter is rescued from jail by an angel, and escapes to the house where the disciples were gathered. When he knocks on the door a maid answers, who upon recognizing his voice becomes so excited that instead of letting him in she runs to tell the others that he is outside. It can't be Peter, they tell her; he is in jail—it must be his angel.

What the Bible did not say about angels, however, later writers did. "I saw a hundred thousand times a hundred thousand, ten million times ten million, an innumerable and uncountable multitude who stand before the glory of God," says Enoch (40:1). He also describes how two hundred of them made a pact and descended to earth to take human wives, and to teach man many arts and sciences (1 Enoch 6-9). A few centuries later, Pseudo-Dionysius divided the angels into nine categories according to their type.[4] And along the way someone decided that although there were millions of angels, one third of them fell from grace to join Satan.

The ancient idea of guardian angels still exists. It appears that most people do believe that even if they may not have a full-time guardian, something divine occasionally intervenes on their behalf in times of need. Numerous books have been published purporting to relate stories of magical, unexplainable encounters with the unknown. Sometimes these involve supernatural rescues from life threatening situations, and sometimes they are just simple acts of grace. Whatever they are, they are still miracles, and to me angels and miracles go hand in hand. The first may cause the second, or they may both be parts of the mystical world around us that is not perceived directly through our senses.

MIRACLES

Yesterday...
all my troubles seemed so far away,
Now it looks as though they're here to stay,
Oh I believe...in yesterday.

Suddenly...
I'm not half the man I used to be,
There's a shadow hanging over me,
Oh yesterday...came suddenly....

Why she had to go I don't know.
She wouldn't say.
I said something wrong,
now I long for yesterday.

Yesterday...
love was such an easy game to play,
Now I need a place to hide away,
Oh I believe...in yesterday....

("Yesterday," by John Lennon and Paul McCartney)

Not being a Beatles aficionado, I had never heard this song, or at least remember hearing it, until one morning many years ago. It was two days after my wife's sudden death, and I was driving to the local supermarket with my two daughters to buy supplies for the luncheon that would follow the funeral, when the radio played it. The song's timeliness was so obvious, that my eleven-year-old daughter turned around and looked at me intently. Being a macho man, I of course kept my eyes straight ahead and showed no evidence of having heard anything special. Thinking back at the event, I suppose that it would have been much better if I had hugged her instead. A year went by before I heard this song again. This time it was the anniversary of my wife's death and I had sneaked out from work to go buy a red rose. I was on my way to place it on her grave when the radio played that song again. I didn't hear it again until twenty years later when it was included on a Carreras tape that I received as a present.

A miracle is a coincidence that defies all probability expectations. To me, hearing that song on the two occasions that I did, was a little miracle. I am still not sure what it meant, except perhaps that someone, somewhere, cared for me and wanted me to know it. In his *Moments of Grace,* Neal Walsch writes: "God talks to us in many ways every day. God is shameless and will use any device to communicate with us. The lyrics to the next song you hear on the radio..." (Walsch 2001, 40). I believe that if you examine your own life carefully, you will identify many instances when something extremely fortuitous or just plain wonderful happened to you, seemingly by chance. Some people say nothing happens by chance; that everything is planned. Perhaps.

The New Testament is full of miracles. But most people doubt them, even religious people. They say miracles don't happen; everything must have a scientific explanation. Perhaps they are right, but I don't think that our science has even started realizing the immense potential that is hidden in our human being, the seeming miracles that our unconscious can perform. We have already talked about people with multiple personalities who require different prescription glasses for each persona; where one personality may be diabetic and the other not, something that causes severe medical problems when the diabetic personality that just received an insulin injection switches to the non-diabetic personality and goes into an insulin coma.

In Jesus' days, prophets and men of God were expected to perform miracles in order to prove that God had empowered them. Some of the miracle workers of those days were Hanina Ben Dosa in Galilee (c. 70 CE), Eliezer ben Hyrcanus (c. 90 CE), Eleazar (c. 68 CE),[5] and Apollonius of Tyana (c. 96 CE), whose life story greatly resembles that of Jesus. Furthermore, many of the miracles that Jesus performed were similar, though greater, to those that prophets had carried out in the Old Testament. In the eyes of his contemporaries, this proved that Jesus was a prophet (though not God). So did Jesus really perform the miracles described in the Bible? Unless you attend a Baptist or an evangelical church, you will probably never hear a sermon that explains a miracle as a real event; preachers always pussyfoot around it. Yet that does not have to be the case.

As I said before,[6] it is true that Jesus could have never made a congenitally blind person see; it is a biological impossibility. We should probably ascribe this miracle to the imagination of the evangelist. It is found only in the gospel of John, whose historical veracity is always suspect, and the miracle is justified by the most uncharitable reasoning found in the New Testament:

> His disciples asked him, "Rabbi, who sinned, this man or his parents, for him to have been born blind?" "Neither he nor his parents sinned," Jesus answered, "he was born blind so that the works of God might be displayed in him." (John 9:2-3 JER)

Condemning a person to spend his childhood and young adulthood blind so that he himself may be exalted does not describe any God that I would care to venerate. I expressed a similar feeling when talking about Job, but at least there we were dealing with the God of the Old Testament. I would like to believe that the God of the New Testament would show more love and empathy for a poor human being.

It is also probable that Jesus did not walk on water, although that may not be as impossible as one might think. Anybody who has visited Wisconsin Dells in the summer has probably seen people water skiing on their bare feet. It is true of course that they are pulled by a motorboat at great speeds, but consider also the movie *The Karate Kid,* where the old pro teaches the young kid to step so quickly on the water that he does not have time to sink. I have never seen any real humans do this, but PBS used to show on television a long-legged bird doing this exact quick-step to skip over the surface of the water.

The miracles of the multiplication of the loaves of bread and fish are usually explained in non-metaphysical ways, by assuming that when the disciples started sharing their food with the people, they in turn took out their own food, which they had hidden away, and shared it with their neighbors. After all, the argument goes, no peasant ever leaves his house for a trip without packing something to eat. These miracles then are not ones of physically multiplying the available quantity of food, but of opening the hearts of people.

All the other miracles, however, could be literally true. Maybe those people who were actually physically sick believed enough for their subconscious to tell their body to respond as ordered—we have already discussed the power that the mind has over the body. Perhaps some of them suffered from hysterically induced sicknesses. Paralysis, deafness, and blindness can all be induced by stressed minds, and in that society everyone believed that sins caused diseases. A person who believed he had sinned could make himself sick, and Jesus was able to cure him by telling him that his sins had been forgiven. Furthermore, remember that Jesus always told the person that it was his own faith that had cured him. This applies particularly to the woman who had hemorrhaged for twelve years, who sneaked close to touch his cloak and immediately became well. Interestingly, this is the only place in the Bible where Jesus' power is quantified: "Someone has touched me; for I sensed that power went out from me," he said (Luke 8:46).[7] But his power by itself was usually not sufficient to perform a cure; the sick person also had to believe. This is why he had so little success in his own home town "where people had little faith."

THE DEVIL IN THE BIBLE

What about the angels' opposite, the demons, and their boss, Satan? What does the Bible have to say about them? In the beginning, when written with a small "s" the word "satan" was a descriptive noun, not a semi-divine being. In its first use in the Old Testament it meant "adversary," and it was applied to an angel who opposed the seer Balaam.[8] (Remember that the Bible does not mention Satan in connection with Eve.) It was used in a similar manner in the book of Kings and in the Psalms.[9] Another meaning for the word "satan" is "accuser," such as a prosecuting attorney at court. And this is how Satan appears in most of the Old Testament—as a loyal member of God's court acting as a prosecuting attorney. Such actions are described in Job and in Zechariah 3:1. Even well into the Talmudic period (third century CE) he was commonly called *Satan mekatreg* (Satan the accuser).[10]

The word devil derives from the Greek *diabolos*, which also means accuser or slanderer.[11] It is a synonym for Satan, and the two words are used interchangeably in the Greek translation (Septuagint). Half of the time it is a descriptive noun, but the other half it refers to semi-divine enemies of God and humans. These first make their appearance somewhere between the writing of the second book of Samuel and of the first book of Chronicles.[12] Both books describe the same event: David orders that a census be taken in order to determine his military strength; God then punishes him, or rather punishes the Hebrews, for placing their trust in their own abilities rather than in their God. In Samuel we read: "The anger of the Lord was kindled against Israel and he incited David against them to say, Go number Israel and Judah" (2 Sam. 24:1). In the Chronicles, which were written at a later time, we read: "And Satan rose against Israel, and provoked David to number Israel" (1 Chr. 21:1).

Sometime between the writing of the two books, the idea was born that satans were semi-divine beings, independent and hostile to God. Up to that time all events, good or bad, had to be attributed to God, perhaps to his dark side. But with the introduction of the satan concept, it became possible to clear God from all responsibility for the bad things that happened to man; they could all be blamed on Satan or his followers—the satans, the demons. This explained away God's seemingly unjustified behavior of first inciting man to do something, and then punishing him for doing it.[13] We must remember that it was during their Babylonian exile that the Hebrews first became acquainted with the concept of a sentient evil power opposed to God. Satan with a capital "S" entered Hebrew literature in the second century BCE, with Enoch and the Jubilees. By the time of Jesus, he was well established and had his own proper name, Beelzebub (or Beelzebul in the original Greek).

In the New Testament, Satan is one of the protagonists. Although he rarely makes an appearance, his power is always apparent. He leads a demonic kingdom in a continuous effort to divert the veneration of humans away from God. And the Bible pictures Jesus as fighting a continuous battle against him, from his first temptation following his baptism to his death upon the cross. Did Jesus win?

Jesus attributed the success of the exorcisms he performed to the fact that the master of the demons had been defeated.[14] "But if I with the finger of God cast out devils, no doubt the kingdom of God has come upon you" (Luke 11:20). But driving demons out of a person was a small battle, a minor skirmish. As he himself pointed out:

> When the unclean spirit is gone out of a man, it walks through arid places seeking rest, and finding none, it says, "I will return to my house from where I came." And when it comes it finds it swept clean and put in order. Then it goes, and brings back seven other spirits more wicked than itself and they enter in and dwell there, and the last state of that man is worse than the first. (Luke 11:24-26)

Displacing an occasional demon, who might come back anyway, hardly seems a great victory. It is apparent that Jesus could not beat Satan; what he could do, if it was indeed true, was to show a person the way to God's kingdom, inside which Satan had no power. So when the people in the cities of Chorazin, Bethsaida, and Capernaum refused to accept his message, he could only voice his frustration, telling them that they will "go down to the netherworld" (Luke 10:15). Even after Jesus' death on the cross and subsequent resurrection, an event that Paul described as a victory over death, Satan remained basically untouched.

> Put on the whole armor of God [so] that you may be able to withstand the wiles of the devil. For we wrestle not against flesh and blood but against principalities, against powers, against the rulers of the darkness of this world, against the spiritual wickedness in the heavens. Therefore, put on the whole armor of God, that you may be able to withstand in the evil day and, having done all, to stand [your ground]. (Ephesians 6:11-13)

Satan's power on earth is acknowledged in all the gospels. The Lord's Prayer that Jesus recites in Matthew's gospel, includes "and do not subject us to the final test, but deliver us from the evil [one]" (Mat. 6:13). Interestingly, John's gospel does not recount any exorcisms, perhaps because in that gospel Satan is never shown to lose his power. In his discourse at the Last Supper, Jesus tells his disciples that "the ruler of the world is coming" (John 14:30). In the later letters of John, we read: "We know that we are of God, and the whole world is under the power of the evil one" (1 John 5:19 AMP).

The Devil in Modern Life

It used to be that most Christian religions believed in the existence of the devil, and some even performed exorcisms to fight him off. I am not sure to what extent they still do, at least in this country. I do know that many of them do not believe in the existence of an independent evil, sentient power, but instead call the absence of God evil. Modern scientists, of course, do not believe in anything that cannot be measured (unless, as we have seen, it is related to their various evolution theories) so they do not accept the existence of either angels or devils or, for that matter, God.

Yet all properties in life have their opposite, like the Chinese principle of duality, yin and yang. Can we think of something without thinking of its opposite? There is happiness and sorrow, light and darkness, goodness and evil. Are there angels and demons? If we believe in angels who catch us when we fall, should we disbelieve the existence of devils who counsel us to jump when we are at the edge of a precipice?[15]

I recently met an Anglican missionary from Madagascar[16] who told a story about some of his students who went into the rain forest to preach the gospel to people who had never heard it before. They set up a tent and talked about the God who had sent his son Jesus to save the world, and how Jesus would intercede with God to save anybody who believed in him. Since the natives there are animistic, believing that everything has a soul and already pray to their ancestors to intercede for them with God, they could easily relate to the Christian Jesus. At the end of the service, or its equivalent, the students asked anybody who wanted to change his life and accept Jesus to step forward. One woman did, saying that she wanted to be relieved of her job; she was a witch doctor, one possessed by a demon.

The students placed their hands on her head and started praying and exorcising the demon. Suddenly the woman started screaming, her eyes rolled up, and she went into convulsions. "What did you do then?" the missionary asked his students, when they were later recounting their story. "We just prayed louder!" they replied. The woman recovered her senses and eventually went back home. At two o'clock the same night, however, the students were awakened by the villagers shouting that the woman was experiencing screaming convulsions again. The students ran to her hut and repeated the prayers and exorcisms. This went on intermittently for three days before the demon apparently left her for good. Afterwards, the woman became a member of the new church, happily helping to spread the word. In the process she lost about thirty pounds, which the missionary attributed to her decreased income, for she could not practice her previous trade any more. "So why did she appeal for a change that would take away her livelihood and her status in the tribe?" I asked. "Because she felt captive and oppressed," explained the missionary.

Perhaps there is no better example of a satanic influence in our everyday lives than the hacker who unleashes a virus on the Internet. His only reward is the satisfaction of knowing how much grief he has caused to his fellow Internet users. If this is not malevolent, what is? To disbelieve the existence of powers that try to take us over, body and soul, is to remain utterly unprepared for them, if they exist. Compare it with a homeowner who does not believe in the existence of burglars and leaves his home completely unprotected; if burglars do visit his neighborhood, his house will surely be burglarized. In a similar vein, yoga teachers instruct us in transcendental techniques that relax and empty our minds from extraneous thoughts so that our freed-up consciousness can spread out and become more aware of the real world around us. But allowing our minds to become blank is precisely what opens them wide to invasion by evil spirits—assuming, of course, that such spirits do indeed exist.

Just as literature is full of stories about angels, so it is also full of stories about demons and their possessions.[17] But let us assume that angels and demons are not separate semi-divine creations. If we believe in life after death and in the existence of spirits who love us and sometimes help us, what do you suppose happens to the spirits of the evil and wicked people who hate us?[18] If the good spirits are free to roam and occasionally interfere in our affairs, the same is probably true for the evil ones also. As Scott Peck quipped, evil is the reverse of live; it seeks our spiritual death. But Jesus has given us the magic amulet against it: love. A heart full of love, clear of bad and envious thoughts, provides no hooks for evil to attach itself.

SOME QUESTIONS TO CONSIDER

1. What do you think of the concept that God governs the world with the aid of a council of angels and a prosecuting Satan?

2. If Satan is God's enemy, isn't God powerful enough to vanquish him?

3. Is it possible to believe in the existence of angels and not of miracles? What would be the job of angels then?

4. If Jesus' miracles were based on the faith of those affected, then Jesus did not possess personal miraculous powers. Is this true?

5. Many religions believe in the existence of Satan as the opposite of God, because everything in life consists of opposite pairs. In that case Satan would not be a fallen angel, but would have existed from the beginning of the world. What do you think?

6. Do you believe in guardian angels? Why?

SELECTED REFERENCES

Fox, Matthew, and Rupert Sheldrake. *The Physics of Angels: Exploring the Realm Where Science and Spirit Meet.*

Martin, Malachi. *Hostage to the Devil: The Possession and Exorcism of Five Contemporary Americans.*

Peck, M. Scott. *People of the Lie: The Hope for Healing Human Evil.*

Sarchie, Ralph, and Lisa Collier Cool. *Beware the Night: A New York Cop Investigates the Supernatural.*

Walsch, Neale Donald. *Moments of Grace.*

CHAPTER 19

THE DEATH OF A REVOLUTIONARY

Returning our attention to the Bible story, Jesus starts his last week riding on a high. He enters Jerusalem riding a donkey, with a huge throng welcoming him like a long-awaited king. He immediately proceeds to the Temple where he starts an uproar, overturning tables and chasing away merchants, and engages the Pharisees in heated debates. Yet no one dares to touch him until Thursday night, when suddenly everything changes. Then he is arrested, his friends abandon him, his enemies become vocal, and it seems as if someone turned off his power. The synoptic gospels next show us a beaten and dazed man led to a cruel death.

Christianity has always tried to explain the ending of Jesus' life as a long-planned sacrifice necessary to magically redeem humanity from some mythical original sin. Yet the similarities between Jesus' actions and the book of Zechariah have been well noted, even in the Bible text itself. And what Christian historians consistently fail to mention is that Zechariah's story not only describes the final battle that ushers in the Kingdom of God, but locates it at the Mount of Olives, at the exact place where Jesus went on that fateful Thursday night. Why should we assume that he had been unaware of it? Jesus did everything else to match Zechariah's narrative; he surely must have also expected the prophesied ending. He may well have thought that he was going to die, but in a majestic battle, not on the cross as a common criminal.

The Start of the Journey

Jesus asks his disciples, "Who do you think I am?" Peter replies, "You are the Messiah," the anointed one. And Jesus agrees. This makes him either the high priest or the king of Judea, since only these were anointed. But there was already a high priest in Jerusalem and, besides, Jesus had never shown much inclination to follow the details of the Judaic laws. Food, purity, Sabbath laws—he broke them all whenever it suited him. It is thus safe to assume that neither Jesus nor his disciples thought that he was the new high priest. The great epiphany at the mountaintop six days later clinched the conclusion. Standing between the greatest past lawgiver and the greatest past prophet[1] of the Hebrew nation, he heard God thundering: "You are my son!" A God's son, a king. So "he sets his face towards Jerusalem." The time to fulfill his mission had come.

But what was this mission? The gospels want us to believe that his mission was to be denounced, to suffer, to be killed, and to be raised up again. Yet his disciples did not understand. And who can blame them?

> I have come to bring fire to the earth, and how I wish it were blazing already! There is a baptism I must still receive, and how great is my distress till it is over! Do you suppose that I am here to bring peace on earth? No, I tell you, but rather division. (Luke 12:49-51 JER)

These are the words of a fighting man who heads to battle, not those of a condemned man proceeding meekly to the cross. As he approaches Jerusalem, increasingly larger crowds come out to meet him. A blind beggar cries, "Jesus, son of David, have pity on me." The son of David, the long-awaited leader, is finally here.

His entry into Jerusalem is that of a king. True, he rides on a lowly donkey, not a commanding stallion, but notice that this is the first time that Jesus is not described walking. A multitude assembles. Some people spread their cloaks on the road, and others put down leafy branches cut from the field. The crowd that precedes him and the one that follows him keep shouting, "Hosanna to the highest!" When some Pharisees ask him to restrain his disciples, he replies, "I tell you, if these keep silence the stones will cry out" (Luke 19:40 JER).

His first task was to go to the Temple to drive out those engaged in buying and selling. "He overthrew the tables of the moneychangers, and the seats of them that sold doves" (Matthew 21:12). Day after day at the Temple he performed miracles and taught the crowd; and the children cried out, "Hosanna to the son of David." The chief priest and the scribes were indignant, but they could do nothing against him because of the people.

PREPARATION AND FAILURE

The gospels tell us what follows. Jesus has a Passover supper with his disciples: he blesses and breaks the bread, passes around one or two glasses of wine, and in the process performs the first communion observance. He announces that he will be betrayed by one of those present, and then tells Judas to go and do what he must do. Although Judas immediately gets up and leaves the group, nobody understands its significance. Jesus and the rest of his disciples then go to the Mount of Olives, to a place called Gethsemane. He tells his disciples to stay awake and pray, while he himself moves apart to pray alone. He returns twice to find the disciples asleep. The first time he awakens them and scolds them lightly, but the second time he lets them be. Then Judas arrives with a crowd, temple guards, and soldiers.[2] He kisses Jesus to identify him to the guards, who then immediately arrest him. All gospels relate that somebody from Jesus' party, usually Peter, takes out his sword and cuts off the ear of the high chief's servant. Jesus, however, stops the fight and cures the servant. He is taken to the chief priest's house where the Sanhedrin meets and decides that he should die.[3] The next morning he is taken to Pilate, the Roman prefect, who was in Jerusalem because of the festival.

There are many problems, with this story. The most minor is that it could not have been the Passover meal that Jesus shared with his disciples, although it could have been the preparation meal on the day before. According to Moses' instructions, the Passover lamb is killed during evening twilight, a little before Passover. Passover starts at sunset and is a day of rest; only the work needed to roast the lamb and prepare the memorial feast is allowed. In the morning everybody goes to a sacred assembly, but does no other work except prepare the food for the feast (Ex. 12:6-16). Obviously holding a trial on that day is out of question. Not to mention the implausibility of Simon the Cyrene returning from work at his field, or of Joseph the Arimathean finding an open store to buy a burial linen cloth.

What follows Jesus' arrest is equally improbable. Mark and Matthew say that he was taken to the high priest's house, where the Sanhedrin met to condemn him. There are two things that make this event highly unlikely: 1) The Sanhedrin could only meet in a specific location, not in the high priest's house; 2) It was Passover, so holding court was forbidden. In Luke's gospel we are told that Jesus was held at the chief priest's house until the morning, and then taken to the Sanhedrin. Again, however, this would have been on the Passover day, and so forbidden. In John's gospel the Sanhedrin is not involved. Jesus is taken to the house of Annas, the chief priest's father-in-law, who questions him. He is then sent to the chief priest's house where he is kept until he is taken to Pilate in the morning. There is no Sanhedrin trial and no witnesses.

John's gospel is more believable. Holding an inquiry in private would have avoided arguments in the Sanhedrin. There, the Pharisees, who were probably in the majority, could well oppose the chief priest, as they did later when Peter was brought in front of that group (Acts 4:6). Furthermore, since the action did not take place on the day of Passover but one day earlier, everything was permissible as long as the Judeans did not enter into Pilate's house and polluted themselves. Although usually John's gospel does not even make an attempt to appear historical, in the matter of Jesus' arrest and trial it does seem to be more accurate and to provide details missing from the other three gospels. Remember also that it is the only gospel that claims that one of Jesus' disciples provided first-hand information while it was being written;[4] perhaps it was the "disciple [who] was known to the high priest, and [who] went with Jesus into the palace of the high priest" (John 18:15). We will examine who that might have been in chapter 26.

Regarding the supper itself, only Luke has the details correct. The Jewish festive meal starts with a blessing over a glass of wine. Then follows a second blessing and breaking of the bread, and after the meal and the singing of some ritual psalms, a thanksgiving over a second glass of wine.[5] Both Mark and Matthew neglected to mention the second glass. But to say, "Eat, this is my body; drink, this is my blood," is something that no pious Jew would have ever said, let alone done.[6]

It is obvious, however, that from his very entry into the city Jesus acted in a strange manner. He appeared to be in control of things well beyond the understanding of his disciples. He first sent them into the city and instructed them where to find a tied-up colt, to untie it and to bring it back after telling its owner that their master needed it. Later he told them to go into the city again and follow a man carrying a pitcher of water who would lead them to a house with a prepared upper room in which they would celebrate the Passover. These things surely suggest that he had made some secret preparations of which his disciples were unaware. During the meal itself, he told Judas to leave immediately and go do what he had to do. The gospel says that he knew that Judas was about to betray him, but this also seems strange. Judas was a zealot, a patriot not an informer;[7] he carried the group's money purse, so he must have been considered to be both trustworthy and levelheaded. Obviously Jesus knew what Judas was going to do; perhaps he had even directed him to do it. Unfortunately the gospels don't reveal any information.[8]

Before we continue, let's look at a strange exchange that takes place at the end of the supper, as related by Luke and only by him: Jesus tells his disciples, "He who has no sword, let him sell his garment and buy one. For I say to you that this that is written must yet be accomplished in me, 'And he was reckoned among the transgressors [lawless]'" (Luke 22:36-37). When they tell him that they have two swords he replies, "It is enough."[9]

And now we turn our attention to what Jesus probably thought would be the climax of his story, the events at the Mount of Olives. The assertion of the gospels that Jesus, while riding a high crest in power and popularity, went to the Mount of Olives to allow his covert arrest seems completely counterintuitive. Although Jesus often surprises us by his actions and words, such an act is beyond understanding, especially when one considers his later uncharacteristic dejected appearance as described in the first three gospels. To get a hint of what must have been in his mind, we must look into the handbook he was following—the book of Zechariah. Let's look at some pertinent excerpts.

The book starts by locating the period of the events: "In the eighth month, in the second year [of the reign] of Darius, came the word of the Lord to Zechariah…the prophet" (Zech. 1:1 AMP). Then it introduces Jesus' chief adversary, Satan (in one of his few appearances in the Old Testament): "Then [he] showed me Joshua the high priest standing before the Angel of the Lord, and Satan standing at Joshua's right hand to be his adversary and to accuse him. And the Lord said to Satan, The Lord rebuke you, Oh Satan! Even the Lord Who [now] chooses Jerusalem, rebuke you! Is not this [returned captive Joshua] a brand plucked out of the fire?" (Zech. 3:1-2 AMP).

It is then the Messiah's turn to be introduced by the Lord: "I will bring forth my servant the Branch.[10]…I will remove the iniquity and guilt of this land in a single day. In that day, says the Lord of hosts, you shall invite each man his neighbor under his own vine and his own fig tree" (Zech 3:8-10 AMP).[11] So it was expected that the act of salvation would be completed in one single day, in the presence of the Messiah. Note also the fig tree, whose importance appears later. Further on, the book predicts Jesus' triumphant entry into Jerusalem riding on an ass:

> Rejoice greatly, O daughter of Zion,
> shout [for joy], O daughter of Jerusalem!
> Behold, your King comes to you;
> he is just and having salvation,
> lowly, and riding upon an ass,
> and upon a colt, the foal of an ass.
> (Zech. 9:9)

Still later, it alludes to Judas's betrayal of Jesus for thirty pieces of silver, and how (according to Matthew's gospel) he later repented and returned the money to the priests who used it to buy a burial field for foreigners. "I then said to them, 'If you think it right, give me my wages; if not, never mind.' And they weighed out my wages: thirty shekels of silver. But Yahweh told me, 'Throw it into the treasury, this princely sum at which they have valued me.' Taking the thirty shekels of silver, I threw them into the Temple of Yahweh, into the treasury" (Zech: 11:12-13 JER).[12]

Eventually Zechariah turns to the final battle between God and the enemies of the Hebrews, which is to take place at the Mount of Olives, exactly where Jesus went to pray: "Then shall the Lord go forth, and fight against those nations, as when he fought in the day of battle. And his feet shall stand in that day upon the Mount of Olives, which is before Jerusalem on the East, and the Mount of Olives shall cleave in the midst thereof toward the east and toward the west, and there shall be a very great valley; and half of the mountain shall remove toward the north and half of it toward the south" (Zech. 14:3-4). The battle brings forth the kingdom of God: "The Lord shall be king over all the earth; in that day shall there be one Lord, and his name one" (Zech. 14:9).

But the new day does not arrive without problems: "And in that day there shall be a great confusion...among them from the Lord: and they shall seize each his neighbor's hand, and the hand of the one shall be raised against the hand of the other" (Zech. 14:13 AMP). Compare with Jesus' words: "Do you suppose that I have come to give peace upon the earth? No, I say to you, but rather division" (Luke 12:51 AMP).

Finally the long-awaited state becomes established: "And it shall come to pass, that every one that is left of all the nations which came upon Jerusalem shall even go up from year to year to worship the King, the Lord of hosts, and to keep the feast of the tabernacles" (Zech. 14:16). Notice the prediction that the battle is to occur on the feast of Booths (Tabernacles); why else would all nations be celebrating the New Jerusalem on that day? "There will be no more traders in the Temple of Yahweh Sabaoth [The Lord of hosts], when that day comes" (Zech. 14:21 JER). As we saw earlier, cleaning up the temple was one of Jesus' first acts upon entering Jerusalem.

Some of Zechariah's predictions had already come to pass (even though they were forced) and Jesus must have expected the others to be also fulfilled. If he was the Messiah and there was a battle against Judea's oppressors, then God would intervene, destroy the oppressors, and introduce his new kingdom. Perhaps this is why Jesus wanted some swords—to put up a nominal fight against the Romans; God would intercede and finish it. And maybe this was Judas's assignment: to tell Jesus' mysterious contacts in Jerusalem to start an uprising, or perhaps to bring the Roman soldiers to the Mount.[13] But no uprising took place, and when the soldiers arrived, two swords were not enough. God did not intervene; Jesus was defeated. Obviously all this is a completely hypothetical story.[14] Yet it makes far more sense to me than all the tales the early Christians thought up in an effort to give meaning to the unhappy ending of the Jesus story.

There remains one little inconsistency. According to Zechariah, the last battle was to occur in the fall, during the feast of the Booths, not in spring during the Passover. Jesus was too intelligent to not have followed the script properly. So if the earlier theory is correct, Jesus must have entered the city for the feast of Booths, not the Passover.

This would explain two other problems with the Biblical account: the insufficient time for all the events described, and the matter of the fruitless fig tree. The Bible maintains that within less than sixteen hours all the following took place: Jesus was arrested, he was taken to the chief priest's house to which the seventy members of the Sanhedrin and witnesses had been called in the middle of the night, and a trial was held. Then Jesus was taken to the residence of Pilate, who must have dropped everything else he was doing to hold another trial. Afterwards, Jesus was mocked by the soldiers, scourged, forced to stagger to his execution place carrying his cross, and finally crucified.[15] It is more reasonable to postulate that Jesus was arrested in the fall and executed during the Passover period as an example to the Judeans. Consider that John the Baptist remained in jail for months before his execution, and that Paul was kept in jail for two years before Governor Felix sent him off to Rome for trial. Then there is the strange incident with the fig tree. Jesus is hungry, walks over to a fig tree, but finding it without fruit, he curses it and it withers. Jesus must certainly have known that fig trees give fruit between August and October. Why would he have looked for fruit in the spring?[16]

THE TRIAL

Matthew's gospel follows Jesus' arrest with the most anti-Jewish polemic in the entire Bible. This is reinforced by the fact that our standard texts translate the appellation "Judeans" as "Jews".[17] The story starts with the chief priests, the elders, and the people bringing Jesus to Pilate for trial. But according to the Bible, Pilate is not convinced of Jesus' guilt, so the gospel introduces the strange tale of Barabbas. It was customary, we are told, for the authorities to release a criminal on feast days, so Pilate asks the crowd whom they want him to release: Jesus or Barabbas. Just like most names, the word Barabbas has a meaning: *Bar* (*ber*) for son and *abba* for father, or better yet, daddy. This was the term used by Jesus when praying to his father, and what he instructed us to use in the Lord's Prayer. Furthermore, Barabbas' first name was Jesus.[18] So, in the story, Pilate asks the crowd: "Whom do you want me to release, Jesus the son of the Father, or Jesus *who says he is* the son of the Father?" The crowd responds: "Release to us Jesus the son of the Father!" When Pilate, who is invariably pictured in the gospels as a decent man, loath to do evil, asks, "What evil has this man done?" the people shout, "Let him be crucified!" When Pilate says, "I am innocent of this man's blood," the crowd responds, "His blood be upon us and upon our children!"[19]

What an intense scene! Give the author an A for drama, but an F for historical accuracy. For there is no record outside the gospels that there ever existed such a custom of releasing a prisoner during the feasts, nor would it make any sense. From the Judean point of view, mercy belonged to the province of God; man judged according to the Law. From the Roman perspective, on the other hand, one never showed weakness; certainly one never let criminals and insurrectionists go free. The Roman authorities in the provinces had only two objectives: to keep order and to ensure that the taxes were collected.[20]

Furthermore, everything we know about Pilate portrays him as a strict soldier, not an inquiring philosopher. It is highly unlikely that he would worry about Jesus' theological ideas or ask philosophical questions such as "What is truth?" (John 18:38) He was a man of action, quick to react to any perceived threat.[21] Had he learned that a charismatic leader had entered the city during festival time at the head of a sizable procession and had been greeted by the population with cries of "Hosanna to the Son of David"—a great past king of the Jews—he would have responded immediately. And if he had not heard about it, the chief priests, upset by Jesus' actions and teachings, would have brought the matter to his attention.[22] And this is exactly what happened and what he did. In fact, according to John's gospel, he dispatched 600 soldiers to capture him.[23] He then imposed the standard Roman punishment for insurrection by a non-Roman: flogging, followed by crucifixion. And just so that Jesus' followers, if they were still around, would not miss the point, he had two common criminals crucified next to him, and then placed on top of his cross the reason for his punishment: King of the Jews. "This is what I did to your leader," he was telling the Judeans. "Anyone want to share his fate?"

A short digression might be in order here regarding the way we should be interpreting the gospel stories of Jesus' crucifixion. The simple truth is that just as there was no Barabbas to let free, there was no crowd to demand Jesus' death. The Romans hardly needed outside instigations for their actions. Can you imagine any conquered and oppressed people asking their masters to execute one of their own? So during the Easter season, when this story is read, a sensitive priest at the pulpit will try to remove some of its sharp edge by pointing out that the gospels really mirror the times when they were written, not the actual events which they are supposedly relating.

We will discuss the gospels in chapter 23, but let us take a quick look at how the perspectives of the evangelists affected their crucifixion stories. Mark, who initiated this account, was probably writing his gospel for the Jews of the Diaspora around 68 CE, a few years after the chief priest had ordered the killing of James, the brother of Jesus.[24] We would thus expect Mark to excoriate the chief priest and his followers, but to not make any offensive remarks about his fellow Jews.

The second evangelist, Matthew, wrote around 80-90 CE. The Romans had destroyed the Temple, and he was unsuccessfully competing against the emerging rabbis for the support of the now leaderless Jews. Little wonder that he lashes out at the rabbis and those who will not join his side, telling them that they are sinning and will lose their souls. Finally, John was writing after his group had been kicked out of the synagogues. Having lost the protection of their Jewish identity, they were now fair game to the Romans for refusing to sacrifice to the idols, a refusal permitted in the empire only to the Jews because of their older tradition. It has been even suggested that the Jews of that time occasionally denounced the new Christians to the authorities. In John's gospel we can almost feel the seething hatred he directed towards the Jews.

So if we are fortunate to have an enlightened person on the pulpit, we may be told in church that the gospel does not always mean what it says.

SOME QUESTIONS TO CONSIDER

1. If Jesus had been involved in clandestine relationships in Jerusalem, why do no hints of it appear anywhere else in the gospels?

2. It has been argued that the upper room prepared for Jesus in Jerusalem was really the covered area that many Judeans would put up on their roofs during the feast of the Booths. Does that sound reasonable to you?

3. Consider the following three possibilities regarding Jesus' actions during the final week: 1) He was acting independently, as he felt was appropriate; 2) He was consciously copying the Zechariah story; 3) His reported actions were fabricated by the evangelists, to parallel the Zechariah story. Which do you think is more probable and how would that affect the way you look at the events of the Holy Week?

4. The gospels clearly quote Jesus as saying that the Son of man was to suffer and die in Jerusalem. Doesn't this contradict any revolution-type theories?

5. How would you explain Judas's thinking process during the final days?

6. Do you find it surprising that the disciples of a man of peace, one who preached turning the other cheek to an assailant, would be carrying weapons?

7. What do you think of the Barabbas story in the gospel?

SELECTED REFERENCES

Maccoby, Hyam. *Revolution in Judea: Jesus and the Jewish Resistance.*

Wilson, A. N. *Jesus: A Life.*

CHAPTER 20

CRUCIFIXION AND RESURRECTION

Jesus' execution by the Romans must have shocked his followers, but probably not surprised them. After all he was a prophet, and prophets were often killed by the authorities or even by the people. But when Paul tried to elevate Jesus to the status of God's son, he must have run into a wall of disbelief. What kind of god would allow his son to be killed?

Then Paul came up with his most brilliant idea: Jesus' death was necessary for mankind's redemption from sin. Only the Son of God could be a proper atoning sacrifice. Jews could understand this, but to Gentiles, who had no history of original sin and atoning sacrifice, he presented a different argument: Sin was a malevolent power that controlled death. By dying, Jesus placed himself under the power of Sin; through his resurrection he proved his mastery over Sin. And through baptism and our belief in him, we also can master Sin and escape death. Paul thus invented the new religion of Christianity, one based on both the death and on the resurrection of Jesus.

But did Jesus really rise from the dead? On one hand, every surviving record presents a different story, a very strong indication that these stories are not true. On the other hand, the resurrection idea would have been too outrageous at that time for the evangelists to make it up. In addition, we witness a complete change in the disciples' behavior. When Jesus was arrested they all ran away, but after he was killed they returned and started fearlessly campaigning in his name. Something must surely have happened in between, and whatever it was it gave birth to the resurrection stories that were written down.

CRUCIFIXION IN HISTORY

The origins of crucifixion are not known. Although Deuteronomy mentions "hanging on a tree," the person was first stoned to death, and only the dead body was placed on what was probably a simple pole. Death by crucifixion was most likely invented by the Persians, and then passed on to the Carthaginians and Phoenicians. During his invasion of Persia in 333 BCE, Alexander crucified 2,000 defenders of the city of Tyre as a successful message to other cities not to put up such a spirited defense. Later, the Hasmonean king of Judea, Alexander Jannaeus (103-76 BCE), crucified 800 Pharisees who had opposed him, and in addition had their wives and children killed in their view while they were still alive. In a victory celebration in 71 BCE, the Romans crucified 6,000 followers of Spartacus along the Appian Way. During the 67-70 CE siege of Jerusalem, the bored Roman soldiers amused themselves by crucifying as many as 500 captured Judean escapees a day, in front of the walls of the city. As a result of the timber required for these crucifixions and for the construction of the assault ramps, all trees within ten kilometers of Jerusalem were cut down. In the Roman world, death by crucifixion was the common penalty for thieves, slaves, insurrectionists, and generally the lower class population; it was usually not applied to Roman citizens except for army deserters. It was always meant to be a humiliating, horrific, exemplary punishment, sending viewers a clear message to abstain from similar behavior.

It appears that the exact crucifixion technique varied greatly. The hands were attached by nails through the wrists—not the palms as had been believed until recently—or simply tied to the horizontal cross bar. Occasionally they were stretched vertically up. The legs were nailed with one or two nails, sometimes in the extended position, and sometimes bent up. Often a short block was placed on the upright member so that the victim could rest his torso, or push up with his feet, enabling him to survive and suffer longer. The final death occurred from asphyxiation, when the victim was too exhausted to push up with his feet to release the weight on his diaphragm and thus be able to breathe. This is why the executioner would break the victim's legs when he wanted to hasten his death. It was the Roman practice to first flog the person almost to death, using leather whips with pieces of bone and stone attached along their length.[1] This of course served to weaken the victim and shorten the duration of the final execution. Despite their wide use of this execution method, Romans thought "that crucifixion was a horrific, disgusting business....[and] there is hardly any mention of it in inscriptions."[2] Most writers, including officials who ordered it, avoided discussing this subject.

In Jerusalem, the condemned man was usually given a drink of strong wine mixed with myrrh to help deaden his consciousness; it was paid for, if not administered, by a group of women. As you probably remember, Jesus refused such a drink when it was offered to him. The practice of crucifixion ended in 311 CE with Emperor Constantine's "edict of tolerance," and was replaced by the somewhat more humane execution on the gallows.

The very early Christians used the sign of the fish, not the cross, as their symbol. The reason for this is unknown. Some relate it to Jesus' multiplication of the fish and loaves at the Sermon on the Mount. Others attribute it to his promise to make his disciples "fishers of men" (Mat. 4:19). Still others point out that the letters in the Greek word "fish" (ΙΧΘΥΣ) are the initials in the expression "Jesus Christ God's Son Savior." By 200 CE, however, according to Tertullian, Christians traced the sign of cross on the foreheads before eating or participating in other important events.

Around 225 CE the symbol of the cross first appeared in the Roman catacombs disguised as a ship's anchor. A sarcophagus built around 350-450 CE shows scenes from the Lord's Passion, but "the shame and horror" of the crucifixion was concealed in the representation of a lamb sitting at the bottom of an anchor. A vase made around 550 CE shows the two bandits with their hands extended but without a cross, and Jesus ascending to heaven. In a seventh-century crypt of St. Valentine's Catacomb in Rome, Jesus is shown in a long robe nailed on the cross with four nails, between Mary and John. The tenth century brought more realism, and artists started to portray Jesus with a drooping head and a crown of thorns.[3]

THE THEOLOGY OF CRUCIFIXION

"For the Jews require a sign, and the Greeks seek wisdom. But we preach Christ crucified, to the Jews a stumbling block, and to the Greeks foolishness" (1 Cor. 1:22-23). The greatest difficulty that the early Christians had in spreading the gospel was the idea that God's son was crucified. It was not just that he had died, because many of the pagan gods had also died: Osiris, Dionysus, Mithras, to mention just a few. They died and were resurrected, more or less in the manner of Jesus, and every year the cycle of their death and life was re-enacted. But they were all killed by other gods, not puny humans, and none of them had died in a manner as humiliating as Jesus had. How could God allow his own son to be killed in this manner? It became an endless source of material for pagan satirists.

Even some Christians refused to believe it. Some of the gnostic gospels discovered near the town of Naj Hammadi (religious writings that the nascent Christian Church had attempted to suppress) describe Jesus switching places with Simon the Cyrene to escape crucifixion.[4] Similarly, Muslims could not believe that God would allow mere humans to kill his own prophet and explained away the crucifixion as an apparition and not the real thing.[5] The Christians faced this problem by borrowing a page from the Hebrews. Just as throughout all Hebrew history, people's defeats were reinterpreted as God's victories, so the Christians used the resurrection story to transform the crucifixion into a triumph. The crucifixion, together with the resurrection that followed it, became the way to deliver humanity from sin and death. As such, it had been God's plan all along.

Paul, who started all serious Christian thinking, presented two different models, depending on how one defined sin.[6] The argument with which we are most familiar is the sacrificial atonement. According to Deuteronomy, murder and the incitement of people to follow other gods were not forgivable; they were punished by death through stoning. But for the myriad other transgressions in their everyday life, Hebrews atoned through guilt and sin offerings. For minor sins they would kill a pigeon or two; for more serious transgressions they would offer a lamb or a goat. And for the really fundamental sins, say those of the entire nation of Israel, they would sacrifice a couple of oxen. The bigger the sin, the bigger the offering. So how could they seek atonement for the sins of the entire human race? What was the ultimate offering by which to atone for Adam's original sin? Paul's, and later, John's answer was the sacrifice of God's son. According to them, Jesus offered his death on the cross as the ultimate and perfect sacrifice for the redemption of mankind's sin. "For God so loved the world, that he gave his only begotten Son, so that whoever believes in him should not perish, but have everlasting life" (John 3:16).

This argument, however, depends on the assumption that the Hebrew beliefs are correct: that one can indeed atone for sins by slitting the throat of an innocent animal; that by wantonly destroying something from God's creation a man can remedy his standing in the eyes of God! It makes no sense. But even if it did, the Hebrews had strict rulings about the procedure. The killing had to be done by the assigned Hebrew priests, in a specific manner, and only in the Temple. When the Romans finally destroyed the Temple, the sacrifices were stopped. A new Talmudic theology had to be developed to explain the God-human relationship in a manner that excluded the taking of animal life. It is obvious that the sacrifice of Jesus did not fulfill any of the Hebrew requirements—it was not performed by priests, it was not done in the Temple, and it was not carried out in the ordered manner (the sacrificial animal had to be killed in such a way that all its blood drained out on the earth). So if Jesus' sacrifice was not based on the Hebrew religion, on what was it based?

Paul's second argument centers on his definition of Sin (with a capital S) as Satan. When Adam sinned, Satan acquired dominance over man, and death entered the world. Thereafter, when a person dies, Satan removes him from the world of God. So when Jesus died as a human, he fell into the hands of Satan. But as his eventual resurrection proves, he conquered death and, hence, Satan. And we too can participate in this victory by being united with Christ in his death and resurrection. According to Paul, this happens when we become baptized.[7] But although during baptism we die with Christ, and thus to sin, we obviously do not immediately acquire physical immortality; we still face a bodily death. Resurrection and attainment of immortality are thus events that will take place when Christ returns at the end of time.

In Paul's first salvation model, sin is a disobedient act that people commit. In the second model, Sin is a cosmic force that works to enslave man. Another concept evolves when we call Jesus our redeemer. In Hebrew law, a poor man might sell himself to another Jew or foreigner in order to satisfy his debts. It was then the responsibility of his closest relative to repay his financial obligations and redeem him. So Jesus, having paid with his blood for our sins, became our redeemer. Always assuming, of course, that there had been an original sin and that we are all contaminated by it.

The Message of the Cross

> Recently I attended a memorial service at a prestigious, mainline Protestant church. Beyond the altar was an immense pentagonal brick wall. The single object on the wall was a large crucifix—Jesus Christ spiked to a cross.
>
> I wondered, is that the message? Jesus Christ violently spiked to a cross to die hideously? Is the message man's vile inhumanity? I thought, why wouldn't the church instead depict Jesus resurrecting, rising to God with his arms spread wide ready to envelop all? Does it not make sense to symbolize church by love and mercy rather than hate and murder? —George E. Noyes

This brief letter was sent to a local newspaper by someone who visited the church I attend, and whom I later came to know. It describes a sight to which many have become so accustomed that they don't really see it in its fullness. It is true, of course, that the antiseptic environment of the church covers up the real ghastliness of what it represents. But in its day, this execution method was so ghastly that even its perpetrators did not talk about it. It was so ghastly that ten generations had to pass between the abolition of its practice and its first tentative representation. It was so ghastly that another ten generations had to pass before the figure on the cross was portrayed in an ever-so-slightly suffering pose, instead of in glorious radiance. Crucifixion was a monstrous thing to do to any human being, let alone to the object of our reverence. So why is the crucifix standing in the front of our church? Why is it the center of our religion? This was my answer to George.

> Imagine if you will, that a powerful and cruel enemy conquers our country. To nip any thoughts of rebellion, and to completely dispirit the population, the conquerors decide to crucify one person from each household. They go house to house picking up people, and they finally reach your door. "Whom shall we crucify," they ask you. "Your wife, your son, your daughter?" "No, no!" you shout. "Take me! Take me!" And they do indeed take you and crucify you. And to insure that the lesson is remembered, they make a picture of you hanging on the cross, and send it to your family.
>
> Guess what your family does with the picture. They frame it, and hang it in the most prominent part of the house, where everybody can see it all the time! Do they do this in order to keep up their hatred for the conqueror? Perhaps to a small extent. The main reason, however, is to remind them of your great love for them, your willingness to suffer for their sake. Whenever they see you hanging on the cross, they think of your incredible love for them, and how they should try to share it with each other. In the end it is love that conquers, not hate.

And here lies the main difference between the Old and the New Testaments. In the Old Testament, when Job complains to God about the evil in the world, God shouts him down: I am superior, he yells; you are too insignificant and stupid to understand. In the New Testament God says instead: Yes, I see the evil, and I share its pain with you.[8]

THE RESURRECTION STORIES

If the Jesus story had ended with his crucifixion, there probably would not have been a Christian religion today. There were numerous charismatic preachers, before and after Jesus, whose disciples scattered and whose teachings were forgotten soon after their death. Certainly Jesus' disciples ran away when he was arrested. Yet only a few weeks later, they bravely stood up and faced the Judeans, castigating them for killing Jesus; or at least this is what Luke says in the Acts. It is obvious that something must have happened in between. It is generally assumed that this was their belief in Jesus' resurrection after he became visible to his disciples and spoke to them. What do we know about it?

The first recorded information about Jesus' resurrection is found in Paul's letters, this one dating from 56 CE:

> For I handed to you first of all that which I also received, how Christ died for our sins according to the scriptures. And that he was buried, and that he rose again on the third day according to the scriptures. And that he was seen by Cephas,[9] then by the twelve. After that he was seen by more than five hundred brothers at once, most of whom remain to this present, but some have fallen asleep. After that he was seen by James, then by all the apostles. And last of all, he was seen by me also, as one born abnormally. (1 Cor. 15:3-8)

It is not clear when, and under what conditions these appearances took place. With the exception of the appearance to the disciples or apostles, as the case may be, the gospels that were written later do not corroborate these sightings.

The next surviving source that discusses Jesus' resurrection is Mark's gospel written around 68 CE:

> When the Sabbath was over, Mary Magdalene, and Mary the mother of James, and Salome had bought spices, that they might come and anoint him....And entering into the tomb, they saw a young man sitting on the right side, clothed in a long white garment;

> and they were frightened. And he said to them, "Do not be frightened! You seek Jesus of Nazareth, who was crucified. He is risen; he is not here. Behold, the place where they laid him. But go your way, tell his disciples and Peter that he goes before you to Galilee; there you shall see him, as he told you." And they went out quickly and fled from the tomb; for they trembled and were amazed. Neither said anything to anyone, for they were afraid. (Mark 16:1,5-8)

Note that the first gospel contains no mention whatsoever of any resurrection appearances. Furthermore, the witnesses to the empty tomb did not even tell the disciples about Jesus' resurrection. It is true that the Bible then goes on to describe the resurrected Jesus appearing to Mary Magdalene, then to two disciples leaving the town, and finally to the eleven disciples gathered at the table, but this part is a later addition.

In the next gospel, written around 80 CE, Matthew presents two resurrection appearances of Jesus. Here, "Mary Magdalene and the other Mary went to the tomb." As they arrived, an angel of the Lord came down and rolled away the stone in front of the entrance to the tomb, filling with fear both the women and the guards that had been posted at the tomb. The angel announced that Jesus had been resurrected and that they should tell the disciples to go meet him in Galilee. On the way, the women met the resurrected Jesus, whereupon they "embraced his feet and did him homage." He repeated to them the message to tell his disciples to meet him in Galilee, which they did.

> When [the disciples] saw him, they worshipped him; but some doubted. And Jesus came and spoke to them, saying, "All power has been given to me in heaven and on earth. Go, therefore, and teach all nations, baptizing them in the name of the Father, and of the Son, and of the Holy Ghost." (Mat. 28:17-19)

Although Mark proposes a meeting in Galilee between the risen Jesus and his disciples, only Matthew claims that such an appearance actually took place. (But see the discussion below about Jesus' appearance in Galilee recorded in John's gospel.) What ensues in this meeting, however, seems highly unlikely. In a terse sentence, Jesus announces the complete transfer to him of God's control over heaven and earth. And the instructions in the second sentence regarding inclusiveness of Gentiles and using the triune baptismal formula are foreign to Jesus' teaching as presented by Matthew.[10]

Turning to the third gospel, written around 85 CE, Luke relates that the women who had followed Jesus from Galilee, and specifically Mary Magdalene, Joanna, and Mary the mother of James, took spices and went to the tomb. They found the tomb empty, but two men in dazzling white clothes appeared and told them that Jesus had been raised. So they returned and told the eleven and all the others. "[But] their words seemed to them as idle tales, and they did not believe them. Then Peter arose and ran to the tomb; and stooping down, he saw the linen clothes laid separately, and departed, wondering in himself at that which had come to pass" (Luke 24:11-12).

"On that very day," Jesus appeared to two of his followers who were on their way to Emmaus, but they did not recognize him. He engaged them in a discussion regarding the recent happenings in Jerusalem, and then explained to them "how Christ ought to have suffered these things in order to enter into his glory" (Luke 24:26). Upon accepting their invitation to join them for dinner,

> He took bread, and blessed it, and broke it, and gave it to them. And their eyes were opened, and they knew him; and he vanished from their sight.... And they rose up the same hour and returned to Jerusalem, and found the eleven gathered together, and those that were with them, saying, "The Lord is risen indeed and has appeared to Simon!"
>
> ...And as they were speaking thus, Jesus himself stood in the midst of them, and said to them, "Peace be unto you." But they were terrified and frightened, and supposed that they had seen a spirit. And he said to them, "Why are you troubled? And why do thoughts arise in your hearts? Behold my hands and my feet,

> that it is I myself. Touch me and see, for a spirit has no flesh and bones, as you see me have." And when he had thus spoken, he showed them his hands and his feet. And while they still did not believe [in the] joy and wondered, he said to them, "Have you here any meat?" And they gave him a piece of broiled fish and [a piece] of honeycomb. And he took it and ate before them....
>
> And he led them [out] as far as Bethany, and he lifted up his hands and blessed them. And it came to pass while he blessed them, [that] he was parted from them, and carried up into heaven. (Luke 24:30-51)

He instructed them to stay in the city until they were clothed with power from on high—a quite different instruction than was recorded in Matthew's gospel. Does this story make any sense? Luke's gospel says that in the morning Peter saw the empty tomb and then returned home, puzzled. In the evening the disciples were gathered discussing Jesus' resurrection and his earlier appearance to Peter. Yet when Jesus appeared to them, they were completely unprepared. If they had known that he had risen and had already appeared to Peter, they would have been filled with joy, not fear. When, and under what circumstances did Jesus appear to Peter?[11]

Finally, in John's version, it is Mary Magdalene alone who goes to the tomb and finds it empty. She runs to Peter and the beloved disciple to tell them about it, and these two run to the tomb. Peter enters and sees the burial cloths, "and the cloth that was about his head, not lying with the linen cloths, but wrapped together in a place by itself" (John 20:7). The disciples go home, but Mary stands outside the tomb. She looks in to see two angels, and then she sees Jesus standing behind her; but at first she does not recognize him and mistakes him for the gardener. When he calls her name, however, she does recognize him and answers: "Rabbouni," teacher. He tells her, "Go to my brothers and say to them, 'I ascend to my Father, and your Father, and to my God and your God'" (John 20:17). Mary goes to the disciples and passes on the message.

On the evening of the same day, Jesus appears to the disciples, who have been gathered behind locked doors from fear of the Judeans. He bids them peace and shows them the wounds on his hands and side. But Thomas was not with the others, and when he is told about Jesus' appearance he refuses to believe. "Unless I shall see in his hands the marks of the nails, and put my finger into the marks of the nails, and thrust my hand into his side, I will not believe" (John 20:25), he says. So a week later, Jesus appears again, and asks Thomas to touch his wounds. Thomas replies "My Lord and my God!" Notice, however, that the gospel does not say that he actually touched Jesus.

John's account ends by stating that Jesus did many other signs in the presence of his disciples, but they were not written down. After the end of the gospel, however, there follows one additional, appended story, probably written by someone else, in which Jesus appears to the disciples while they are fishing in Galilee and he calls them out for breakfast. There he tells Peter three times to tend and feed his sheep, perhaps once for each time that Peter had denied him during his trial.

What do we make of all these stories? The resurrection is without doubt the most important part of the Christian story. And yet we are offered five entirely different accounts. One gospel relates no resurrection appearance. Another says that one occurred in Galilee. The other two locate multiple appearances in Jerusalem (although John's gospel does have an appended appearance story placed in Galilee). There are only two points of agreement: all four gospels refer to an empty tomb, and in all Mary Magdalene is among those who first discovered it. Furthermore, in the two of the three gospels that do recount appearances, she is also the first one to see Jesus.

The problem starts with Paul's list of those who saw the risen Christ. Not only was Paul the first to write about these things, but he had also been in Jerusalem at the time of these presumed sightings. We might expect, therefore, that he would have more correct information than the gospel writers. Yet his list does not correspond with anybody else's. He neglects Mary Magdalene completely, although she was almost certainly the first witness. We can explain this away, perhaps, by the fact that he was probably a misogynist and that since in those days women could not be legal witnesses what they had to say was irrelevant.

The rest of Paul's list, however, presents greater difficulties. He identifies Cephas as the first person to see the risen Christ. But the four evangelists don't agree with him, except for Luke's one strange sentence.[12] He lists an appearance to James, Jesus' brother, which is also not found in the four canonical gospels, although it is described in the mostly lost Gospel of the Hebrews.[13] There is no remaining record of any appearance to five hundred, and of course the appearance to the "twelve" should have said eleven.[14]

So are these resurrection stories true? Perhaps. It is more probable however, that these particular accounts just represent some of the many stories that circulated in the area at that time. Furthermore, it appears that a number of competing Christian groups arose during the first century: the Hebrew Christians, with James as head; Peter's group, supported by the lost Gospel of Peter and to a lesser extent Mark's; the new Christians of Paul; the breakaway Hebrew Christians of Matthew; miscellaneous gnostic groups that fully developed in the second century; and finally the almost gnostic Christians of John. We can assume that each group tried to preserve those stories that gave its side greater pre-eminence. Unfortunately, the emerging Church was quick to destroy the records of those groups with which it did not agree. The Gospel of Hebrews, for example had been completely lost until some fragments were discovered in the nineteenth century. Even less survived from the Gospel of Peter. Some say that the reason that Mark's gospel has such a truncated ending is that Jesus' appearance to Peter at Galilee was deleted very early on.[15]

In chapter 17 we discussed the assertion that dead people appear to relatives and other interested people immediately after their death. Why should this not have also occurred with Jesus? And in the culture of those days, such appearances would have seemed even more real than they do in today's more scientific world. It is interesting to note along those lines that the most reliable resurrection appearance stories describe them as taking place on Sunday, two days after Jesus' death.

Although some gospels stress that the resurrected Jesus appeared as a real body and not a ghost, he may not have been visible to everyone. Peter said: "Him God raised up [on] the third day, and showed him openly, not to all the people, but to us, the witnesses chosen by God in advance, who ate and drank with him after he rose from the dead" (Acts 10:40-41).

Resurrection of the Body and the Empty Tomb

Although pagan Hellenists who agreed with Plato's philosophy believed in the survival of the soul, Hebrew Pharisees believed in the resurrection of the body. Survival of the soul and bodily resurrection are two different concepts: the first claims that one's identity survives the death of his body in some non-corporeal manner; the second argues that sometime after one dies, one's physical body is reconstituted from the physical remains of one's dead body, whatever its condition, and is joined by its soul. Notice that this second concept also includes the first one, since it assumes the survival of the soul during the period between the death and the resurrection of the body. In the previous section, I argued that Jesus' identity could have survived and appeared to the disciples and perhaps some others. But this is not what was understood by Christians or even Pharisees. For them, if Jesus rose from the dead, he would have done it in his own body. So all four gospels pay special attention to the discovery of the empty tomb.

But if we agree that Jesus' tomb was found empty, where did the body go? Was it actually resurrected as the Bible says? If so, it must have acquired some completely new properties. His resurrected body could appear and disappear at will; it could enter through locked doors;[16] it could be selectively recognizable or not. In the following two millennia, there have been heated discussions in the Church about the exact characteristics of the resurrected body, and therefore presumably of Jesus' body. Through most of these years, Christian writers widely insisted that the risen body would consist of the actual specific atoms that had constituted the body at the time of its death.[17] This contradicted Paul's words: "Flesh and blood cannot inherit the kingdom of God, nor does corruption inherit incorruption....The dead shall be raised incorruptible, and we shall be changed. For this corruptible must put on incorruption, and this mortal must put on immortality" (1 Cor. 15:50-53). Will a different body require the atoms of the earlier body?

No current evidence that we have about life after death (however disputed) involves the physical body. Why should Jesus' resurrected body be different? If he rose with his actual body, one that he could teleport at will (a la science fiction) what did he do with it? If he took it with him to heaven, we will have to modify our concept of heaven and make room in it for physical bodies. Then we will have to find an actual physical location for this heaven. But more important, Christians believe that Jesus had existed since the beginning of eternity as the *Logos*, God's Word who helped create the world and was sent to Earth for a specific purpose. He did not have a body before he became man, and he certainly would not need one after he was restored to his previous position. Otherwise one would be forced to say that depending on the worth of an earthly body, either the pre-incarnated Logos was less than the risen Christ, or the reverse.

But if Jesus did not take his body out of his tomb, who did? Could someone have stolen it? If so, it was not the Judeans, and it was not the Romans, since neither group would have any reason to do it. It was certainly not the disciples, because if they themselves had stolen the body they would have neither believed nor been affected by the resurrection stories. So that leaves only someone from Jesus' family. Although his brothers later became leaders of the church, they never believed in his divinity. They probably came to believe in his teachings, but only as those of a saintly man, a philosopher. So this is one possibility, probably the best one.[18] But why would they have removed it?

On the other hand, is there any scientifically explainable miracle that would cause the disappearance of Jesus' body from a guarded tomb? Not really, unless one chooses to believe in the theory of spontaneous combustion. There have been improperly documented cases of people, who for no apparent reason, suddenly burn completely to ash—something that requires temperatures in the hundreds of degrees—while sitting in a chair or lying in bed, without the heat affecting their clothes, or their surroundings. Improbable? Yes, but no more improbable than the creation theories presented by our physicists.

SOME QUESTIONS TO CONSIDER

1. What do you believe was accomplished through Jesus' crucifixion?

2. Why does the author assume that the differing descriptions of the resurrection appearances indicate their falsehood, rather than normal changes of orally transmitted stories that occur with the passage of time? Do you think any of these stories sound more plausible than the rest?

3. Which one of Paul's explanations for Jesus' death do you find more believable: the substitute sacrifice or the overpowering of Sin?

4. The Acts records that the resurrected Jesus was only visible to some of his close followers, not to everyone. Remembering our discussion from chapter 17, what can we conclude from this?

5. Was Jesus' resurrection due to his own power or due to God's power? Does it matter?

6. It was the standard Roman practice to let crucified people remain on the cross for days after their death, as an example to others. Why would Jesus' body have been removed immediately after his death?

7. What do you think when you see a reproduction of the crucified Jesus?

SELECTED REFERENCES

Bynum, Caroline Walker. *The Resurrection of the Body in Western Christianity, 200-1336.*

Davis, Stephen T., Kendall, Daniel SJ, O'Collins, Gerald SJ, eds. *The Resurrection: An Interdisciplinary Symposium on the Resurrection of Jesus.*

Ehrman, Bart D., *The New Testament: A Historic Introduction to the Early Christian Writings.*

Hengel, Martin. *Crucifixion: in the Ancient World and the Folly of the Message of the Cross.*

Kessler, William Thomas. *Peter as the First Witness of the Risen Lord.*

PART III

WHAT MAN MADE OF THE TEACHING—THE CHURCH

Introduction to Part III

In the last few chapters we discussed the life and teachings of Jesus. We saw that, in essence, he was a traveling preacher, calling on people to open their hearts to God to prepare for his coming reign. But that was two-thousand years ago. The only way to know what he said and did is to read what had been written about him. And this is where the Church comes in.

Just as the Old Testament writers interpreted God's works and deeds according to their own beliefs and interests, so also did the New Testament writers with the life and work of Jesus. At first there were many conflicting versions, but as the years went by, an organization developed that took upon itself the task of deciding on the accuracy of the writings. It placed the writings it considered true on an approved list, called the canon. The others were anathematized, and when the organization grew in strength and received political support from the Roman Empire, they were proscribed and burned.

This organization that grew from Jesus' work we call the Church. Jesus himself had not thought that it was needed, but Jesus also thought that God's reign was around the corner. When he died and the reign failed to materialize, his followers needed guidance, and this is what the Church set out to provide. In the process, the Church discovered that Jesus had failed to discuss many subjects, including even himself. Sometimes he called himself the Son of man, but it was not clear what that meant. He called God his Father, but what did it mean to be the Son of God? The Church set about to resolve these problems, and some of the greatest minds of the first few centuries CE devoted all their lives to their solution.

In the next few chapters we will discuss the writings about Jesus and his work that have survived, who wrote them, and why. Then we will look at the Church's efforts to answer the remaining questions, and how it went about doing it. Finally, we will examine the Church's effect on two burning issues of the last century: the position of women in the world, and the propriety of sexual relationships. We will conclude by trying to answer the original question that started us on this road of discovery: how can we find God? If all our religious writings are man-created, and thus contaminated, who is God?

CHAPTER 21

THE FIRST CHRISTIANS

Today's Christianity consists of many denominations ranging from the "You must follow every single word in the Bible" Seventh-day Adventists, to the "Anything goes if you are comfortable with it" Episcopalians. In a similar manner, Jesus' followers soon separated into numerous sects: James's Jerusalem Christians were regular Hebrews, except that they were expecting the momentary arrival of God's reign and perhaps may have also believed that the Messiah had come bringing new teachings; the Ebionites were very similar except that they believed only in the Torah and abstained from meat; the Nazarenes retained the Hebrew practices but believed in Jesus' divinity and perhaps in the Holy Spirit; Matthew's Christians tried to retain their Hebrew beliefs and practices while declaring that Jesus was God's son, begotten through a virgin birth. Paul's followers believed in the pre-existence of Jesus and that he possessed some kind of semi-divinity but not equality with God. Finally John's group separated from the Jews and proclaimed the absolute divinity of Jesus: he and God were one. Outside of this mainstream were various gnostic groups that believed Jesus was a semi-divine person who had revealed secrets that could save those who knew them. In many ways John's gospel supported their position and had been accepted by some of their groups.

THE CHRISTIANS OF THE ACTS

In his Acts of the Apostles, Luke describes the life of Jesus' followers in Jerusalem immediately after his death. It sounds like the ideal communist society—from everyone according to his abilities, to everyone according to his needs.

> Now the company of believers was of one heart and soul, and none of them claimed that anything which he possessed was [exclusively] his own, but that everything they had was in common and for the use of all. And with great strength the apostles delivered their testimony to the resurrection of the Lord Jesus, and great grace rested richly upon them all. Nor was there a destitute or needy person among them, for as many as were owners of lands or houses proceeded to sell them, and one by one they brought the amount received from the sales. (Acts 4:32-35 AMP)

It isn't clear who, if anyone, still worked—certainly not the disciples who spent all their time teaching and praising God. On the other hand, we can assume that a poor person gained much by joining the group. But just as in all other communistic societies, a little persuasion was occasionally needed. When Ananias and Saphira, a married couple in the group, sold a piece of property they owned, they kept some money for themselves before bringing the rest to the apostles. But Peter realized it and accused them of letting Satan fill their hearts and of lying not to humans but to God. Whereupon they immediately fell dead at his feet. I imagine that this successfully dissuaded others from following their so-called selfish example.[1]

The growing community contained two groups: the Hebrew-Christians, who spoke Aramaic and followed the strict Torah rules, and the Hellenists, who spoke Greek and who believed that Christianity had abrogated some of the Torah rules.[2] Some members of this second group complained to Peter that they were being discriminated against during the food distribution. So Peter called the congregation together and told them that he and the other apostles were too busy proselytizing "to serve at tables." He asked them to select seven reputable men, to be called deacons, to perform the various managerial tasks needed in the community. Even these people, however, quickly became involved in preaching the word, and Stephen in particular being "full of faith and power, did great wonders and miracles among the people" (Acts 6:8).

Unfortunately, this idealistic life did not last long. Stephen raised the wrath of the religious authorities, who brought him to trial and condemned him to death by stoning, thus turning him into the first martyr of the Christian Church.[3] A general persecution of the commune followed; "All of the Lord's followers, except the apostles, [abandoned Jerusalem and] scattered everywhere in Judea and Samaria" (Acts 8:1 CEV). Actually, despite Luke's words, it is probable that only the Hellenists were chased out, since both the Acts of the Apostles and Paul's letters report that the Church of Jerusalem remained the main headquarters of the sect at least until 60 CE.

Ebionite and Nazarene Jews

Escaping the persecution that followed Stephen's execution, Philip, another of the seven elected deacons, found refuge in Samaria. There he preached, chased away demons, and performed other miracles, so that many believed and were baptized in the name of Jesus. For some reason however, he did not pass on the power of the Spirit, so Peter and John were sent from Jerusalem to lay their hands on the people.[4]

This group eventually came to be called Ebionites, meaning the "poor ones," probably in reference to those poor who had been blessed by Jesus in the Beatitudes. They believed only in the Torah, ignoring all the other books of the Old Testament. Although they esteemed James, they held Peter in much higher regard. They practiced vegetarianism, rejected the divinity and virgin birth of Jesus, and abhorred Paul and his religion.[5]

Some of the other early Christians, perhaps the ones who remained in Jerusalem, became known by the name of Nazarenes, or Nazoreans, probably because they were followers of Jesus who was called Nazorean.[6] Although Acts presents Peter as making the first proselytizing speeches to the people, and leading the defense against the Sanhedrin, it does not appear that he was their formal leader.[7] That one was James the

Just, the brother of Jesus, who apparently joined after the Easter appearances and took over the group. When he was executed by the high priest in 62 CE, he was replaced by Simon ben Clopas, a cousin of Jesus. He in turn was replaced upon his death by Jesus' brother Judas. Does this sound a little like nepotism, similar to that of the Maccabees two centuries earlier, or like the return of the old Israelite kingdom? The Church in Jerusalem disappeared when most of its inhabitants were slain by the conquering Romans in 70 CE. It is said that the few survivors found refuge in the gentile city of Pella.[8]

The Nazarenes believed in the complete Old Testament and all its laws, but not in the *Halakah*, its interpretation by pharisaic scholars. They followed circumcision, Sabbath rest, and all other Hebrew practices. There exists some confusion, however, regarding their early beliefs about Jesus. Since they were led by James, his brother, we can assume that they would not have believed in either the divinity of Jesus or his virgin birth. I say this, because the gospels make it clear that during his lifetime, Jesus' family thought him to be confused, if not outright crazy, and tried to stop him from preaching. Nevertheless, the great Church historian Eusebius (260-339 CE) wrote that the Nazarenes accepted Jesus' Sonship and his virgin birth, but not his pre-existence before the beginning of time.[9]

I believe that the Nazarenes had their roots in the Jerusalem Christians. They used the Gospel of Hebrews that resembled a little the Gospel of Matthew, but which held James in special esteem and described how Jesus had appeared to him after the resurrection. They even had some rudimentary beliefs about the Holy Spirit. Because their beliefs were very similar to those of the later Church, they were not considered heretics until well into the third century when it was their Jewish practices, such as circumcision, and not their beliefs, that placed them on the proscription list. The Nazarenes considered themselves to be Jewish but believed that Jesus was the Messiah. They attended synagogues and tried to proselytize other Jews. The two sects separated, however, when the Jews instituted the *Birkat ha-Minim*, a curse upon the Nazarenes that was pronounced by all attending synagogue services. That was aggravated during the final (132-135 CE) Jewish revolt against the Romans, when the Nazoreans failed to support the Judean acclamation of Bar Kochba as the Messiah. They were never too numerous, however, and they disappeared completely by the fourth century.

MATTHEW'S JEWISH CHRISTIANS

We don't know who exactly were the people for whom Matthew wrote his gospel. Nevertheless, some authors have tried to deduce their possible characteristics by working backwards from the gospel's content. They concluded that the group was essentially Jewish, except that it believed that the long-expected Messiah had come in the person of Jesus. Although the group considered Jesus to be the Son of God, this was in the same way that Israel was considered to be the son of God. But Jesus himself was not God, since that would obviously conflict with Hebrew monotheistic beliefs.

The group was small, perhaps restricted to only one synagogue. This conclusion was derived from Matthew's instructions that people should solve their problems by first dealing with each other on a one-to-one basis, and that only if they were unsuccessful they should bring the matter to the attention of the Church. Since they were all brothers, they were not supposed to have any leaders or teachers.

> But you are not to be called rabbi (teacher), for you have one Teacher and you are all brothers. And do not call anyone [in the church] on earth father, for you have one Father Who is in heaven. And you must not be called masters (leaders), for you have one Master (leader), the Christ. He who is greatest among you shall be your servant. Whoever exalts himself shall be humbled, and whoever humbles himself shall be raised to honor. (Mat. 23:8-12 AMP)

Because the Temple in Jerusalem had been destroyed a few years earlier, the Torah's Deuteronomic instructions could not be followed any longer. As an alternative, the surviving Pharisees developed a new rabbinic-based religion to draw the allegiance of the Jews. Matthew's group positioned itself against them and tried to reform Judaism by preaching that the Messiah had arrived. At first, both groups worshipped in the same synagogues. But sometime around 80 CE the rabbis got the upper hand and introduced in the Jewish service the *Birkat ha-Minim*, a blessing/curse against heretics.[10] As was the case with the Nazarenes, this would have forced the attending Jewish-Christians to curse themselves aloud, something that was not taken lightly in those days. As a result, Matthew's group withdrew into their own synagogues.

JOHN'S COMMUNITY OF THE BELOVED DISCIPLE

The last gospel, written about 90 CE or a little later, describes a community that is under continuous attack from those around it. Its first members were probably past disciples of John the Baptist, which is why this gospel places so much emphasis in minimizing his importance compared to Jesus.[11] Perhaps they brought with them to the group the Essenes' concepts of Good and Evil, Light and Dark. The community was later joined by anti-Temple Samaritans, which explains why this is the only gospel that talks about Jesus entering Samaria and making converts. It has been suggested that the Samaritans introduced to the group the high Christology that changed Jesus from a Messiah to someone who is one with God, perhaps because many famous Gnostics of those days were Samaritans.

The proclamation of Jesus' divinity incensed the Jews in the synagogues, who were not content with just expelling John's followers, as they had done with Matthew's group, but tried to execute them as blasphemers and heretics; they often betrayed them to the Roman authorities, causing their martyrdom.[12] As a result, John's followers stopped thinking of themselves as Jews and started preaching the gospel to the Gentiles. Some of these joined the group, but many refused. So although this gospel starts by considering only the Jews as tools of Satan—all the Jews, not just their leaders, as was the case with Matthew's gospel—it later expands the hated group to include all outsiders.

The most intriguing feature of this gospel is the character of the Beloved Disciple whose identity is never revealed, but who was claimed to have been alive during the time the first version of the gospel was being written.[13] His presence during the writing was used to justify the greater inherent reliability of this gospel compared to the others, and to increase the gospel's importance through the numerous preferential comparisons of the beloved disciple versus Peter.

Sadly, the closely knit community was soon broken apart by disagreement on Jesus' exact nature. John's group raised Jesus to the status of God, but in the process caused him to lose his humanity. In John's gospel, Jesus constantly proclaims his divinity: God and he are one. Even when he exhibits human emotions towards Mary or Lazarus, it is from a superior, condescending position. It was an easy step to go from John's gospel to Docetism, the belief that Jesus was God only disguised as human—one of the most prominent gnostic heresies of the second century. (Remember that Gnostics believed that human salvation could only be attained by acquiring secret knowledge, knowledge that Jesus had shared with a few select persons.) The two epistles written in the gospel author's name reveal the unsuccessful efforts made to mend the growing rift. And because the gnostic group also adopted the gospel as its own, it took some time before the growing orthodox Church adopted it.

PAUL'S FOLLOWERS OF THE WAY

Paul became a believer of Jesus in 37 CE after a vision he had while going to Damascus to arrest Christians. Twenty-five years later, he was probably executed in Rome during Nero's persecutions. In between he proselytized in Middle Eastern and Greek cities, and during the last ten years of his life he wrote half a dozen letters to the communities he had founded to help them with their problems. He also wrote one very important letter to the believers in Rome in which he explained carefully and at great length his religious beliefs. These letters, as well as another half-dozen letters written by his followers in his name, are the foundation upon which Christianity was built.

Upon arriving in a new city, Paul's *modus operandi* was to go to the Jewish synagogue and preach about Jesus to the gentile God-fearers—Gentiles who frequented synagogues but had not converted to Judaism. The Jews would invariably kick him out of the synagogues, but he usually managed to win over some Gentile converts. He would then settle in the craftsmen's neighborhood, and talk to people while he practiced his tent-making profession. More often than not, his radical views would infuriate the Jewish inhabitants of the city and they would rise up against him and chase him out of town. According to his own words, they scourged him five times with the standard thirty-nine strokes of the lash.[14] Understandably, the authorities usually considered him to be a troublemaker and an undesirable.

Nevertheless, in each city he managed to gather together a small group (*ecclesia*) of believers who practiced the new religion with baptism, breaking of bread, and preaching the word of God. The group was probably composed of a general representation of city inhabitants: mostly slaves and laborers, but also some artisans and tradespeople. Once a week they would gather at the house of their richest member and celebrate Eucharist in remembrance of Jesus' Last Supper. To outsiders, the group probably appeared like any other voluntary association—a gathering of people with some common interests, whose main purpose was to insure the proper burial of its members—very common in the Roman era, except that in this case it also included women. Because the basic society unit in those days was the family, a new convert would bring with him into the group his entire family, servants, and slaves, if any. Obviously not all of these were equally enthusiastic about the new ideas they were expected to embrace.

One group of people that was particularly attracted to Paul's religion consisted of the God-fearers. These were Gentiles who went to synagogues, believed in the Jewish God, but did not convert because they were unwilling to follow all the Hebrew laws, especially those regarding circumcision and clean foods. Since Paul prohibited circumcision, and dropped food proscriptions so long as the food did not originate from sacrifices to pagan gods, these people were a perfect fit in the new Christian gatherings. As we will see in the next chapter, Paul's religious ideas formed the basis for the new Christian religion.

SOME QUESTIONS TO CONSIDER

1. The Early Christians in Jerusalem cooperated in a communistic type of society because they thought that the

end was near. What do you think they would have done if they had known that the second coming was going to be delayed?

2. Why do you think there arose so many different Christian sects after Jesus' death? Which one do you think represented better the ideas he preached?

3. To what do you attribute the survival of Christianity in the first century?

4. In Matthew's gospel John the Baptist says: "I am baptizing you with water, for repentance, but the one who is coming after me is mightier than I....He will baptize you with the Holy Spirit and fire" (Mat. 3:11, similar to Luke 3:16). How can you explain that the first three gospels don't describe Jesus performing any baptisms.

SELECTED REFERENCES

Brown, Raymond E. *The Community of the Beloved Disciple.*

Ehrman, Bart D. *Lost Christianities: The Battle for Scripture and the Faiths We Never Knew.*

Goulder, Michael. *St. Paul versus St. Peter: A Tale of Two Missions.*

Pritz, Ray A. *Nazarene Jewish Christianity: From the End of the New Testament Until its Disappearance in the Fourth Century.*

Saldarini, Anthony J. *Matthew's Christian-Jewish Community.*

Stegemann, Ekkehard W. and Wolfgang Stegemann. *The Jesus Movement: A Social History of the First Century.*

CHAPTER 22

THE INVENTION OF CHRISTIANITY

Neither Jesus nor his disciples started any new religions. Instead they considered themselves pious Hebrews and worshipped at the Temple and the synagogues. The only thing that may have distinguished them from the others was their belief in the imminent arrival of God's reign. It was Paul, a self-converted outsider, who set out to logically and methodically construct a new religion, based entirely on Jesus' death and resurrection.

He reasoned that if Jesus had been raised from the dead, he must have had a special relationship to God. Since he beat death, that is Satan, Jesus must have been stronger than him and hence must have predated him. And through his resurrection and escape from death's grip he must have cancelled Adam's sin that had resulted in man's mortality. Thus Jesus the man became Christ the *Logos* through whom all things were made according to the old philosophers. Then came Paul's crowning assertion: if Jesus was man and beat death, then mankind can do the same, if it only believes and suffers with Jesus Christ. And these are accomplished through the sacraments of baptism and Eucharist. He asserted that salvation is achieved not through man's actions, but through man's faith and God's grace.

Paul

In spite of what one would have thought, Jesus did not start a new religion. His message, as described in the three synoptic gospels, was simple: repent, because the kingdom of God is coming. Jesus never said that he was God; he never even mentioned his unusual birth, if he indeed had an unusual birth. He did say that God was his father, but he also said that God was our father. Besides, in those days the Hebrews considered that God was the father of all of Israel. It was not Jesus, but Paul who first laid the foundations of Christianity in his letters, and later John who finished the edifice in his gospel.

Paul was born in Tarsus, in Cilicia—just north of Syria at the northeastern corner of the Mediterranean—of parents who had bought Roman citizenship.[1] His original name was Saul, and he claimed that he had been sent to Jerusalem to study under the famous Pharisee Gamaliel. He insisted that he had lived as a Pharisee, but never said that he had been one. By profession he was a tentmaker, something which involves working with hides, a rather questionable practice for a pious Jew. This, however, would have brought him in close contact with the Temple authorities who accumulated huge quantities of hides from the daily sacrifices.

It is also probable that he was somehow involved with the Temple guard, since he admitted to vigorously suppressing the early followers of Jesus in the city. Luke tells us that Paul had also been present at Stephen's execution, watching the cloaks of the people who were involved in his stoning (Acts 22:20). Afterwards he was supposed to have asked the chief priest for letters of recommendation to the Jewish establishment in Damascus so that he could go there to locate the Christians and bring them back in chains for punishment.[2] But why was he so enraged against the new sect? The explanation that is usually given is that, as a pious Jew, Saul considered the crucified Jesus to have been cursed in accordance with the Law: "He that is hanged is accursed of God" (Deut. 21:23). Hence, his followers must have also been cursed.[3]

On the way to Damascus, a most extraordinary thing happens to Saul. He sees a bright light that blinds him, he is thrown off his horse, and he hears a heavenly voice. He himself does not describe this event. Instead he just says that, "It pleased God, who separated me from my mother's womb, and called me by his grace, to reveal his Son in me that I might preach him among the heathen" (Gal. 1:15-16). Luke, however, relates the event three times. The stories differ slightly, but the last one is the most detailed, with Paul describing the event to King Agrippa II:

> About noon I saw a light brighter than the sun. It flashed from heaven on me and on everyone traveling with me. Then I heard a voice say to me in Aramaic, "Saul, Saul, why are you so cruel to me? It's foolish to fight against me!" "Who are you?" I asked. Then the Lord answered, "I am Jesus! I am the one you are so cruel to. Now stand up. I have appeared to you, because I have chosen you to be my servant. You are to tell others what you have learned about me and what I will show you later." The Lord also said, "I will protect you from the Jews and from the Gentiles that I am sending you to. I want you to open their eyes, so that they will turn from darkness to light and from the power of Satan to God. Then their sins will be forgiven, and by faith in me they will become part of God's holy people." (Acts 26:13-18 CEV)

Recollecting this event, Paul asserted that Jesus appeared to him and gave him the gospel that he was teaching; he claimed that no human had instructed him. It is obvious, however, that there was no Jesus sighting of any kind; it was at most a bright light and a voice, and the voice said very little. The rest of Paul's assertion is even more important; did Paul really make up his gospel independently from scratch? This is difficult to believe, since he had numerous opportunities to exchange data with Jesus' followers. He started his discipleship by spending three days with Ananias in Damascus. Three years later, he went to Jerusalem and spent fifteen days as Peter's guest, a time that must have been devoted to exchanging information and theologies. Later he spent most of his proselytizing travels in the company of people who had been in Jerusalem during the days of Jesus: Barnabas, John Mark, and Silas.

But perhaps this is something that Paul was making up to shore up his arguments. He had major disagreements with the Jerusalem church regarding proper teachings. If he had learned the gospel from the leaders of that church, then he would not have been in a position to disagree with them; he would have had to follow their interpretation. But if his source of knowledge was independent, then he could teach whatever he wished, claiming superior knowledge. Peter did the same thing when he argued that he had a vision that permitted him to baptize Cornelius, an uncircumcised Gentile, and to stay and eat in his house (Acts 10:9-16).

It is strange, however, that when we compare Paul's letters with the gospels, we find only three points of agreement between them: Jesus was crucified; he was resurrected and seen by various people (although there is disagreement about who exactly these people were); and on the night of his arrest there was a supper where the wine and bread were blessed in a manner agreed upon almost identically by all sources. With the exception of this Last Supper, which we will examine in in detail chapter 25, his letters don't seem to have much in common with the first three gospels.[4] What we have instead is a set of assumptions, followed by the logical, philosophical conclusions that could be derived from them, at least by him. In the final analysis, it didn't make any difference what, if anything, he knew about Jesus. Starting from scratch, he built a theological system that became the basis of Christianity as we know it today.

PAUL'S CONVERSION AND TRAVELS

Much has been written about Paul's conversion incident. He might have been a diabetic or had some other congenital disease—he himself refers to bearing the signs of Jesus on his body—so he might have sustained a sudden attack of some kind. But, with one exception, nothing we know could have precipitated such a sudden and complete reversal of his beliefs. Just a bright light and a voice would not have done it, but a near-death experience could have. Remember our discussion in chapter 17 about how people's attitudes change instantaneously and completely after such an event? Paul actually mentions that somebody he knew had such an experience during which that person visited the third heaven.

> I know a man in Christ who fourteen years ago—whether in the body or out of the body I do not know, God knows—was caught up to the third heaven. And I know that this man—whether in the body or away from the body I do not know, God knows—was caught up into paradise, and he heard utterances beyond the power of man to put into words, which man is not permitted to utter. (2 Cor. 12:2-4 AMP)

He says that it was not he, and we can calculate from known dates that the event he describes would have occurred about six years after his conversion. Nevertheless, it is highly probable that something similar must have happened to him. We know of nothing else that is powerful enough to have changed his beliefs instantaneously.

Our knowledge about Paul derives from two sources: his seven authentic letters (1 Thessalonians, Galatians, 1 and 2 Corinthians, Romans, Philippians, Philemon), and Luke's Acts of the Apostles, which was written perhaps forty years later. It is believed that Luke traveled with Paul towards the end of the latter's career, so we can assume that his stories regarding this part of Paul's life are fairly accurate. But since Luke had no firsthand information about Paul's earlier life, he might have made errors when relating events that occurred during that earlier period.

After the incident on the road, Paul entered the city of Damascus where a man named Ananias was directed by God to go lay his hands on him and cure him. Paul spent the next two or three years in Arabia, the Gentile-inhabited region southeast of the city. Some say that he used this time to put together his newly discovered theology; others say that he actually preached there. The truth is that we have no way of knowing for certain. In any case, he returned to Damascus afterwards and started preaching that Jesus was the Son of God. But when the city Jews threatened to kill him, the Christians helped him escape in the middle of the night. He next visited Peter in Jerusalem, where presumably they compared theologies for fifteen days. But his presence there became known to his former friends, now opponents, so the disciples had to spirit him out of town and he returned to his hometown Tarsus.

It is assumed that Paul proceeded to preach in Cilicia and nearby territories, but we have no specifics. We do know, however, that when a new church was started by the Hellenists in Antioch, the third largest city of the Roman Empire, Barnabas fetched him from Tarsus to help, since presumably he could communicate better with the Gentile inhabitants.[5] According to Luke's Acts it was there that Jesus' followers were first called Christians.[6]

With this one exception, we know nothing about Paul's actions in the period between 33 and 46 CE. He founded no surviving churches, and he never referred to those years in his letters. Judging by later events, however, one can perhaps assume that the Jerusalem church was dispatching teams to all the places Paul went to try to bring any converts he had made back to the proper Torah-following orthodoxy. (Remember that the believers in Jerusalem considered themselves Hebrews, followed the instructions of the Torah, and sacrificed in the Temple.)[7] In any case there are no results to show for his efforts during this period, prompting many historians to assume that he did nothing in these so-called "hidden years." It was then that Agrippa I started his persecution of Jesus' followers in Jerusalem. He had James Zebedee killed and Peter imprisoned, although as we saw previously, Peter was helped to escape from jail, and he then fled Jerusalem (43 CE).

Paul spent the next eleven years traveling and preaching in northwest Asia, Macedonia, and Greece. This is a fairly well documented period, since during some of that time Luke, the author of the Acts of the Apostles, presumably traveled with him. Furthermore, we also have letters that Paul wrote to some of the churches he had founded in that area. In 49 CE, he went to Jerusalem to discuss his status as a preaching apostle at the so-called Apostolic Council, with James, Cephas,[8] and John. At James's direction, it was decided that Paul would preach to the uncircumcised (Gentiles), and Peter to the circumcised (Jews).

Despite this agreement, however, Paul eventually entered into a confrontation with Peter, which ended their friendship, assuming, of course, that there had ever been one. As we will see below, Paul took a rather cavalier attitude about Torah rules such as circumcision and clean foods. So while he was in charge of the Antioch church, both Gentiles and Jews ate the same food at the same table. When Peter came in town, he went along with the arrangements, but then a delegation was sent from Jerusalem by James to check out the church's practices. The members of the delegation refused to eat at the same table with the Gentiles but ate separately, and Peter joined them. The rest of the Jews then followed Peter's example, prompting the furious Paul to engage Peter in an angry dispute and break off with the Antioch church.

Paul's preaching travels ended in 57 CE, when he returned to Jerusalem, bringing a collection for the church that was in a bad financial condition. Despite attempts at reconciliation, the Jews arose against him, accusing him of bringing a Gentile into the Temple, and only the intervention of the Roman garrison, which arrested him as a troublemaker, saved his life. When a Roman officer ordered him flogged, Paul revealed his Roman citizenship that prohibited such punishment. Not knowing what to do with him, the Roman commander sent him to Felix, the governor of the province, at Caesarea. According to Acts, Paul was protected from the aroused population by a force of two hundred soldiers, seventy horsemen, and two hundred auxiliaries, something that gives us an idea of how much upheaval he must have caused in Jerusalem.

Felix didn't know what to do with Paul either, so he kept him in jail for two years before finally sending him to Rome for trial. It seems that in Rome he remained under some kind of house arrest for a couple of years, during which time he apparently did some preaching and wrote his last two letters. Luke's narrative ends at this point. It is assumed that he was executed together with Peter in 62 or in 64 CE during Nero's persecution of the Christians. At least this is the story told by Eusebius (Maier 1999, 85), although today many doubt that Peter had actually ever been in Rome. Others think that maybe Paul did fulfill his ambition to go and preach in Spain, the furthest part of the world, and that he was executed in 64 CE after his return to Rome.[9] Although Luke, writing thirty years later, probably knew the answer, he did not reveal it. We can assume that he did not want to antagonize his Roman readers by recounting Paul's execution, but I see no reason why he would not have mentioned Paul's trip to Spain, if he had indeed made one.

Paul's Beliefs

Since Christianity depends so heavily on Paul's beliefs, we should take a little time to summarize them.

Paul's gospel was inspired: Paul believed that what he was teaching had been directly revealed to him[10] by Jesus. "But I certify to you, brothers, that the gospel which I preached is not man's. For I neither received it from man, nor was I taught it, but by the revelation of Jesus Christ" (Gal. 1:11-12).

Jesus annulled the laws of the Torah: The old Hebrew law commanded by the Torah has been superseded by Jesus' death and resurrection. Note that Jesus' teachings are not involved; in fact Paul never mentions them. This nullification of the Torah through Jesus' death and resurrection is Paul's basic assumption from which all else follow logically. But on what did he base it? Some suggest Isaiah:

> Therefore I will divide him a portion with the great, and he shall divide the spoil with the strong; because he has poured out his soul unto death; and he was numbered with the transgressors; and he bore the sin of many, and made intercession for the transgressors. (Isa. 53:12)

Paul believed that nobody could be saved by following the Law (Torah) because it was impossible for man to do so. Only the elimination of sin itself could save man from becoming prey to sin, and this is what Jesus did; his death resulted in our justification: "As one man's fall brought condemnation on everybody, the good act of one man brings everyone life and makes them justified" (Rom. 5:18 JER). He argues that otherwise, "if the Law can justify us there is no point in the death of Christ" (Gal. 2:21 JER). In other words, if we could have been justified by following the Law, then Jesus' death would not have been necessary. But according to Paul, man's physical nature does not allow him to follow the Law, so however much we try, it is impossible for us to be justified by it. The death of Jesus provides an alternate possible solution. Paul bases the rest of his theology on this one assumption.

Jesus' role in creation: Since, according to Hebrew beliefs, the Torah laws had existed before man was created, Jesus would have been able to annul them only if he had existed before them. "He is the image of the invisible God, the first-born of every creature" (Col. 1:15) wrote one of his followers. Not only that, but he must have been above the Law; in order to be able to destroy the Law, he must have created it himself: "For us there is one God, the Father, from whom all things come and for whom we exist; and there is one Lord Jesus Christ, through whom all things come and through whom we exist" (1 Cor. 8:6 JER). His followers added: "All things were created by him, and for him" (Col. 1:16).

Jesus became man: But apparently the Law could not be destroyed from outside, but only from within. So "God sent his son, made from a woman, made under the law" (Gal. 4:4). He became like one of us by God "sending his own son in the likeness of sinful flesh" (Rom. 8:3). Only a man could undo the damage done by another man: "For if by one man's offense death reigned by one; much more they who receive abundance of grace and of gift of righteousness shall reign in life by one, Jesus Christ" (Rom. 5:17). Jesus had to become a man not only to save us from sin, but also to bring us eternal life: "For since by man came death, by man came also the resurrection of the dead" (1 Cor. 15:21).

Humans cannot follow the Law: As much as we may want to, our human nature makes it impossible for us to follow the laws of the Torah. "For we know that the law is spiritual: but I am carnal, sold under sin.... For the good that I would do, I do not, but the evil that I would not do, that I do" (Rom. 7:14,19).

The Law's annulment destroys sin: If there is no Law, you cannot break it: "[To be sure] [S]in was in the world before ever the Law was given, but sin is not charged to man's account where there is no Law" (Rom. 5:13 AMP). Note the difference here between Sin as an evil power, and sin as the result of an evil deed. Man cannot sin until prohibitions are put in place: "The sting of death is sin; and the strength of [S]in is the law"

(Cor. 15:56). (But consider question #6 at the end of the section.)

Faith substitutes for the Law: Paul looks back in the scripture to Abraham, who was justified solely through faith in God: "[Abram][11] believed in the Lord; [who] counted it to him for righteousness" (Gen. 15:6). In the same way, we will be justified not through our actions, but through our faith in Jesus. "For by the works of the law no flesh shall be justified" (Gal. 2:16). "For if righteousness comes by the law, then Christ died in vain" (Gal. 2:21).[12] Later on in the letter, however, Paul seems to say that our actions still matter: "Whatever a man sows, that (and that only) is what he will reap. For he who sows to his own flesh will from the flesh reap decay..., but he who sows to the Spirit will from the Spirit reap eternal life" (Gal. 6:7-8 AMP). Nevertheless, our righteous actions by themselves cannot save us, because not one of us is righteous. We can only be saved through unconditional trust. Not through our own efforts, but through complete trust in God. An example of this would be a drowning man whom somebody is attempting to save. He can only help by letting himself go completely and allowing his savior to pull him to shore; if he starts thrashing his arms wildly trying to stay afloat, both he and his rescuer will go under.

Baptism substitutes for circumcision and becomes a saving action: Just as circumcision used to mark a Hebrew as one of God's own, so baptism marks a Christian as a follower of Christ. It is the representation of sharing Jesus' death and resurrection: When "we were baptized into Jesus Christ [we] were baptized in his death. Therefore we are buried with him by baptism into death, so that as Christ was raised from the dead by the glory of the Father, we also should walk in newness of life" (Rom. 6:3-4).

Our resurrection will be in a heavenly body: In the years that followed Paul, there was much discussion in the growing Church regarding the characteristics of our resurrected body. After a few centuries it was finally agreed that the resurrected body would consist of the identical atoms that had been present in the earthly body. Paul had thought otherwise: "There are celestial bodies, and bodies terrestrial....So also is the resurrection of the dead. [The body] is sown in corruption; it is raised in incorruption....It is sown a natural body; it is raised a spiritual body....We shall not all sleep, but we shall all be changed, in a moment, in the twinkling of an eye, at the last trumpet" (1 Cor. 15:40,42,44,51,52).

The kingdom of God is imminent: As we saw earlier, Paul expected that the Second Coming would occur as soon as everyone had an opportunity to accept the new faith. His converts probably thought that it would be even sooner, so many stopped working and started spending their savings, something against which he had to admonish them. Similarly, when some people died in Thessalonica before the arrival of God's reign, he had to write to his followers and reassure them that though dead, these people would still be resurrected and enter the kingdom, in fact ahead of the living.

Be satisfied with your status in life: As a result of his belief that the end was imminent, he saw no reason for people to try to change their current status. "Let each one lead the life which the Lord has allotted to him....In whatever station of life each one was when he was called, there let him continue with God" (1 Cor. 7:17,24 AMP). Besides, in the eyes of God, everyone was the same. "There is neither Jew nor Greek, there is neither slave nor free, there is neither male nor female; for you are all one in Christ Jesus" (Gal. 3:28).

Sex and married life interfere with efforts to please the Lord: Married people care for each other instead of concentrating their attention on God. So if people are not married, it is best for them to continue like that. "This I say, brothers, the time is short. In what is left, let those who have wives be as though they had none....For the way of this world is passing away. But I would have you be without worries. He that is unmarried cares for the things that belong to the Lord, how he may please the Lord. But he that is married cares for the things that are of the world, how he may please his wife" (1 Cor. 7:29-33).

SOME QUESTIONS TO CONSIDER

1. Was Paul right in believing that Jesus' death negated the validity of the Torah?

2. Where did Paul find the facts that he preached in his gospel? Did he make them up?

3. Paul said, "If righteousness comes by the law, then Christ died in vain." Is he assuming the falsehood of the first half of the sentence and using it to disprove the second half, or the reverse?

4. How can you justify (or explain) Peter's first sitting for dinner with the Gentiles, and then refusing to do so?

5. What do you think of Paul's interpretation of baptism?

6. Paul maintains that if there were no law there would be no sinners. Could we reduce our criminal problems by rescinding all laws?

7. Did Paul believe that Jesus was divine? How do you define divine?

8. If a person only reaps what he sows, how can we say that he should put his faith in God's mercy and not try to justify himself?

9. Paul's preaching technique caused much upheaval at the cities he visited and resulted in his spending many years in jail. Do you think that a different approach could have been more effective?

10. Genesis describes God creating the world through his commands. How do we defend the statement that it was all created through Jesus, the *Logos*?

11. Is Paul's position that only the creator of the Law can destroy it, logical?

12. Why should the creator of the world, if this is what Jesus was, have to become man in order to eradicate the original sin?

SELECTED REFERENCES

Dunn, James D. G., ed. *Paul and the Mosaic Law.*

Goulder, Michael. *St. Paul versus St. Peter: A Tale of Two Missions.*

Hengel, Martin and Anna Maria Schwemer. *Paul Between Damascus and Antioch: The Unknown Years.*

Seifrid, Mark A. *Christ, Our Righteousness: Paul's Theology of Justification.*

Wilson, A. N. *Paul: The Mind of the Apostle.*

CHAPTER 23

THE GOSPELS

Until recently, Christians had generally ignored Paul's contribution to their religion, concentrating instead in the teachings of the four gospels. Of these, the first three present a fairly similar story. A prophet, perhaps God's son,[1] is traveling through Judea announcing the coming of God's kingdom and the need for people to repent their sins and to love their fellow men. After confronting the authorities, the prophet is executed, but he then rises miraculously from the dead. The fourth gospel, John's, presents a different picture. The protagonist is not a man but an incarnation of God's eternal Son, who had helped create the world, and who is now teaching that not repentance, but belief in his divinity, will allow one to enter into the kingdom of God.

Although named after disciples and followers of Jesus, we don't really know who actually wrote these four gospels. But they are all presenting a Jesus who is not responding to the problems that he himself must have faced in his life, but to the problems that the gospel writers had faced in their own times. As a result the message varies slightly from gospel to gospel, and it is impossible to separate Jesus' actual words from those ascribed to him by the evangelists. In recent years many religious authorities have informally added one additional text, the Gospel of Thomas, to the four gospels in the cannon, and consider it as meriting equal attention.

The Writings and their Corruption

When Jesus walked and preached across the land of Palestine, he talked; he did not write. Nor did any of his disciples take notes. So during the first two decades after his death his teachings spread orally from person to person. But as the followers of "The Way" grew in number and spread out across the entire Roman world, the words became more and more hearsay. On one hand, the original witnesses were few and getting older, and on the other, the churches were becoming increasingly more numerous. So before the old stories were forgotten, some people wrote down what Jesus had said and done. To be more precise, they wrote down what they thought he had said, since memories become fickle with age. We know that Paul's letters to various churches were copied and forwarded to sister churches. Probably some of the writings about Jesus' sayings were also copied and read in churches, forming the basis for the discussions on Sunday meetings.[2]

But records in those days were not what we think today as historical documents. People did not write down what someone had actually said, but what they believed he must have *meant* to say. So the interpretation of the events that took place became even more important than their precise recollection. Think back to discussions you may have had with a long-dead grandfather. I am sure you cannot remember the exact words, but you can probably recollect the sense of the things that he had said. And that recollection is probably affected by your knowledge of his likes and dislikes. You can imagine what he probably would have said in a particular situation, and so you are fairly comfortable quoting him, even if such a situation had never actually arisen in his lifetime.

My grandfather died more than fifty years ago, when I was barely fourteen. Just like most village people in those days, he was prone to quote many sayings, but I can remember word for word only one of them: "Oh, height indiscernible and depth beyond measure!" Perhaps it was his favorite; he certainly used it often enough. But the reason I remember it is that its meaning puzzled me for years. And this is the exact feature that allows today's researchers to separate Jesus' actual sayings from those that have been put in his mouth by the ancient writers: if something did not allow intelligent interpretation in that ancient world, he must have said it exactly that way. Everything else is assumed to be fabrications of what each writer had thought that Jesus would have said, had he experienced the same situation as the writer did in his day.[3]

Numerous versions of what Jesus said and did were written down in the first two centuries. The Christian Church, however, destroyed most of these in the fourth century in its effort to shield the "true beliefs" from heretical writings, defined as anything that disagreed with the Church's point of view. As a result we have been left with four gospels, plus some letters attributed, mostly incorrectly, to Paul, John, Jude, James and Peter. To these we can now add the Gospel of Thomas, which was discovered in Egypt in 1945, although it has not been formally included into the canon yet.

None of the original works written during the first century have survived. What have survived instead, are copies of copies of copies of them. More often than not, the earliest fragments of the surviving writings date back to the third century. By that time they included well-intended corrections by numerous copyists, who were either trying to fix perceived errors, or were purposefully altering the original material so that it could better represent the then-current point of view. But even reading and interpreting the initial writings was not a simple matter. Greek words in those days were written in capitals and joined together without spaces or punctuation. When a word reached the end of the line, usually about three inches long, it continued onto the next line without any indication that a division had occurred.

The crucifixion scene in Luke presents an example of how punctuation can alter the meaning of a sentence. Jesus has been crucified between the two bandits, when one of them turns to him and taunts him: "Are you not the Messiah? Save yourself and us." But the other bandit castigates him. Whereas they themselves were guilty, he tells him, Jesus was innocent. Then he turns to Jesus and says: "Jesus, remember me when you come into your kingdom." And Jesus replies: "Truly I say to you, today you will be with me in Paradise" (Luke 23:43). Those who believe that the final judgment follows immediately upon our death have used this sentence to justify their position. But the sentence had originally been written without the punctuation. Suppose we rewrite it after moving the comma one word to the right: "Truly I say to you today, you will be with me in Paradise." See how the meaning has altered? I imagine that the translator's job must have been incredibly difficult when dealing with Paul's writings, which involve long, complicated sentences with multiple subordinate clauses.[4]

DIFFERENCES AMONG THE GOSPELS

Throughout most of Christian history it had been accepted that the authors of Matthew's and John's gospels were the identically named disciples of Jesus. Indeed, the Church Fathers' chief criterion for deciding which Scriptures to place in the accepted canon was their belief that they had been written by eyewitnesses to the events they were describing. It is now generally accepted, however, that none of Jesus' disciples had the education and expertise to write the gospels, and that, in fact, no gospel was written by an actual eyewitness. Nevertheless, John's gospel does maintain that an eyewitness was involved in its writing. Three of the gospels—Mark, Matthew, and Luke, the so-called synoptic gospels—presume to be biographical histories of Jesus.[5] The fourth one, John, may at first glance also appear to be a biography, but it disagrees so much with the others that it is generally looked upon as a theological exposition, intertwined around the story of Jesus' life but without adhering to the proper timeline. Even so, it appears that John's gospel contains many actual details missing from the other gospels, particularly details of events during Jesus' last days. This seems to confirm the assertion of the writer of John's gospel that he had access to information from an actual observer of Jesus' ministry.

One way to classify the four gospels is by the degree to which they reflect Paul's gentile inclination versus Peter's and James's Jerusalem orientation. Mark's gospel clearly follows Paul's ideas, as shown by his comment that Jesus "declared all foods clean" (Mark 7:19).[6] Matthew, on the other hand, follows the Petrine line: "I have come not to destroy but to fulfill [the Law]" (Mat. 5:17). Although Luke has allocated most of the space in his Acts to Paul's work, he has been very careful to insure that both Peter and Paul are represented as equals, performing exactly the same number of miracles. So we can probably conclude that Luke takes the middle ground.[7] Finally, John expanded Paul's ideas to construct a completely new religion, essentially unrelated to Judaism.

THE FIVE GOSPELS

MARK

Until recently, it had been assumed that Matthew wrote the earliest gospel, which is why it is located first in the Bible. We now know, however, that Mark's gospel was the most primitive one, written sometime around the destruction of Jerusalem in 70 CE.[8] It is generally accepted that the author of this gospel was John Mark, a nephew of Barnabas, who had accompanied Paul in his first travel and who later may have become a disciple of Peter. He was supposed to have written down by memory what Peter had taught him, probably in Rome.[9] He is also supposed to have been the son of that Mary in whose house the early Christians in Jerusalem used to gather and where Peter went after his escape from jail. He would have thus heard all the stories in the growing community firsthand. Others, however, disagree. Theissen, for instance, believes that the author of Mark's gospel was a Gentile, who wrote somewhere in Syria near Palestine (Theissen 1991, 257), and that he mainly pasted together things written down by others in the years between 40 and 50 CE.[10] Some say that Mark wrote his gospel in an effort to protect the Christians by clearly differentiating them from the Jews who had revolted against the Romans and were being prosecuted. If true, it was a futile effort, since it is doubtful that any Roman ever read these writings. In any case, no particular church group is associated with him.

Mark starts his gospel with Jesus already grown up and ready to be baptized by John. He ends it abruptly with the empty tomb, without any resurrection appearances. His Jesus appears to overrule the Torah, declaring all foods to be clean and that "The Sabbath was made for man, not man for the Sabbath" (Mark 2:27). Jesus is here cast as the oppressed, misunderstood Son of God. Nobody understands what he says, including his disciples, who are presented as a bunch of hicks, with Peter being the worst. Even so, although Mark talks at length about Jesus' miracles, he covers very little of his actual teaching.

Permeating the gospel is the so-called messianic secret. Jesus wants to keep his real identity hidden: when Peter declares him to be the Messiah, Jesus warns him not to tell anybody. He gives the same warning to his disciples after his transfiguration, and whenever he cures somebody, he admonishes that person not to tell anyone. Why? Was it that he was afraid that the Judeans might mistake him for the kind of regal Messiah they expected, the one who would liberate them from the Romans? Or did he think that people might consider him as just a healer instead of an inspired teacher? At the end of the gospel, Mark's Jesus dies on the cross, showing as little understanding as his disciples: "My God, my God, why have you forsaken me?" (Mark 15:34) are his last words.[11]

Matthew

We have no idea who Matthew was, but it is assumed that he was the head of a small Christian church who wrote the gospel to prove to his followers that the long-awaited Messiah had finally come, and that this Messiah was Jesus. He lists Jesus' genealogy, starting from Abraham, showing that Jesus was indeed a direct descendant of David, a requirement for the real Messiah. Interestingly, he includes four women in Jesus' ancestry: Tamar, Rahab, Ruth, and Bathsheba. "All were outsiders to Israel, all were sexually suspect; and through all of them God worked surprisingly for the salvation of the people. They prepare for the birth of the Messiah by the Virgin Mary" (Johnson 1999, 139).

Both Matthew and Luke based their gospels on (1) Mark's gospel, (2) writings from a so-called "Sayings Source" usually referred to by scholars as "Q," and (3) personal written and oral sources. It is the Q source that presumably provided those teachings of Jesus that are present in the two later synoptic gospel but are missing in Mark.[12]

Matthew's Jesus has not "come to destroy [the Law] but to fulfill [it]....Till heaven and earth pass, one jot or one tittle shall in no way pass from the law, till all be fulfilled" (Mat. 5:17-18).

Matthew corrects some of the most flagrant deviations from the Law that Mark postulated: it is not man, but the Son of man, who is Lord of the Sabbath; also, although he copies Mark's saying that it is not what goes in the mouth that renders people unclean but what goes out—he does not conclude that Jesus declared all foods to be clean. In Matthew's gospel Jesus does not change the Law, but he does know the real intention of the Father. Again and again we read: "You have heard it said....But *I* say to you...." Matthew presents him as a new Moses: he shows him going to Egypt as a baby, so that he can come out of Egypt later as Moses did; he has him deliver his Sermon from the Mount, just as Moses brought down the Law from the mountain (interestingly, Luke portrays Jesus giving his sermon on a level plain, not on a mount). Matthew's Jesus performs all sorts of actions "so that the scriptures may be fulfilled." He is the person whose coming was foretold by the scriptures.[13]

In Matthew's gospel Jesus' enemies are only the Pharisees, not the scribes and Pharisees as they were in Mark's gospel. This is probably because Matthew's church included scribes among its members but was in direct competition with Pharisees for the support of the Jews. The fight for their allegiance had reached a peak, possibly tearing families apart:

> Think not that I have come to bring peace on the earth. I came not to bring peace but a sword. For I have come to set a man against his father, and the daughter against her mother, and the daughter-in-law against her mother-in-law, and a man's foes shall be they of his own household. (Mat. 10:34-36)

Only Matthew makes this statement (which is almost identical to the verse in Micah 7:6). His gospel was probably written in Antioch, the city where Paul lost his church to the Jerusalem (Petrine) faction.[14] Matthew may have believed that the destruction of Jerusalem was God's punishment of the Jews who had not accepted his son. He most likely wrote the gospel to give his Church an alternative to the teaching of the Torah. Although originally written in Greek, it was almost certainly translated into Aramaic and edited versions (downplaying Jesus' divinity aspirations) were used by the Ebionites and the Nazarene Christians.

LUKE

We don't know if Luke, the author of the third synoptic gospel, was the person whom Paul called "his beloved physician," the fellow traveller mentioned in Colossians, Philemon, and 2 Timothy. Similarly we don't know whether the "excellent Theophilus" to whom Luke addresses his gospel was an actual Roman noble. It is assumed that Luke was a Gentile who wrote his gospel to show that Christians were a peaceful group that did not present any threats to the Roman Empire. But is it reasonably to believe that a highly placed Roman officer would have taken time to read such a long treatise about an obscure Judaic sect? A diferent explanation has also been proposed: Theophilus means lover of God; was Luke addressing the gospel to his fellow-Christians, hoping to provide them with arguments they could use to justify their innocence when accused by the Romans?[15]

What is obvious is that Luke took great pains not to blame the Romans for anything bad that may have been done to the Christians. He represents Pilate as an almost saintly figure—little wonder that the Coptic Church canonized him. Luke's Jesus did not come just for the Jews, but for the entire world, so his genealogy starts with Adam, not Abraham.

This gospel differs from the previous ones in that Jesus did not die for the expiation of anyone's sins. He was falsely accused and died as a martyr. Furthermore, since it was obvious by this time that the end of the world was not imminent, Luke turns his attention to the correction of social injustices. To him, the blessed ones were the poor in goods, not the poor in spirit as Matthew had said. The wicked people were not the Pharisees, but the rich. The good news was that God's standards are different from those of man (Johnson 1999, 230); even a prostitute could be saved before a Pharisee, if she repented.

One of the characteristics of Luke's gospel is the importance of women and how he tries to include in his teachings both sexes. The man who searches for his lost sheep (15:4-7) is followed by the woman who searches for her lost dowry gold coin (15:8-14). A male householder pounds on his neighbor's door for some flour (11:5-8), and a widow pesters a judge for justice (18:2-8). He is the only one to talk about the Galilean women following and helping Jesus (8:2-3), the penitent woman who bathed his feet with her tears (7:38), and finally the story of Mary and Martha (10:38).[16] Furthermore, most scholars believe that the story of the adulterous woman related by John (John 8:1-11) has been misplaced and really belongs in Luke's gospel. Because of his inclusiveness and his kinder social standards, some people have suggested that his patron might have been a woman. Another unique detail in this gospel is its mention of Jesus' dispatch of seventy-two disciples to go preach across the countryside.[17]

JOHN

The fourth gospel, written perhaps around 90-100 CE, is quite unlike the other three. It starts with an extraordinary statement:

> In the beginning was the Word, and the Word was with God, and the Word was God. [He] was in the beginning with God. All things were made by him; and nothing was made without him.
>
> And the Word was made flesh, and dwelt among us, (and we beheld his glory, the glory as of the only begotten of the Father,) full of grace and truth. (John 1:1-3,14)

From the very first chapter, John's gospel shows its difference from the previous three. Those said nothing about Jesus being God, that he had existed before he was born, or that he had been instrumental in the creation of the world.[18] Now, barely a short ten years after Luke's gospel, Jesus has become the Word, and the Word is God.[19] But how did we get here? Let's read about the Hebrew Wisdom in Proverb 8:

> From the beginning, I was with the Lord.
> I was there before he began to create the earth.
> At the very first the Lord gave life to me.
> When I was born there were no oceans or springs of water.
> ....

> I was there when the Lord put the heavens in place
> and stretched the sky over the surface of the sea.
> I was with him when he placed the clouds in the sky,
> and created the springs that fill the ocean.
>
> ....
>
> I was right beside the Lord, helping him plan and build.
> I made him happy each day, and I was happy at his side.
> I was pleased with his world and pleased with its people.
>
> ....
>
> By finding me, you find life,
> and the Lord will be pleased with you
> But if you don't find me, you are in love with death.
> (Proverbs 8:22-36 CEV)

Much of what the proverb says about Wisdom, John attributes to the Word (who he assumes is Christ); through him all things were made.[20] Look at the last stanza above. This idea is restated again and again throughout John's gospel:

> For God so loved the world, that he sent his only begotten Son, so that whoever believes in him should not perish, but have everlasting life....He that believes in him is not condemned; but he who does not believe is already condemned, because he has not believed in the name of the only begotten Son of God. (John 3:16,18)

And here we have a simple summary of John's gospel: Salvation comes *only* through belief in Jesus, and perdition through disbelief. This is the only criterion, just as it was in Paul's teachings. John describes Jesus saying, "He who does not honor the Son, does not honor the Father who has sent him. Truly I tell you, he who hears my word, and believes in him that sent me has everlasting life, and shall not come to condemnation, but passes from death to life" (John 5:23-24). Little wonder that the Jews responded: "We were right to say that you are a Samaritan and that you have a demon in you" (John 8:48 CEV).

In John's gospel, Jesus starts his mission the way he ends it in the other gospels: by cleansing the Temple from the moneychangers and animal sellers. This brings him into immediate conflict with the authorities, positioning him so that he can rant and rail against them throughout the rest of the story. Most people think that this is the greatest of the gospels. It certainly is a glorious gospel, reaching peaks of rapture. Yet to me, its most important point is Jesus' explanation of the relationship between God and himself, himself and us, us and God; it is a full circle. (See the figure in chapter 15.)

But could Jesus' arguments really convert anybody? Perhaps this is why in this gospel Jesus performs miracles not to heal sick people, but to manifest the glory of God: "Unless you see signs and wonders, you will not believe" (John 4:48). A man was born blind so "that the works of God would be made visible through him" (John 9:3). Apparently, however, that was not enough: "Although he had done so many miracles in front of them, yet they did not believe in him" (John 12:37). Perhaps this depressed statement mirrors the growing failure of John's church.

In any case, I admit to my own lack of understanding also. The same Jesus who had said that the only thing that counts for salvation is belief in him, and hence in the Father, tries to atone for the human race by becoming the Passover lamb in some ancient Hebrew rite; a rite that for many centuries had failed to save the Hebrew nation both from its sins and from its enemies. Jesus' merciful Christian God moves backstage for a day, letting the vengeful Hebrew God take over. And the God who had said, "I desired mercy, and not sacrifice" (Hos. 6:6), the God who had stopped Abraham from sacrificing his son, that same God can now find no other way for human salvation than the sacrifice of his own son!

Yet John presents Jesus as being always in control, always knowing the future, as well as his next actions. There is no Gethsemane experience here, no "Take this cup away from me" prayer. Instead, we have his final soaring prayer to the Father:

> I have brought glory to you here on earth by doing everything you gave me to do. Now, Father, give me back the glory that I had with you before the world was created. You have given me some followers from this world, and I have shown them what you are like....My followers belong to you, and I am praying

> for them. All that I have is yours, and all that you have is mine, and they will bring glory to me.... I am also praying for everyone else who will have faith because of what my followers will say about me. I want all of them to be one with each other, just as I am one with you, and you are one with me. I also want them to be one with us. (John 17:4-21 CEV)

Jesus remained in control through his last seconds of life on the cross. He said, "'It is finished.' And he bowed his head, and gave up the spirit" (John 19:30).

THOMAS

You will not find the Gospel of Thomas in your Bible. No Christian denomination accepts it yet, and none even knew about it sixty years ago. Yet it may well have been written, in its earliest form, even before John's gospel.[21] In 367 CE, however, Athanasius, the bishop of Alexandria, ordered all religious writings to be burned, except those on the approved list, the canon. This resulted in the destruction of much of the early Christian literature, including most heretical, gnostic, and other writings that presented unapproved ideas.[22] Fortunately for us, some monks in the Egyptian desert chose to hide some of their banned books in nearby caves, where they were found sixteen centuries later. Included among them was a Coptic translation of the Gospel of Thomas.

Even though he almost certainly did not write the gospel himself, its authorship was attributed to Jesus' disciple Judas Thomas the Twin. Whose twin was he? According to the early Syrian Church, he was Jesus' twin.[23] According to the Gospel of Thomas, however, the term could apply to anyone who heard and understood Jesus' words: "He who will drink from my mouth will become like me. I myself shall become he, and the things that are hidden will be revealed to him" (Thomas 108).

This gospel does not present a history of Jesus' life as the others purport to do, but only a selection of his sayings. So it is much like the Gospel of Q that has never been found, but whose existence is assumed by scholars. It even contains some of the same sayings and parables found in Matthew and Luke and hence also in Q, although some of them appear to be earlier renditions. Like other gnostic works, it proposes that we can achieve salvation through knowledge, but it dispenses the knowledge in a concealed manner. "Whoever finds the interpretation of these sayings will not experience death" (Thomas 1).

Perhaps the main characteristic of this gospel is its inclusiveness. We don't have to believe in Jesus or even in God to be saved. The required knowledge is hidden in our hearts waiting for us to discover it: "The Kingdom [of the Father] is inside of you. When you come to know yourselves, then you will become known, and you will realize that it is you who are the sons of the living Father" (Thomas 3). Again: "The Kingdom of the Father is spread out upon the earth, and men do not see it" (Thomas 113).

In this gospel, as in John's, light plays a central part. Jesus says: "It is I who am the light which is above the All. From me did the All come forth, and unto me did the All extend. Split a piece of wood, and I am there. Lift up the stone and you will find me there" (Thomas 77). Yet Thomas' Jesus is more inclusive; we are all part of the light: "There is light within a man of light, and he lights the whole world. If he [or: it] does not shine, he is darkness" (Thomas 24). Again: "If they say to you 'Where did you come from?' say to them, 'We come from the light, the place where the light came into being on its own accord and established itself and became manifest through its image' If they say to you, 'Is it you?' say to them, 'We are its children, and we are the elect of the living father.'" (Thomas 50).

It is assumed that as with the other gospels, this one was also not written by the disciple after whom it was named. It is also probable that the author did not originate the sayings, but collected them from among those that were circulating around. As a result, we find some inconsistencies among them. For instance, although the gospel says that we all have the light inside us and if we find it we will be saved, it also says that Jesus selected those worthy of salvation: "It is to those worthy of my mysteries that I tell my mysteries" (Thomas 62). Again: "I shall choose you, one out of a thousand, and two out of ten thousand, and they shall stand as a single one" (Thomas 23). The gospel ends with a most enigmatic saying:

> Simon Peter said to them: "Let Mary leave us, for women are not worthy of Life." Jesus said, "I myself shall lead her in order to make her male, so that she too may become a living spirit resembling you males. For every woman who will make herself male will enter the Kingdom of Heaven."[24] (Thomas 114)

As we shall see in another chapter, by the end of the second century women made serious efforts to follow this teaching. They renounced their femininity, dressed like men, and became monks in order to grow closer to God. Is it possible that this saying is a later addition that corresponded to the accepted ideas of the second and third century?

SOME QUESTIONS TO CONSIDER

1. If Mark was Peter's follower, why was his gospel so uncomplimentary to him, and why does he say things that conflict with Petrine thinking?

2. If Christian scribes edited the contents of the gospels to justify the politically correct thinking of their time, how does that affect the way we should now read and interpret these gospels?

3. Pretend that you are a pious Hebrew in old Palestine, following God's commandment in the Torah. You come across a wild-eyed charismatic who tells you that you will be saved only if you believe that he is the Son of God. What are your thoughts?

4. Mark and John give completely different descriptions of Jesus' last words on the cross. Which do you think are more correct?

5. The synoptic gospels present us Jesus as a Hebrew teacher, concerned about the coming of God's kingdom. Paul and John, on the other hand, somehow derive a theology that in many ways is completely opposed to Judaism, but which has become the basis of Christianity. How can you justify this difference between the works of the evangelists?

6. How can you explain that the God, who had forbidden human sacrifices, asked for the sacrifice of his own son?

7. The Hebrew status of a person depended on the status of his mother, not of his father. (Because one could always be certain who his mother was, but not his father.) Why then are we following the male ancestors to determine someone's lineage?

8. Since Joseph was not Jesus' real father, how can we trace Jesus' lineage back to Abraham or Adam?

9. In Chapter 19, the author proposed that Jesus was following Zechariah's text when he said, "I have not come to bring peace but division." Now he states that Matthew made it up to describe the disputes that arose between his followers and those of the new rabbis. Which is true?

SELECTED REFERENCES

Barnstone, Willis, ed. *The Other Bible.*

Brown, Raymond E. *The Community of the Beloved Disciple.*

Dunn, James D. G., ed. *Jews and Christians: The Parting of the Ways AD 70 to 135.*

Ehrman, Bart D. *The Orthodox Corruption of Scripture.*

__________. *The New Testament: A Historic Introduction to the Early Christian Writings.*

Fowler, Robert M. *Let the Reader Understand. Reader-Response Criticism and the Gospel of Mark.*

Funk, Robert W., Hoover, Roy W., and The Jesus Seminar. *The Five Gospels: The Search for the Authentic Words of Jesus.*

Griffith-Jones, Robin. *The Four Witnesses: The Rebel, the Rabbi, the Chronicler, and the Mystic.*

Metzger, Bruce M. 3d ed., *The Text of the New Testament: Its Transmission, Corruption, and Restoration.*

Pagels, Elaine. *The Gnostic Gospels.*

_________. *Beyond Belief: The Secret Gospel of Thomas.*

Theissen, Gerd. *The Gospels in Context: Social and Political History in the Synoptic Tradition.*

CHAPTER 24

TRINITY: POLYTHEISM, SPLIT PERSONALITY, OR LEGISLATED MYSTERY?

What makes one a Christian? Strangely enough, following Christ's teachings does not do it, though it may make a difference on how good a Christian one is. It is not what you do that makes you a Christian, but what you believe in. According to the Church, a Christian is one who believes in the God that was defined by one or two thousand clergymen sixteen hundred years ago in the Councils of Nicaea, Constantinople, and Chalcedon. Commonly known as the Nicene Creed, its hundred or so words define the Christian God. The problem is that in all the years since then, nobody has ever understood these words well enough to explain them.

God is three in one. All three are composed of the identical particles of substance or whatever, yet they are separate and distinct, so that one of them can send another somewhere else. Yet where one is, the others are also. Take one away and you have nobody left. Although one brought the other into being, there never was a time when one was not there. Do you want to see your clergyman squirm? Ask him to explain the Trinity, and insist that he explains what he says so that you can understand him.

A CHALLENGING CONCEPT

The seventh Sunday after Easter is the day the assigned church preacher dreads the most. It is Trinity Sunday, so his sermon should properly deal, and hopefully explain, the concept that summarizes the basic Christian belief. So why is this so difficult? The Church itself developed this concept that represents its doctrinal foundation; many were those who were burned at the stake for challenging one or more little details of this doctrine. Yet it appears that we cannot always explain what we declare. In his book on heresies, Harold Brown says: "Because the doctrine of the Trinity deals with the very nature and inner relationships of the godhead, it will never be fully understood by human beings in any full sense" (Brown 1984, 91). Now I agree there are many things in the universe that human beings cannot understand, like the creation process, for instance, but most of these are not of our own doing. The Trinity, however, is a doctrine that we developed ourselves after years of discussion—sometimes very violent—among the greatest theologians of antiquity; yet we are told that nobody can understand it? How did this strange state of affairs come about?

John's gospel summarized Jesus' message clearly. "For God so loved the world, that he gave his only begotten Son, so that whoever believes in him should not perish but have everlasting life" (John 3:16). "If you do not believe that I AM (he), you shall die in your sins" (John 8:24). In other words, our sins will be forgotten, and we will have everlasting (eternal) life, if we just believe in Jesus, God's son, who is also God. The goal is clear, as well as very desirable—yet the methodology is mystifying. The one and only God has a son who is God? How many Gods make one God? Forgiveness of which sins will make us eternal, and how is that accomplished? The greatest minds of the early Christian community were applied to these problems, directed and commanded frequently by the Roman emperors. Clerics who disagreed were often banished, although sometimes they were subsequently recalled and installed in high positions. The result is the well known Nicene Creed, developed in a series of councils, starting in 325 at Nicaea[1] and ending in 451 at Chalcedon.[2]

The sin that Jesus cancelled was, of course, Adam's sin, his attempt to make himself an equal to God; mysteriously this sin is somehow relentlessly transmitted from father to son (and daughter). It is argued that before he sinned, Adam was immortal, and that through Jesus' actions, we have now also become immortal, but only after we die.[3] Adam had been disobedient, and Jesus atoned for it through his own perfect obedience. Note that after Paul and John, Christianity separated from Judaism. We don't view Jesus' act of atonement any more as his sacrifice on the cross, but as his complete obedience to the Father, which leads him to willingly accept even death on the cross. (This, of course, raises the question of why the Father would ask him to do it.)

Since Adam was man, Jesus also had to be man. But in addition, Jesus had to be fully God, because the job was presumably too big for a mere man to accomplish. And to resolve the problem of polytheism, he had to be indistinguishable from God; the two were one. But since John had said that Christ was our advocate to God, he also had to be distinct from him. Add the Holy Spirit, and now you have a Trinity: three distinct personas in one entity. Yet not quite distinct, because you cannot remove one and have the others left. Where one is, the others also are. Like…in all these years nobody has ever come up with a satisfactory example.

FIRST CENTURY: JUDAISM AND JESUS' TEACHING

As we saw in the last two chapters, soon after Jesus' death, substantial disagreements arose in the nascent Christian community. These were mostly related to what Jesus had said, or perhaps meant to say, and to a much lesser extent, to who he was. On one side was the Jerusalem faction, consisting of Jesus' original disciples, led by his brother James; it may have viewed Jesus as the Messiah, and tried to merge his teachings with the Hebrew theology. On the other side was the newly converted Paul, who claimed that he had received a revelation directly from the risen Christ; he preached that Christ was "the firstborn among many brothers" (Romans 8:29), and that his teachings had superseded the Mosaic Law. The middle ground was occupied by the new Hellenistic converts in Jerusalem, who were persecuted in the early forties and had to leave the city; they also believed that Jesus had repealed some of the Hebrew rules, something that helped them proselytize the Gentiles in the area.

The main argument in those early days was whether the old Hebrew laws were still valid. Did new converts have to follow rules on circumcision, food, and all other Deuteronomy regulations? This made little difference to the Jews in Asia since they followed these rules anyway, but affected greatly the new Gentile proselytes in the West. There was one regulation, however, to which no one objected: the required rest on the seventh day. This was the one Jewish law that everybody loved to follow; at least as far as its basic work-for-pay requirement was concerned.

In the heated debates that took place in the early period, Paul fared by far the worst. The Jerusalem faction had much greater prestige and membership, and did not permit any of Paul's work in Asia to survive. It was only in Europe that Paul finally succeeded in holding onto the churches he had started, probably because of the larger number of Gentiles that were involved there. The Jewish revolt in 67 CE, however, with the resulting destruction of Jerusalem, changed all this: the Jerusalem faction was largely wiped out and the Gentile church gained the upper hand.

The three synoptic gospels, which were written towards the end of this period, deal with the relationship between Judaism and Jesus' teachings, sometimes at considerable length. This was completely neglected in the fourth gospel, which deals almost exclusively with the sonship of Jesus. By that time, the separation between the Jews and Jesus' followers was complete, and the mostly gentile Christians happily dumped all Hebrew Law requirements. The question that rose to prominence then was who was Jesus? From that time on, Christianity became more concerned about Christ's being than about his teaching.

THE GNOSTIC INFLUENCE

Throughout John's gospel, Jesus keeps proclaiming that he and the Father are one. Towards the end, Thomas addresses Jesus as "My Lord and my God" (John 20:28). What does it all mean? If Jesus is God, do we then have two gods? If, on the other hand,

Jesus and the Father are one, what kind of God is this? Was Jesus man or God? If he was God, did God die on the cross? In that case, did he then resurrect himself?

The first group to start asking these and similar questions was the Gnostics, those who believed that salvation comes from knowledge. In the previous two centuries they had developed complicated theologies about a totally good and spiritual God who had created a whole series of semi-divine entities, one of which, in turn, created the world—God himself, being remote and completely immaterial, was incapable of directly creating a material world. As we saw earlier, the world's creator was called the *demiurgos* and was usually assumed to be an evil deity, something that explained the evil that exists in the world. According to the Gnostics, the human souls had been created before the world, and became enslaved in life, captured inside human bodies. Some souls, however, could hope to escape and return to their original place in heaven, provided they attained the proper knowledge.

And this knowledge was what Jesus, one of the various created divinities, had come to bring to selected people. Notice how well it fits the description in the gospels. Jesus was the world's savior, but he spoke in parables so that only the few select could understand. In John's gospel he tells his followers: "You shall know the truth, and the truth shall make you free" (John 8:32). Later he refers to the *demiurgos*: "The ruler of the world is coming" (John 14:30), although he had no power over him. The Gnostics did not believe that Jesus suffered on the cross, but that at the last minute he switched bodies with Simon the Cyrene. This group generated a large number of writings, and although the emerging Church destroyed most of them, many survived to our day, either whole or in segments.

SECOND CENTURY: JESUS' DIVINITY

At first, Jesus' followers believed that his return was imminent, so they did not bother to record their doctrine and beliefs. But as it became apparent that the Second Coming was being delayed, the Church Fathers were forced to start combating the teachings of the Gnostics. They first turned their attention to Jesus' divinity, to oppose the Gnostics' Docetic ideas.

Docetism, derived from a Greek word that means "to seem," maintained that Jesus was fully divine and only appeared to be human.[4] This was the first heretical Christian belief to appear—defining as heresy any belief that does not agree with the final accepted dogma of the Christian Church. It must have appeared very early, because Ignatius, the bishop of Antioch, wrote extensively against it before his martyrdom around 110 CE. During this century the Church accepted Christ as the *Logos*, although the exact meaning of the term remained unclear.

Irenaeus (125-202), one of the great theologians of this period, described the tasks that God had assigned to the other members of the triad: The Word (Logos) caused all things to be made and the Holy Spirit inspired the prophets to speak wisely. Strangely, almost as soon as the Holy Spirit was brought into the Trinity, it was forced into retirement. The Church decreed that the age of direct revelation had ended with the death of the apostles.[5] By freezing the selection of canonical books that were accepted as revealing the true word of God, and by refusing to accept any new ones, the Church was essentially stating that God had finished talking to humanity.

But not everyone agreed. Montanus, the founder in 156 CE of a new ecstatic movement in Phrygia, considered himself to be a vessel for the outpouring of the Holy Spirit. Together with his two female disciples, Maximilla and Prisca, he preached that the Holy Spirit spoke directly through them. Much like the millenarians of later eras, he also believed that the end of the world was at hand, so single people should not marry, married ones should separate, and they should all gather at an appropriate place to await the descent of the heavenly Jerusalem. Because of his great piety and asceticism, many followers joined him, including the famous theologian Tertullian. The Church, however, succeeded in preventing him, and others like him, from putting forth new ideas, by constructing the canon of accepted books. God's word had been fully revealed.[6] It was now up to the Church to interpret it. But first the Church had to explain how a man could be God.

THIRD CENTURY: REFUTATION OF POLYTHEISM

Origen, a great thinker of those days, had an opinion. Human souls, he suggested, were pre-existent, just as the Neo-Platonists had maintained, and each soul was assigned a body according to the moral quality of the free-will decisions that it had made relative to God. One soul was perfect, was merged with the *Logos*, and was assigned to Jesus' perfect non-corrupt body that God created especially for it. After the resurrection, the human soul remained merged with the divine *Logos*. But there was still a Father-person, a Logos-person, and I suppose a Holy Spirit-person. Too many persons, too many gods! How could the Church get back to one God? Although monotheism had been around for centuries, the Christian theologians gave it a new name, monarchianism: one monarch in the heavens. They then proceeded to devise two ways to implement it, both of which were almost immediately rejected by the Church.

Adoptionism considers Jesus to be a human being, albeit a morally perfect one, who was adopted by God at his baptism and given special powers to perform miracles. Remember the dove that descended upon him at his baptism?[7] Because of his moral perfection and the magical powers granted to him, he was able to stay in constant communion with God. As you can probably imagine, the idea did not go too far, and the Church excommunicated its proponents. The problem with the proposition was that the Church believed that Jesus had to be God in order for him to be able to atone for mankind. Only a God could atone for humanity's sin. Yet Paul didn't think so: "Just as by the offense of one all men were condemned, so by the righteousness of one justification of life (came) to all men" (Romans 5:18).[8] Paul was too pious a Hebrew to even consider the possibility that Jesus was God. Son of God, yes, but we have already seen that in his time this did not entail any divinity.

Modalism was perhaps the exact opposite. There was only one God, and he was Jesus. Jesus, God, and the Holy Spirit were all one single entity, which, however, had three separate, distinct modes of manifestation, or perhaps personalities. At any given time God appears to mankind under one of these guises—something like a person being simultaneously a parent, a spouse, and a son or daughter. Although one is all of these things, at any single instant one behaves from only one of these positions. Or we can compare it to the father, adult, and child personas that we all possess according to the principles of transactional analysis (discussed in chapter 16).

This is indeed the easiest way to explain the Trinity, and you may even hear it expounded in church on Trinity Sunday. It is certainly supported by John's gospel: "I and the Father are one. Whoever has seen me has seen the Father." Unfortunately for us and for the Sunday preacher, the Church condemned this concept also as heresy. If Jesus was not a separate entity, then he could not atone for us to the Father. Or looking at it from the other side, how could the Father accept his own sacrifice?

Sabellianism, named after its originator Sabellius, took modalism one step further, adding the Holy Spirit to make God a triad instead of a dyad. The Son, however, was restricted to his earthly work. Once he returned to heaven, God made himself visible through the Holy Spirit. Interestingly, two popes, Victor (189-199 CE) and Zephyrinus (199-217 CE), accepted this concept as preferable to Adoptionism, but it was eventually repudiated by their successors.

FOURTH CENTURY: THE PRE-EXISTENCE OF THE LOGOS

Everybody agreed that the Son pre-existed creation; after all, the Son took an active part during the process. But when did he come to be?

Arianism proposed that the Son was not begotten by the Father (not procreated, generated by his own being) but created by him (fashioned using other matter). He is therefore inferior to him and not fully divine. Although he came to be before the creation of time, when "before" and "after" have no meaning, he had not always been with the Father. The statement above is fairly clear, but the resulting controversy with the orthodox (i.e., the winner) group, was far from clear.

The orthodox faction maintained that the Son was begotten, not created, and that in addition, God had never been alone. But the very word "son" presupposes a time dependency; sons always come after the father. How could it be different? This seeming paradox of the son's eternal existence was resolved by accepting Origen's idea that the son had been eternally begotten. I am not sure what that means, but I suppose it was defined as fulfilling the "Son was not after the Father" requirement.[9] Logically the statement makes no sense. If there is no timeline, there is no before and after.

In any case, the dispute became a dividing issue among Christians in the Roman Empire. And Emperor Constantine, who had backed the Christian religion in an effort to unite his domain, became very unhappy. So in 325 CE, he called together a council of about three hundred bishops at Nicaea, a town near Constantinople (now Istanbul) Turkey.

The council rejected the doctrine of Arianism, "that our Lord was a pure creature, made out of nothing, liable to fall, the Son of God by adoption, not by nature, and called God in Scripture, not as being really such, but only in name".[10] A huge argument followed, regarding the similarity between God's and Jesus' make-up: whether to use the word *homo-ousion* or the word *homo-iousion*. The former word meant "of the same substance", the second "of similar substance." The problem with the first term, was that its meaning was not clear. Did it mean "of identical-type substance," or "of numerically same, identical substance?" There was disagreement, but Constantine told the bishops to use the word—probably because the bishops could then agree on the word, but disagree on its interpretation—so they did. Two bishops refused, and Constantine banished them, together with Arius and some of his followers. Within a year, however, Constantine recalled them and reinstated them, and when he moved his capital to the new city of Constantinople, he made one of them its bishop. For the next fifty years, Arianism was in full control, until the council of Constantinople, in 381 CE, when the Nicene Creed was reaffirmed and Arianism was banned.

FIFTH CENTURY—JESUS' HUMANITY

If Jesus was God, was he then also a complete man? He had a human body, but did he also have a human soul? If he was both a complete God and a complete man, shouldn't the Virgin Mary be called "mother of God and man" instead of just "mother of God"? By this time the theological arguments had become too complicated to understand, let alone explain. Did Jesus, the incarnate Word, have one or two *physes*? But what did this word mean? To the Latins it meant person, to the Greeks, nature. There followed a comedic series of accusations and counteraccusations, aphorisms, and exiles, with popes and emperors becoming involved.[11]

The disagreement between the pope in the West and Emperor Theodosius in the East was providentially resolved when the emperor fell from his horse and died. Marcian, his successor, was a simple soldier who just wanted unity throughout the empire. So in 451 CE he called another council, in Chalcedon, and ordered the five hundred or so bishops to reach an agreement that everybody could sign. As a result it was finally decreed that Jesus had not only been consubstantial (*homo-ousion*) with God, but he had also been consubstantial with man, except, of course, for the original sin. His distinct human and divine natures were separate but coalesced in one, indivisible nature.[12]

The Bible says that God made man in his image. On the other hand, many philosophers and theologians say that it was man who made God in his image. From what we have seen above, it appears that man also made religion to suit his own political and social needs. It is possible that the creed mirrors the truth, but if so, it must be through sheer chance or divine providence. Little wonder that the unhappy preacher has problems on Trinity Sunday.

SOME QUESTIONS TO CONSIDER

1. If Jesus was God and he died on the cross, does this mean that man can kill God?

2. Christianity claims that it is a revealed religion. How can you justify this when its main doctrine is based on documents that were produced after years of theological, but mostly political, strife?

3. In putting together its canon of accepted wisdom books, the Church proclaimed that with the apostolic era over, the Holy Spirit had finished communicating with mankind. Why would this be so?

4. Did Jesus have to be God, or just perfect, in order to atone for mankind's sins through his sacrifice at the cross? After all, Hebrews atoned for their sins by sacrificing all sorts of petty animals. Why must a God be sacrificed to atone for man?

5. Could Jesus have been "fully human" if he had no human father?

6. If Jesus was consubstantial (of the same numerically identical substance) with the Father, and he was similarly consubstantial with man, then it follows that man and God are consubstantial also. If God is a house made of stone, and mankind is a house made of bricks, you cannot build a house made of stone and bricks and call it either a house of stone or a house of brick. What do you think?

7. We are told that upon his incarnation, the *Logos* became joined indivisibly with the human *nous* (*pneuma*) of the man Jesus. Does this mean that the pre-incarnation *Logos* differs from the post-incarnation *Logos*? Did human *physis* become a part of God?

8. In Matthew, the resurrected Jesus tells his disciples at Galilee: "All power in heaven and on earth has been given to me" (28:18). But if Jesus Christ was the divine *Logos*, he was the creator of the world, so he presumably already had this power. What does this statement say about equality among the three God aspects?

9. If Jesus proves his complete obedience to the Father by accepting death on the cross, is this death a test? When God tested Abraham's faith in a similar way, he stopped before the act was completed. Was he less merciful towards his own Son, than towards Abraham's son?

SELECTED REFERENCES

Brown, Harold O. J. *Heresies: Heresy and Orthodoxy in the History of the Church*.

Hughes, Philip. *The Church in Crisis: A History of the General Councils: 325-1870 CE.*

Kelly, J. N. D. *Early Christian Doctrines.*

Rusch, William G. *The Trinitarian Controversy.*

CHAPTER 25

THE EUCHARIST

A long time ago I read a science-fiction story, whose name and author I have since forgotten, about a magic-performing prophet who had started a religion based on love. When he died his disciples put him in a big pot, cooked him, and then sat around the table and ate him.[1] Gory? Gross? Compare it with John's gospel: "Unless you eat the flesh of the Son of man, and drink his blood, you have no life in you. Whoever eats my flesh and drinks my blood has eternal life; and I will raise him up at the last day. For my flesh is truly food, and my blood is truly drink" (John 6:53-55).

The Eucharist is the Church's most solemn ceremony and certainly the most misunderstood one. An action that is supposed to exemplify our oneness with God has been interpreted literally by the Church because it is presented literally in the Bible. But is it also the literal truth? The Hebrew Law was clear: "Make sure that you do not eat the blood; for the blood is the life, and you may not eat the life with the flesh" (Deut. 12:23). As an allegory, the gospel's words fit their intent perfectly, but taken literally, how many of us can think of the Donner Party story without shuddering? Would Jesus, a pious Hebrew, have really said these words? No wonder "Many of his followers said: 'This is intolerable language. How could anyone accept it?'" (John 6:60 JER).

The Last Supper

Our first reference to Jesus' Last Supper with his disciples comes from Paul.

> For I received from the Lord that which I also delivered to you, that the Lord Jesus, on the night he was handed over, took bread, and having given thanks, broke it and said *"Take, eat, this is my body which is being broken for you. Do this in remembrance of me."* In the same manner [he] also [took] the cup, after he had supper, and said, *"This cup is the new covenant in my blood. Do this, as often as you may drink (it), in remembrance of me."* For as often as you may eat this bread and may drink this cup, you proclaim the death of the Lord until he may come. (1 Cor. 11:23-26 Berry)

This is the only event in Jesus' life, other than his crucifixion and resurrection, that Paul ever mentioned. Certainly these are Jesus' only words that he has ever quoted, and his quote is almost identical to those found in the three synoptic gospels. Here is how these gospels describe this event, with the common words in italics.

> And as they were eating, Jesus, having taken and blessed [the] bread, broke it and gave it to them, and said: *"Take, eat; this is my body."* And having taken the cup and given thanks, he gave it to them, and they all drank from it. And he said to them: *"This is my blood of the new covenant,* which is being shed for many. Amen I say to you, that I shall not drink again the fruit of the vine until that day when I drink it new in the reign of God." Then having sung a hymn they went out to the Mount of Olives. (Mark 14:22-25 Berry)

> And as they were eating, Jesus, having taken and blessed the bread, broke it and gave it to his disciples, and said, *"Take, eat; this is my body."* And having taken the cup and given thanks, he gave it to them,

> saying, "Drink from it all [of you], for *this is my blood of the new covenant*, which is being shed for many for the forgiveness of sins. I tell you that I shall not drink anymore from this fruit of the vine until that day when I drink it with you new, in the reign of my Father." And having sung a hymn they went out to the Mount of Olives. (Matthew 26:26-30 Berry)

> Then having received a cup and given thanks [he] said: "Take this and divide it among yourselves; for I tell you that I shall not drink form the fruit of the vine until the reign of God comes." And having taken and blessed the bread, he broke it, and gave it to them saying, "*This is my body*[2] which is being given for you; *do this in remembrance of me*." And likewise the cup after he had eaten, saying, "*This cup [is] the new covenant in my blood*, which is being shed for you." (Luke 22:17-20 Berry)

The common elements are the story of the supper (which is missing in John's gospel) and the passing of the bread and cup of wine: the bread that is his body, and the cup that is his blood in a (new) covenant.[3] We read about the old covenant in Exodus, where Moses confronts the people with the words of God:

> And he sent young men of the children of Israel, who offered burnt offerings, and sacrificed peace offerings of oxen to the Lord. And Moses took half the blood, and put it in large basins; and half of the blood he sprinkled on the altar. And he took the book of the covenant, and read in the audience of the people; and they said: "All that the Lord has said, we will do, and be obedient." And Moses took the blood, and sprinkled it on the people, and said, "Behold the blood of the covenant which the Lord has made with you concerning all these words." (Exod. 24:5-8)

Just as the oxen's blood sealed the old covenant, so also Jesus' blood seals the new one. Looking again at the various descriptions of the Last Supper we find two main differences: only the gospels describe Jesus saying that he would not drink wine again until he was in God's reign (and in Matthew he also specified that it would be together with his disciples) and only Paul and Luke call for a commemoration of the event in his memory (though Luke's reference may have been a later addition). Since Paul believed in the survival of a spiritual, not material, body, we can understand why he did not mention drinking wine in God's kingdom. But why did Mark and Matthew omit the directive to repeat this ritual? Is it perhaps, because the ritual of blessing the bread and wine, and then passing them to those at the dinner table was an already existing tradition followed in all Jewish households, and so Mark's and Matthew's readers knew about it, whereas Paul and Luke were addressing Gentiles who did not have such a custom? Let us look at the Jewish dinner ritual.

THE JEWISH MEAL BERAKOTH

A Jewish ceremonial meal, especially during holidays, progressed through various distinct and required steps:[4]

- First was the ritual hand washing, similar to the one with which every Jew started the day.
- Each person drank a cup of wine, repeating for himself the blessing: "Blessed be thou Yahweh, our God, King of the Universe, who givest us this fruit of the vine."
- The father, or other officiating person, broke the bread and gave it to the participants with this blessing: "Blessed be thou, Yahweh, our God, King of the Universe, who bringest forth bread from the earth." (This was a general blessing for the entire meal, and anyone who came afterwards could not join in.)
- Food courses and cups of wine followed, with each person pronouncing appropriate blessings.
- If the meal was being celebrated on the eve of a holy day, the mother brought in a lamp that she had prepared and already lit. It too was blessed.

- A second general hand washing took place, with the father receiving the water from a servant, and passing it on to the others at the table. (Although John does not include a meal in his gospel, the foot-washing ceremony he describes may fit in at this point.)

- The father, with the cup of wine mixed with water in front of him, would then invite those assisting to join him in a thanksgiving. "Let us give thanks to the Lord our God," he would say. "Blessed be he whose generosity has given us food and whose kindness has given us life," would be the response.

- Then the father would sing four *Berakoth* (blessings) psalms.

It is obvious that Jesus' actions at the last supper correspond exactly to the typical Jewish prayers at a meal. Notice, however, that only Luke correctly mentions the two cups of wine, one at the start and one at the end of the meal (except, of course, that this may have been a later addition by the scribes). So the Last Supper, despite what followed, was a typical Jewish ritual meal, including the concluding psalms mentioned in Mark and Matthew. The difference, however, is that when he was giving the blessings, Jesus added, "this is my body...this is my blood." But is this believable? All gospels describe Jesus as a very pious, though perhaps somewhat eccentric, Jew. Drinking the blood of animals was a grave sin among the Hebrews. There were detailed instructions specifying how to kill an animal so as to allow all its blood to drain out and return to earth. The mere idea of eating human flesh and of drinking human blood would have seemed an even greater abomination to the Jews of that day than it does to us today. Would a pious Jew have ever said such a thing? Yet Paul and all four gospels unanimously declare that Jesus did say these things. All our records describe the Early Church as repeating these words in its celebration. All records that is, except one.

THE DIDACHE

The Didache, meaning "the Teaching," was probably written some time during the first century and sets down what the converts to the new religion should know and how they should act. It begins with various directives derived from the Old Testament and then proceeds to describe the Christian rituals, prayers, and expected behavior. Although many of the early Christian Fathers made reference to this document, a complete copy of it was only found in 1873. It includes a copy of the Lord's Prayer that is almost identical to the one we currently use. But what interests us here, is the description of the Eucharist in the Didache:

> Now regarding the Eucharist, give thanks in this way:
> First concerning the cup:
>
> "We thank You our Father,
> for the holy vine of David Your servant,
> which you made known to us through Jesus Your servant.
> To You belongs the glory forever."
>
> And concerning the broken bread:
>
> "We thank You, our Father,
> for the life and knowledge which You made
> known to us through Jesus Your servant.
> To You belongs the glory forever."
>
> And as this broken bread was scattered over the mountains, and was brought together to become one, so let Your Church be gathered together from the ends of the earth into your Kingdom, for the glory and power are yours through Jesus Christ forever.
>
> But let none eat and drink of the [E]ucharist except those who have been baptized in the Name of the Lord. For concerning this did the Lord say: "Give not what is holy to dogs."
>
> But after you are satisfied with food, give thanks this way:[5]

> We give thanks to You, O Holy Father, for Your Holy Name which You made live in our hearts, and for the knowledge, the faith and the immortality which You did make known to us through Jesus Your servant. To You belongs the glory forever.
>
> You, Lord Almighty, did create all things through Your Name, and did give food and drink to men for their enjoyment, that they might give thanks to You, but us have You blessed with spiritual food, drink and eternal light through Your servant.
>
> Above all we give thanks to You, because You are mighty. Yours is the glory forever.
>
> Remember, Lord, to deliver Your church from all evil and to make it perfect in Your love, and gather it together from the four winds, holy in Your Kingdom which You have prepared for it. For Yours are the power and glory for ever.
>
> Let grace come, and let this world pass away. Hosanna to the God of David. If any man be holy, let him come! If any man be not, let him repent! Maranatha! Amen!
>
> But permit the prophets to hold the [E]ucharist as they see fit.[6]

Notice the close correspondence between these prayers and the Jewish prayers at mealtime. Although the second cup of wine is missing, the final prayers are still there. But more important, notice the complete absence of any Christological mentions of "this is my body...this is my blood." In the Didache, Jesus is called a servant of God, not his son or a divine being (although the Greek word *pais* can be translated as either servant or son). This document probably describes practices that we would associate with the Ebionite and Nazarene segments of Christianity.

ORIGIN OF EUCHARIST

Clearly, Jesus' first-generation followers were divided into two camps: the first considered him to be saintly, the second semidivine. The two disagreed on Jesus' exact words at the Last Supper. They all believed that he had been crucified and was raised, a basic Christian principle, but not that he had said that the bread and wine at the supper were his flesh and blood, an equally basic part of today's Christian belief. When and where did these sayings originate?

Since he was the first writer, Paul was also the first to report them. But did he report an accepted belief, or did he originate the idea? He himself says that the second is true: "For I received from the Lord what I also handed on to you." He takes personal credit for the information. As we have already seen, his teachings deviated greatly from those of the Jerusalem faction. He believed that Jesus was the pre-existent *Logos*, whereas the others remembered him as a plain man, the brother of James. He believed that the Law of the Torah had been cancelled, and along with it the requirements for eating clean foods; the old covenant was overwritten by the new covenant. And since the old covenant had been sealed in blood, the same should also hold true with the new one. But how could this be done years after Jesus' death, especially since it was essentially a bloodless death? What better way than to use the symbolism of the red wine? And what about eating his flesh? After sacrificing an animal, the Hebrews generally ate its flesh, and Jesus was the sacrificial lamb.

It seems reasonable to conclude that the Didache describes the actual practice of Jesus' early followers in Palestine and that Paul himself invented the words attributed to Jesus in the Last Supper. Mark became acquainted with Paul's version of the events either in Rome, if he had indeed been there, or in Antioch, where Paul spent many of his early years and where some believe that Mark wrote his gospel. Matthew and Luke then just copied the story from Mark. This explains the close similarities in Jesus' purported words in all four documents. If true, Jesus' quoted words were a complete fabrication, a possibility supported also by the fact that Paul shows little knowledge of anything that Jesus had actually said or done.[7]

The Church's Interpretation

Nevertheless, by the beginning of the second century, the Church Fathers apparently believed in the current interpretation of the Eucharist. In 110 CE, on his way to martyrdom, Ignatius, bishop of Antioch, wrote to the Smyrnaeans: "Who abstains from the Eucharist and prayer, because they refuse to acknowledge that the Eucharist is the flesh of our Savior Jesus Christ, who suffered for our sins and whom the Father by his goodness raised up...will perish in their contentiousness."[8] This sentence, of course, also indicates that there existed at the time a group of Christians that did not accept this interpretation of the Eucharist.

The Church went a step beyond what Paul and the gospels said. It drew the magical conclusion that one becomes what he eats. So if we eat something spiritual, we will also become spiritual. (It is well known, for instance, that in many native cultures the hunters eat the heart of the powerful animal they have just killed in order to acquire its strength.) So when during Eucharist we eat Christ's flesh and drink his blood, we are joined with him and we gain some of his spirituality. I am not going to discuss here the theology regarding the transubstantiation of the communion bread and wine into the actual body and blood of Christ, a theology that has evolved painfully through the centuries. Is the change in the bread and wine real or allegorical? If it is real, at what point during the ceremony does it take place? There was a period in the Middle Age when the laity was only shown the host but not allowed to partake of it. Later, the priest placed it directly in the mouth of the faithful, in case they dropped it. Some Christian faiths offer only the bread, some both bread and wine, some together, some separately. I will simply conclude with the transubstantiation theology of monk Paschasius Radbert composed around 833 CE.

> The body of Christ, which is present in the Eucharist, is the very body born of Mary. What is present in the Eucharist is the physical flesh of Christ, which is as it were "veiled" by the appearance of bread and wine, so that believers will not feel horrified at eating it: The appearance of bread and wine are nothing but a

garment to deceive the sense. If it were possible to remove this veil, the flesh and blood of Christ would appear in their natural form: not only is this possible, but it occurs in eucharistic miracles. (Mazza 1998, 183-184)

SOME QUESTIONS TO CONSIDER

1. In the first covenant Moses sprinkled the blood from the sacrifices onto the people. In the second covenant we are asked to drink the blood, much like it was done in older pagan and aboriginal religions. Why the difference?

2. In the *Berakoth*, Yahweh is blessed for giving us wine, the fruit of the vine. Does this render the wine holy?

3. In the Didache, the wine in the Eucharist is specifically called the holy vine of David, made known to us through Jesus. Is this a reference to Jesus being in the wine?

4. In the Didache the bread is compared to the Church, not to Jesus, who is called the teacher of faith and knowledge. But we now often say that Jesus is the Church, or the other way around. Does this make the bread Jesus?

5. The fact that the Jerusalem church knew Jesus to have been a normal human being like the rest of them does not take away from the possibility that he could have been God incarnate. Who do you think were correct, the Jerusalem faction or Paul's followers and on what do you base your opinion?

6. Jesus' divinity resided in his combined soul and *Logos*, not in his body which was completely human (since God does not have a material body). When in communion we partake of Jesus' body and blood, are we not receiving his earthly presence rather than his divine one?

7. For two thousand years Christians have argued and disagreed about the exact significance and substance of the Eucharist. Yet it remains the central part of many denominations' religious services. How can something so important be so misunderstood, and what makes us think that *we* understand it better than all the others, both now and in the past?

8. Deuteronomy says that blood is life, and Jesus said that whoever drinks his blood has eternal life. Doesn't this make perfect sense? Why does the author maintain that Jesus, a pious Hebrew, would not have said it?

9. It has been argued that the importance of the Eucharist lies in the belief of the participant that he is partaking in a holy action—precisely what he is physically consuming is immaterial. Do you agree? Justify your position.

SELECTED REFERENCES

Aquilina, Michael. *The Mass of the Early Christians.*

Bouyer, Louis. *Eucharist: Theology and Spirituality of the Eucharistic Prayer.*

Cooper, D. Jason. *Mithras: Mysteries and Initiation Rediscovered.*

Mazza, Enrico. *The Celebration of the Eucharist: The Origin of the Rite and the Development of its Interpretation.*

Witt, R. E. *Isis in the Ancient World.*

CHAPTER 26

WOMEN

Up to the 1970s, no Christian denomination had ordained women as priests. This was justified by two main arguments: Jesus did not have any women disciples, and priests were stand-ins for Jesus, who obviously was a man. It is true, however, that throughout their histories, both the Christian Church and the Hebrew religion had held women in contempt, considering them to be tools of the Devil, easily manipulated, and too flighty to be trusted. The situation has started to change in recent decades, ever so slightly, one denomination at a time. But, just as is the case in the business world, what is probably holding women down is the good ole boy club and the refusal to consider women as equals. What will change it is the eventual realization that there are too few men to do the job and that women have special contributions to make.

Yet there are many indications that Jesus himself had treated women very similarly to the rest of his disciples. He taught them, and they followed him. They certainly were the only ones who did not abandon him at the end. We can probably attribute the general absence of women in the gospels to the misogynist attitude of the period and to the Church's expurgation of the surviving writings. After all, there is no reason to expect that Christian attitudes would have been different from those of the Hebrews, Greeks, and Romans of the day.

ATTITUDES IN OLD TESTAMENT DAYS

The Old Testament is not only about men. It also contains many stories of women who are virtuous, powerful, or even acting as God's messengers. In what is a very strange episode (Exod. 4:25), Zipporah, Moses wife, protects him from God who was about to kill him, by circumcising their son on the spot. Miriam, Moses' sister, was a prophetess who led the people in a triumphal song after the Pharaoh's army was destroyed. The well that provided Hebrews with water during their forty years in the desert was named after her. Deborah was a prophetess and a Judge of Israel who led the Hebrew army to a victory that brought them peace for forty years. When the priest Hilkiah discovered the Book of Law in King Josiah's days, they called for Huldah, a prophetess, to explain to them God's intentions. Esther saved the Jews in Persia from the powerful vizier of King Xerxes. In 2 Maccabees, a nameless mother encourages her seven sons to keep their faith and withstand their tortures by the soldiers of the Hellenistic king Antiochus IV and then dies herself in support of God's laws.

Yet the Hebrews ignored all these stories about women and instead looked back to the Genesis tale: God made man first, and only later did he make the woman as his helper, after none of the animals proved acceptable. Thus the woman was made for man, a point used by both Jewish and Christian men to justify their superiority over women. Then Eve was tricked by the snake's guile and induced to break God's commandments. Worse yet, she convinced Adam to do the same thing, bringing forth death where previously reigned eternal life. The conclusion was obvious: women were foolish, could not be trusted, and should be subservient to men. Since they could be easily fooled they should not be allowed to testify in court.[1]

But Hebrew women were also physically challenged. For one week every month, while they had their menses, they were ritually unclean. They were also considered to be ritually unclean for forty days after giving birth to a son or eighty days after giving birth to a daughter. (Notice the ever-present gender double standard.) During such times a woman had to stay away from men[2] and obviously could not take part in religious ceremonies. It was decreed, therefore, that none of the time-related requirements of the Torah applied to women. Women were also exempt from saying the *Shema*,[3] wearing *phylacteries*,[4] and staying in the Booths (*Sukkahs*).[5] Of course, you will not find these decrees in the Bible. Instead, we are told that in addition to the written law there were oral laws. But since the oral laws were not written down, we can't tell when they were developed. Two centuries after Christ, the rabbis recorded them in the Talmud, but we have no way of knowing whether they really had existed before or whether the rabbis inferred them. In any case, these rabbis concluded that the main purpose of women under the law was to help the men follow the law.[6]

Since, according to men in those days, a woman was incapable of proper judgment, she always had to remain under the supervision of a man: first her father, and later her husband. If her husband divorced her—and simply displeasing him was a sufficient reason for that—she returned to her father's home. Wedding papers were very similar to bills of sale. In the early years a father sold his daughter much like he would sell his cow. (Remember the story of Jacob in the Genesis chapter?) By the days of the Romans, things had changed a little. The husband kept the bride-money (*ketubbah*) in the name of the bride, and the father added a dowry to it. But since the husband controlled this money, it did not have to actually change hands until he died or divorced his wife. This made divorce a very expensive proposition. Although Hebrews could have more than one wife, only the very rich could afford them. Herod the great, for instance, had ten wives.

A Hebrew was not enthusiastic about having daughters, for "the birth of a daughter is a loss" (Sirach 22:3 NRS).

> A daughter is a secret anxiety to her father,
> and worry over her robs him of sleep;
> when she is young, for fear she may not marry,
> or if married, for fear she may be disliked;
> while a virgin, for fear she may be seduced
> and become pregnant in her father's house;
> or having a husband, for fear she may go astray,
> or, though married, for fear she may be barren.

> Keep strict watch over a headstrong daughter,
> or she may make you a laughingstock to your enemies,
> a byword in the city and the assembly of the people,
> and put you to shame in public gatherings.
> See that there is no lattice in her room,
> no spot that overlooks the approaches to the house.
> (Sirach 42:9-11 New English Bible)

As a result, girls were betrothed (engaged to be married) as soon as possible. As long as she was less than twelve and a half year old, a girl could not object to any marriage arranged by her father. Thereafter, however, she could not be forced into a marriage against her will. Deuteronomy specified that all children should be taught the Law (Deut. 11:19) but, as we discussed earlier, the oral laws insisted that only boys should be taught; not girls. Even so, some rabbis had been known to teach the Torah to their daughters.

It was desirable to keep the women at home working at suitable tasks like grinding flour, baking bread, washing clothes, and most important, spinning. Even if there were servants in the house to do all the work, a woman should still spend her time spinning and weaving, since these occupations kept her away from mischief.[7] If she had to go to the market or the temple, she should do it when the streets were empty. When outside, she should keep her head covered. Loose hair, in particular, was considered to be the sign of a prostitute. Strangely, the only occasion when a Jewish girl did not cover her head in public was when she walked to her bridegroom's house. It follows from all this that it was up to the husband alone to do all the work needed to earn a living. Since most of the Hebrew people were poor, however, many of these rules could not be followed: the wives often had to work at the store and, sometimes, even in the field.

WOMEN IN THE GRECO-ROMAN WORLD

Hebrews classified women as spiritually and mentally inferior. In the Classical world, on the other hand, they were considered to be physically deficient and incomplete. According to Aristotle, the procreating power lay in man's semen; the woman just provided growth material. It was something like a farmer's seed and field: man provided the seed, the woman the field.[8] The Romans accepted this and extended it further. In a culture that glorified strength, a woman's inferiority was unquestionable. But she was also an incomplete human being. First, she did not have all the physical parts of the body, and second, her body shape kept changing, proving that her growth had not been completed. It was all because the female fetuses had not "amassed a decisive surplus of 'heat' and fervent 'vital spirit' in the early stage of their coagulation in the womb."[9] Although these things may sound ridiculously stupid to us today, they were considered facts of life in those societies.[10]

An Athenian woman who was restricted to her own house quarters and whose husband spent all his time talking with his friends and dining with them and with prostitutes (*heterai*) would have been envious of the Hebrew wife. So would the Spartan wife whose son was taken away from her at the age of six to live with other boys and whose husband had to stealthily visit her on an occasional night. Spartan wives, however, had considerable independence since there were no men at home to order them around.

Romans had the unique institution of *paterfamilias*, whereby the father was lord and master of all members of his family, including adult children. In the early days, he even had the right of life and death over all family members, but that changed by the end of the millennium. Even so, because a woman was considered to be weak and lightheaded, she always had to have a guardian: first her father, and then her husband. If the latter were to die, the court would need to appoint another guardian, usually a relative. Although in theory a Roman matron did not have many more rights than the Athenian wife, in reality she had greater independence. Her husband was often away for extended periods, during which she managed the finances of the estate. Furthermore, Roman girls were better educated, since they were usually sent to school at the forum and the richer ones had private tutors.

Roman girls were married off when they were twelve to fifteen years old, but some even earlier. Childbirth at such an early age was a factor influencing the high female mortality of the period. But even if they survived to adulthood, women still died ten to fifteen years younger than men did, excluding, of course, men killed in the wars. The law required wives to have children, and being barren was suitable grounds for divorce. After three or four children, however, if the husband died, a wife was considered sufficiently mature to handle her own affairs without a guardian. Because Roman women inherited from both husband and father, there were many wealthy widows.

Throughout the empire, a woman was considered property: first of her father,[11] and then of her husband. Only after her husband's death would a Roman woman acquire her independence and some control of her husband's estate, although in theory she was still subject to a legally appointed guardian. We read in the Bible that many such Greek and Roman women made their houses available to the early Christian churches, thus becoming their patrons.

JESUS AND WOMEN

There is nothing in the gospels to indicate that Jesus treated women as second-rate people. Some might even say that he was perhaps a little more tender with them, though never condescending. It is true that in John's gospel he deals with his mother in what appears to be a cold manner. When at the Canaan wedding she asks him to resolve the problem of insufficient wine, he replies coldly: "What [concern] is it to me and to you, woman?" (John 2:4, my translation). At the end of his life, when he looks down from the cross and sees his mother next to the beloved disciple, he says: "Woman, behold your son" (John 19:26). We have to remember, however, that to John, Jesus was the divine *Logos*, God's Word. It would have been quite incongruous for the *Logos* to address Mary as his mother. For some unexplained reason, however, John always refers to Mary as Jesus' mother and never mentions her name. From the previous argument we would have expected the opposite.

All the gospels specify that women were the only followers of Jesus who were present at his crucifixion. This is reasonable, because if Jesus was being executed as a revolutionary the authorities would have been keeping an eye out for any of his followers; women, however, would have been safer. But two of the gospels go even further, telling how women traveled with

Jesus and supported his mission financially. This must have been true, because in Judea it would have been quite shocking for women to travel through the country with a group of young men so no one would have written about it unless it had actually occurred.[12]

According to the gospels, Jesus selected twelve male disciples. No woman disciple was chosen, something that the later Church used to justify its position that only male priests were acceptable in Christianity. This ignores the fact that in his day Jesus could have hardly invited women to go traipsing around the countryside with him, although some of them apparently did. Besides, the twelve disciples probably represented the twelve tribes of Israel, none of which had ever been represented by a woman. But Jesus had more disciples than the twelve listed in the gospels. In Luke's gospel, for instance, he appoints seventy-two of them to go out and preach in all the towns that he expected to visit later.

There is little doubt that some of Jesus' disciples were women. Certainly the women who followed him to Jerusalem must have been his disciples.[13] Furthermore, Luke describes Mary, Martha's sister, as sitting at Jesus' feet listening to him. In the Judean community, "sitting at the rabbi's feet" was a common way to describe a disciple's position. And let's not forget that after his resurrection, Jesus appeared to women first and to men later. Of course, it depends on how we define the word "disciple." The gnostic Gospel of Philip says, "There were three who always walked with the Lord: Mary, his mother, and his sister, and Magdalene, the one who was called his companion. His sister and his mother and his companion were each a Mary." Obviously someone who always walked with Jesus would be called a disciple.

There is no doubt, however, that later writers objected to the inclusion of women among Jesus' chosen. Peter, in particular, is often presented as one of the chief opponents. We have already seen that the Gospel of Thomas concludes with a puzzling discussion about women and the kingdom of heaven:

> Simon Peter said to them, "Let Mary leave us, for women are not worthy of life." Jesus said, "I myself will lead her in order to make her male, so that she

> too may become a living spirit resembling you males. For every woman who will make herself male will enter the kingdom of heaven." (Thomas 114)

And now we are ready to look at the identity of the beloved disciple in John's gospel.

The Beloved Disciple

Just as he never used the name of Jesus' mother, John also never identified by name Jesus' favorite disciple. For years it had been assumed that it was John himself, the gospel's author, who did not use his name out of modesty. But just as John's authorship of the gospel is not accepted anymore, there is also no reason to identify him with the beloved disciple who, according to the gospel, was the one who had testified to the gospel's accuracy. In none of the four gospels is Jesus shown to have been partial to any single disciple. Peter stands out, but he was treated as the most mature, not as the most beloved. Only Lazarus, the brother of Mary and Martha, was said to have been loved by Jesus. Nevertheless the identity of the disciple is clearly revealed if one reads the Crucifixion story and restricts himself to the factual descriptions without adding any outside assumptions. Examine the pertinent sentences:

> Now there stood by the cross of Jesus his mother, and his mother's sister, Mary the wife of Cleopas, and Mary Magdalene. When Jesus saw his mother and the disciple whom he loved standing by, he said to his mother... (John 19:25-26)

Who was standing by the cross of Jesus? Some say four women, and others say three, depending on whether Mary the wife of Cleopas was the sister of Jesus' mother. But only these were standing there. There were no men. The other three gospels also agree that none of the twelve disciples were present at the Crucifixion. So when Jesus looked down from the cross whom did he see next to his mother? His mother's sister, Mary the wife of Cleopas, and Mary Magdalene. Which one was the disciple he loved? Mary Magdalene! We will see below what some gnostic gospels say about Mary. But first we have to explain the next verses in the gospel story regarding what Jesus said to his mother:

> "Woman, behold, your son," Then he said to the disciple, "Behold your mother." And from that hour the disciple took her into his home. (John 19:26-27)

The words here clearly refer to the disciple being male, and this is why most people assume that John was also standing there. If that were the case, why didn't the gospel mention him? But on the other hand, if no male had been beneath the cross, to what can we attribute the discrepancy in the wording? We can ascribe the discrepancy to changes made by the editor, or the redactor, as Bible scholars usually call such people. As we will see in chapter 27, by the end of the first century, the Church had become strongly misogynistic and was starting to preach sexual renunciation. To let Jesus single out a woman for a special honor would have been unthinkable. But by that time it was well known that from among Jesus' followers only women had witnessed the Crucifixion. So the editor could not openly place a man by the cross. But he could change the word "Mary" to "the beloved disciple," which she apparently was. Then he changed a couple of pronouns from feminine to masculine, the word "daughter" to "son," and he accomplished his goal in a fairly surreptitious manner.[14]

The redactor also had to remove Mary's name from all other places in the gospel where she appeared to be close to Jesus. So in the Last Supper it was not Mary, but the disciple whom Jesus loved who was reclining on his bosom.[15] The story at the empty tomb is more complicated. Mary finds the empty tomb and runs and tells Peter and the disciple whom Jesus loved. Those two run to the tomb, see it empty, and leave. Mary remains outside. But we were not told that Mary had returned to the tomb. (In Luke's gospel, for instance, Peter alone investigated the empty tomb.) Remove the second disciple from the story and it becomes smoother. Mary finds the empty tomb, she runs and tells Peter; she and Peter run back to the tomb, Peter leaves, Mary remains.

The only gospel incident that would contradict the thesis that Mary was the beloved disciple is Jesus' appearance to the seven disciples at the sea of Galilee. Here the beloved disciple is shown as fishing with the others, something that Mary would certainly not have done. It is generally accepted, however, that this story was added later by the redactor, probably to bring it in correspondence with Matthew's gospel that described an appearance by Jesus in Galilee.

Except for being the first person to see the resurrected Jesus, the canonical gospels don't say much about Mary Magdalene. In the Gospel of Thomas quoted earlier, however, and some other gnostic gospels, Jesus appears as being closer to Mary than to any of the other disciple.[16] The Greek Orthodox Church venerates Mary Magdalene as a saint. According to its tradition, she left Palestine after Jesus' death to teach in various places, including Rome.[17] (It has been argued that she is the Mary whom Paul thanks in his letter to the Romans for having worked hard for them there.) Later Mary went to Ephesus where she helped write the fourth gospel. Presumably she was the beloved disciple who testified to the veracity of things in John's gospel (John 21:24) and who supplied details about incidents missing from the other gospels.

In contrast to the Orthodox Church, which holds Mary in high regard, the Roman Catholic Church changed Mary into a prostitute, identifying her with the sinful woman whose sins Jesus forgave. Pope Gregory the Great, in a homily that he gave in 591 CE pronounced that Mary Magdalene, Luke's unnamed sinner, and Mary of Bethany were all the same person.[18] Unfortunately, the Internet is almost the only place where you can find information about Mary Magdalene today. Her repute, however, has followed her even there. If your computer has viewer-discretion controls activated, it may very well refuse you access to the sites because "it includes sex."

Paul and Women

In his letters, Paul gives much credit to the many women who helped him, in particular, to the wife-and-husband team of Priscilla and Aquila who even "risked their necks" for him (notice that he lists the woman's name first). He refers to a Phoebe as both a deaconess and a protector. He calls Junia, his kinsperson and fellow prisoner, an apostle. He refers to the mother of Rufus as his own mother and, finally, as we have already mentioned, to Mary who had worked so hard spreading the gospel in Rome. At one point he exhibits zero discrimination:

> For as many [of you] as were baptized into Christ have clothed yourselves with Christ. There is neither Jew nor Greek, there is neither slave nor free, there is not male nor female; for you are all one in Christ Jesus. And if you are Christ's, then you are Abraham's descendant, and heirs according to the promise. (Gal. 3:27-29 AMP)

But above all, Paul was a "keep the status quo" guy; "don't rock the boat" was his motto. He wrote, "Let every person be loyally subject to the governing authorities. For there is no authority except from God, those that exist do so by God's appointment....For this reason you pay taxes, for [the civil authorities] are official servants under God" (Rom. 13:1,6 AMP). In other words, don't start any revolutions. Similarly, don't start doing strange things in church, which will shame us in front of the heathens. "Let your women keep silence in the churches, for they are not permitted to speak; but should be obedient [subordinate], as the law says also" (1 Cor. 14:34).[19] Women should know their place: God was the head of Christ, Christ the head of every man, and a husband the head of his wife (1 Cor. 11:3). He describes and justifies the proper husband-wife relationship in his letter to the Ephesians:

> Submit yourselves one to another in the *fear* of God. Wives submit yourselves to your own husbands as to the Lord. For the husband is the head of the wife just as Christ is the head of the Church; and he is the savior of the body. Therefore, as the Church is subject to Christ, so let the wives be to their own husbands in everything. Husbands, love your wives....So ought men to love their wives as their own bodies. He who loves his wife, loves himself....Let every one of you love his wife as himself, and the wife should respect [*fear*] the husband. (Eph. 5:21-32 emphasis added)[20]

Note the clear double standard: the man loves, the woman respects. Actually the Greek word is not "respect" but "fear." The very same word that was used at the beginning of the quotation relating to God. Whoever wrote these words, erroneously attributed to Paul, felt that wives should look at their husbands as God. Conversely, husbands should love their wives the way God loves us. One of Paul's followers, writing later under Paul's name, was even harsher to wives:

> Let the woman learn in silence with all subjection. I do not permit a woman to teach, nor to usurp authority over the man, but to be in silence. For Adam was first formed, then Eve. And Adam was not deceived, but the woman being deceived was in transgression. (1 Tim. 2:11-14)

For the next nineteen centuries the Church accepted the attitude towards women put forth by Paul and his followers. It was not until the twentieth century, when women began battling for political rights, that Church attitudes started to change, ever so slowly. Numerous (but not all) Christian denominations now have women deacons, priests, even bishops.

SOME QUESTIONS TO CONSIDER

1. In Genesis God made man and, only after all the animals were paraded in front of the man and he did not find a suitable companion, did God make woman to be his companion. Does this tell you anything regarding the way the writer of this book thought about women?

2. Were Hebrew women better or worse off for not having to follow all the religious requirements of the Torah?

3. Young girls in today's society have greater freedom than in the old days. Would you attribute it to the fact that their parents don't care?

4. Where do you think women were better off: Greece, Rome, or Judea?

5. From what the author has said, what do you think was the relationship between Jesus and Mary Magdalene?

6. In Galatians (3:27-28) Paul says that in Christ Jesus there is neither male nor female. Then why did both he and the Church treat women differently than men?

7. How were the rights of women different in the nineteenth century from the days of Jesus?

8. A question arose in Germany in 2003 on whether to permit Muslim women to wear scarves while working in government offices. Was this a gesture of religious nondiscrimination or an act of religious oppression of Muslim women? What do you think?

SELECTED REFERENCES

Cantarella, Eva. *Pandora's Daughters: The Role and Status of Women in Greek and Roman Antiquity.*

Ilan, Tal. *Jewish Women in Greco-Roman Palestine.*

Kraemer, Ross Sheppard. *Her Share of the Blessings: Women's Religion among Pagans, Jews, and Christians in the Greco-Roman World.*

Mortley, Raoul. *Womanhood: The Feminine in Ancient Hellenism, Gnosticism, Christianity, and Islam.*

Pomeroy, Sarah B. *Godesses, Whores, Wives, and Slaves: Women in Classical Antiquity.*

Witherington III, Ben. *Women and Genesis of Christianity.*

CHAPTER 27

SEX

You can find these days on the shelves of your local book store cooking manuals right next to sex manuals. It was not always this way. It is true that the old Hebrews thought that sex should be kept inside the bedroom, but they still considered it a sacred act that fulfilled God's command to multiply themselves. This changed, however, when the Early Christians associated the sex act with Adam's fall. It is not clear why this came about, but the idea became entrenched with Augustine, who first tried sex with a woman, then with a man, and finally decided to give it up altogether.

Furthermore, in the early days of Christianity, women recognized that the only way they could achieve equality with men was to renounce not only sex, but also their sex. They took an oath to remain virgin, got dressed in men's clothes, and hid themselves in monasteries in the middle of the desert. There they abstained from men and, as much as possible, from food. Men did the same thing. They figured that since they would shed their bodies in the next world, they would be one step ahead if they gained control over them in this one.

Today's generation does not question the propriety of the sex act anymore, but the sex of those engaged in the sex act. Homosexuality has become the new dividing issue. Is it normal or abnormal? Should the Church accept it or denounce it? Science and government are for its acceptance; most denominations think it is a terrible sin and quote Paul for justification.

The Old Testament Days

After he created man and woman, God charged them to "Be fruitful and multiply, and replenish the earth, and subdue it" (Gen. 1:28). By Noah's day, however, mankind had grown so fast and had become so wicked that God all but exterminated it. Then God became more discriminating: fecundity would be rationed. He told Abraham: "A father of many nations I have made you. And I will make you exceedingly fruitful" (Gen. 17:5-6). He also promised to render fruitful Abraham's first son, Ishmael. Thereafter, fertility ceased to be an unrestricted gift to all mankind, but became something to be controlled by God.

To the Hebrews, procreation was a divine command that affected deeply the way they looked on life. "Concerning the man who loves his wife as himself, who honors her more than himself, who guides his sons and daughters in the right path, and arranges for them to be married around the period of puberty, of him it is written: Thou shalt know that my tent is at peace."[1] "Marriage was regarded as a sacred duty and family life was holy....A woman was commanded to take a ritual bath after the menstrual period (plus an additional seven days), to prepare herself for the holiness of what came next: sexual relations with her husband. The idea that sex could be holy in this way would be alien to Christianity, which would sometimes see sex and God as mutually incompatible."[2]

This attitude was in complete contrast to the way the rest of the world looked at sex. The Baal religion of the Hebrews' Canaanite neighbors included orgiastic sex acts as part of festivities, as well as both male and female temple prostitution. Further away, in the Hellenistic world, the family was more of a political entity than a closely knit group of mutually loving people. Although the population requirements of the Roman Empire promoted and rewarded births and, thus, sex between married couples, most men preferred to stay single and carefree for as long as possible. Furthermore, as we shall see below, Roman society, just as the earlier Greek one, tended to find sexual pleasures outside the family, often with partners of the same sex.

Sexual Renunciation

Paul believed that the cares of this world detracted from people's ability to concentrate on God. "I should like you to be carefree. The unmarried man cares about the things that belong to the Lord, how he may please the Lord. But he that is married cares about the things that are of the world, how he may please his wife" (1 Cor. 7:32-33). Paul felt similarly regarding women, but he did not want to make waves. "So let the married people stay married, and the single ones stay single," he counseled. "But if you marry, you have not sinned; and if an unmarried woman marries, she has not sinned. Nevertheless, such people will suffer affliction in their earthly life, which I would like to spare you" (1 Cor. 7:28).

Then Paul goes into details: it is a good thing for a man not to touch a woman, but because of immorality every man should have his own wife, and every woman her own husband. He wishes that everybody could be like him (presumably not needing women) but understands this cannot be, and he does not want to place people in a situation where they could be tempted. So, although Paul sounds a little unhappy about sexual relations, he does not come out and say that there is anything really wrong with them.

Paul, however, did not represent Jesus' attitude towards marriage. The gospels, which were written a few decades later, generally show Jesus to be supportive of marriage.[3] He tells his disciples: "A man shall leave his father and mother and be joined to his wife, and the two shall become one flesh. So they are no longer two, but one flesh" (Mat. 19:5-6). He applies to his supporters, however, a significant qualification: he asks them to leave their families and follow him. And in Matthew's gospel (and only in that one) he seems to advocate celibacy: "There are eunuchs, who have made themselves eunuchs for the sake of the kingdom of heaven. Whoever is able to receive it, let him receive it" (Mat. 19:12).[4] Today we are not sure whether Jesus actually said that and, if so, what he meant by it. But in the early days of Christianity there was no such doubt: Jesus had said it, and he had meant it; sex and the kingdom of God did not mix. As a result, the great early theologian Origen castrated himself, as did many others also.[5]

Even before the days of Jesus there were groups whose members practiced celibacy. The elite among the community of Qumran, at the northwest shores of the Dead Sea, had sworn to abstain from sex while preparing for the final battle against evil. Similarly, in Egypt, the monk-like Theraputae, described in Philo's writings, also abstained from sex. Although this last group had both men and women members, the two sexes were separated by a partition during the services.

The Gnostics were the first Christians to actively campaign against sex.[6] They believed that the world in which death reigned supreme had been created by an evil *demiurgos* and that through the sexual act man had lowered himself to the status of an animal. But by giving up sex—boycotting the womb, so to speak—this current age would come to an end. Marcion, a Gnostic who separated from the emerging Church in 144 CE, went even further, calling for the complete breakup of families—husbands from their wives, children from their parents. Tatian (d. 180 CE) believed that intercourse entered the world only after Adam lost his immortality, and in order to return to the previous state, married couples should give up sexual intercourse. The Manicheans, a sect still existing today that was founded by Mani (216-277 CE) in Babylon, believed that a very small part of humanity remained beyond the power of darkness and that only a complete renunciation of sex, plus proper diet, would allow these elite to push back these powers.[7]

Soon thereafter, mainstream Christianity also decided that sexual intercourse, even among married people, was an unholy act. It was necessary, of course, for begetting children, but it was sinful to get pleasure out of it.[8] Holy people were not expected to indulge in sex. By the third century, priests and bishops were expected to live in continence with their wives.[9] Theologians started associating the original sin with sex: instead of continuously contemplating God, Adam and Eve had looked at their own bodies. Virginity, both male and female, became a cherished state. The Egyptian desert became filled with monasteries inhabited by thousands of men, and others filled with women who sought salvation through complete disregard of all the needs of the physical body. They figured that since upon death the soul would separate from the body, gaining control over the body while they were still alive would put them closer to their goal.[10]

Women discovered that the only way to attain any kind of parity with men was to become brides of Christ. They were supported by fictional stories such as the *Acts of Paul and Thekla*. Thekla, a beautiful, strong-willed virgin, rebuffs her fiancé in order to live a chaste life and successfully faces all sorts of persecutions. She not only baptizes herself but also preaches and baptizes others. When she is placed on a pyre to be burned, God sends rain to put the fire out. When she is thrown to the wild animals, a protective lioness saves her. When she jumps into a pool of killer seals for baptism, God sends fire to kill the animals. The spectators shower her with flowers and perfumes. Until his own death, Paul encourages and supports her throughout all her endeavors. Eventually she is freed and she spends the rest of her life preaching until her final "noble sleep." The implication of the story was clear. Through complete dedication to Christ and fearless protection of their virginity, women could become equal, or even superior to men. Later theologians, bent on attaining complete subjugation of women, hated this story.[11]

Augustine (354-430 CE) was born in Africa and probably influenced the development of the Roman Catholic dogma more than anyone except Thomas Aquinas (1225-1274 CE) eight centuries later. At nineteen, Augustine took a concubine and lived with her for thirteen years, begetting one son. During that time he studied philosophy and probably joined the gnostic Manicheans, who abstained from sex. He then went to Rome where in 398 CE, together with some philosopher-friends of his, he underwent a group conversion to abstinence. As you might expect, his concubine took his son and left him.

Augustine believed in a completely physical Adam and Eve. He thought that in the Garden of Eden they had sex and that they could have had children. At first they had enjoyed a harmonious unity of body and soul, but that was somehow lost and they fell, carrying down with them all mankind. Augustine believed that the human body had been created to be loved and cherished, yet he found that during the act of marital intercourse one feels the shame of Adam. What really disturbed Augustine was that the human mind could not control the various sexual functions: arousal, orgasm, impotence. Such failure of control meant that instead of being godlike, man was more like an animal. He naturally assumed that in paradise things had been different, but that man sinned, and one result of his punishment was his loss of sexual control. Now, sixteen hundred years later, we know that men and chimpanzees share more than ninety-seven percent of the same DNA structure, so there really is not much physical difference between them.[12] Augustine, of course, had probably never heard of DNA, but there are numerous theologians today who have thought and who still do think like Augustine.[13]

There were others who disagreed with Augustine's position regarding original sin. Pelagius (d. c. 418 CE) and his followers Celestius (dates unknown) and Julian (386-454 CE) questioned the reality of the original sin. They maintained that Adam had been created mortal and that even without baptism his children were eligible for eternal life. Man's free will made him independent of God, and God's goodness would have precluded the punishment assumed by Augustine.[14] They believed that the sexual urge was a God-given drive for the purpose of procreation. How the drive was channeled depended on the choices that man made: using it for marriage and children was a good choice.

Julian accused Augustine of harboring old Manichean thoughts that sex was the tool of the Devil rather than a gift from God to help man with his mortality. Augustine counterattacked, saying that man's inability to control his sexual drive was God's punishment for Adam's disobedience. As it turned out, Pelagius and his followers lost the argument, and their doctrine was anathematized at the Council of Ephesus in 431 CE. Augustine then spent the last years of his life in Africa preaching to married people about the sins of Lenten sexual intercourse.

By the sixth century, most of the bishops came from monasteries, and wives had disappeared from their homes.

HOMOSEXUALITY

Segregating men and women in monasteries did not make their sexuality disappear. Although monks and nuns had no access to the opposite sex, they did have access to each other; directing their sexual urges to those of their own gender soon became a widespread practice.

Furthermore, homosexuality had been a fairly accepted practice in both the Greek and the Roman worlds. Both cultures looked down on women, kept them uneducated, and considered them, at best, inferior, and at worst, half-human. Especially among the higher classes, there was no communion of equals between men and women. When sexual intercourse was described in literature, the gender of the participants was not even mentioned, as it was always assumed that they were both male. The important roles were those of the penetrator and the penetrated: the former was manly, the latter effeminate. Usually young boys started in the latter role and as they grew up they took over the more aggressive position. The fact that there was no clear-cut rite of passage for a growing boy generated confusion. But the act was not sexual as much as it was social; it proved one's manliness. In reality, it was not even a social act but one of outright aggression, generally directed at a helpless young person or slave.

The alert or suspicious reader may have noticed in the previous section that Augustine, after thirteen years of common-law marriage that resulted in only one child, took a vow of abstinence together with some of his philosopher friends. The reader would have been even more suspicious had I mentioned that one of these friends was his lifelong comrade Alypius who, after only one unhappy sexual experience, had decided that sex was not for him. Augustine did confess eventually that a sexual relationship had existed between the two of them, which he ultimately regretted (Kuefler 2001, 199). It is unfortunate that a man who was psychologically unable to resolve his own sexual problems succeeded in impressing his personal ideas regarding sex on the millions of Christians who lived in the next seventeen centuries.

Although acceptable in the Greco-Roman culture, homosexuality was considered sinful in the Judeo-Christian religion. The Bible either describes homosexual events in a critical manner or directly prohibits such actions. Only three such situations are described in the Old Testament: the story of Ham and Noah after the latter got drunk;[15] the tale of Sodom and

Gomorrah, where the Sodomites tried to rape Lot's two angelic visitors (Gen. 9); and the fatal gang rape of the Levite's concubine by the men of Benjamin's tribe after they were unsuccessful in their attempt to rape the Levite himself (Judg. 19). Note, however, that not a single one of these cases deals with homosexual love; the first case was a pure power, and the other two were violent acts performed by violent mobs. In fact, the last, and probably the only historically true story, involved a strictly heterosexual abuse.

Prohibitions of homosexual acts appear twice in the Old Testament: "You shall not lie with a man as with a woman; it is an abomination" (Lev. 18:22); "If a man lies with a male as with a woman, both of them have committed an abomination; they shall be put to death" (Lev. 20:13). The two statements are obviously connected; the second one repeats the first and then sets the penalty for the transgression. They are preceded and introduced, however, by the following verse: "Don't follow the customs of Egypt where you used to live or those of Canaan where I am bringing you" (Lev 18:3 CEV). We don't have any independent sources confirming the existence of homosexual practices in Canaanite society. We can assume, however, that their religious practices involved both male and female temple prostitutes, since the Hebrews often incorporated them in their own temples. "There were also sodomites (male cult prostitutes) in the land. They did all the abominations of the nations whom the Lord cast out before the Israelites" (1 Kings 14:24 AMP).[16] The prohibition, therefore, is simply another order to the Hebrews to not follow foreign gods. But nothing in the Old Testament deals with homosexual practices among mutually loving members of society.[17] This does not mean that such practices, if they had existed, would have been accepted; they were just not mentioned. Lesbianism was also not mentioned in the Old Testament, perhaps because women and their actions were completely unimportant to men of that time.

The four gospels do not contain any references to same-sex relations, perhaps because it was not relevant at the time.[18] This is true even of the gnostic writings.

In New Testament writings, it is only Paul who clearly condemns homosexual acts. In his first letter to the Corinthians he writes: "Do you not know that the unjust will not enter the reign of God? Be not deceived; neither fornicators nor idolaters nor adulterers nor effeminate nor sodomites nor thieves nor the covetous nor drunkards nor revilers nor robbers will enter the reign of God" (1 Cor. 6:9-10, my translation). This, however, is just a list probably copied from the Old Testament, and so it is almost irrelevant. Far more important is what he says in his letter to the Romans:

> For that which is known about God is evident to them, because God (Himself) has shown it to them. For ever since the creation of the world His invisible nature and attributes [of] eternal power and divinity, have been made intelligible and clearly discernible in and through the things that have been made. So (men) are without excuse....For this reason God gave them over and abandoned them to the vile affections and degrading passions. For their women exchanged their natural function for an unnatural and abnormal one. And the men also turned from natural relations with women and were set ablaze with lust for one another—men committing shameful acts with men and suffering in their own bodies and personalities the inevitable consequences and penalty of their wrongdoing, which was their fitting retribution (Rom. 1:19-20,26-27 AMP).

Paul says that it should be clear to all what is a natural action and what is not. And that by performing unnatural acts people have reaped the due penalty. The entire argument, however, hangs on the statement that God's purposes can be deduced by all and that everybody can figure out what is a natural act and what is not. Obviously we only need to define what is natural and get everybody else to agree with us. If a homosexual agrees that he is performing an unnatural act, then he is acting against nature, he is a sinner. But what if he maintains that his actions are not unnatural?

Homosexuality in the Modern Era

My dictionary gives a circuitous definition of the word "natural": "existing in or formed by nature; based on the state of things in nature." The usual argument against homosexuality, of course, is that it is against nature. The proponents of this argument point out that in all animal species there are two sexes and each one has a specific role to play in reproduction.[19] The opponents of the argument point out that an advanced species like mankind has, or should have, other purposes and interests in life besides procreation. To define people by their reproductive capabilities, they say, is to reduce mankind to the level of breeding stock.[20]

A man can fulfill God's "be fruitful" command in less than thirty minutes of his life. That leaves him with another forty million minutes to use in other ways: plant crops, draw pictures, build pyramids, go to the moon. How do we define him? By how he spends the thirty minutes or how he spends the forty million minutes of his life? Furthermore, with the world's population over the six billion mark, procreation has ceased to be humanity's most pressing need.

Natural is whatever we, or better yet our minds, are accustomed to. Almost anyone over fifty years old, black or white, feels strange when he sees an interracial couple kissing. It is something they have rarely seen in their lives; it presents a new image to their brain, so their mind classifies it as different, and hence not natural. The same holds true when most of us see two grown men kissing; it is a new image for most of us, and hence we judge it not natural. Even when I was in the restaurant business and had homosexual couples as customers, I never saw them kissing. Fighting, yes; but kissing, never. Seeing women kissing does not seem as unnatural because they are always kissing each other anyway.

Apparently it is not easy to grow up as a homosexual. Most homosexual adolescents suffer greatly both psychologically and from the actions of their peers. Many attempt suicide, often successfully. So homosexuality is a state that few people seek but one that many attain, whether because of genes, early environmental influence, or combination of the two. In the last two to three decades, U.S. society has started to accept the premise that homosexuality is a natural condition.[21] Laws have been passed to insure that homosexuals are not discriminated against and that they have the same opportunities as heterosexuals.

But just as men and women have different mental characteristics, so apparently do homosexuals. Women are more artistic than men, so, perhaps, it is not surprising to find many gay men attracted to fields that require an artistic flair, whether it is that of an art director, interior decorator, or hairdresser. This, of course, is a personal observation. If one quantified it, however, and proved it to be true, then it would be an indication of actual differences in the genes or brain structure.

People differ in other ways besides sexual inclination. Blood type is an example. There are four blood types, eight if we consider the positive or negative factors. Use the wrong blood type in a transfusion and the patient dies. But if we only started with two people—one really, since Eve was part of Adam—how did we end up with eight blood types? Are some of them unnatural? Should we forbid marriages between people with certain blood type combinations because the second child may die unless it gets an immediate, complete blood transfusion? Perhaps it is time to stop trying to divine what God had in mind when he created humanity and concentrate instead on the marvelous ways in which Nature molded mankind. If we start by assuming the answers, we will never be able to ask the proper questions.

Some Questions to Consider

1. If we are going to shed our bodies in the next world, how does learning to better control them on this earth put us ahead?

2. Could there be a kernel of wisdom in the gnostic ideas? God's creation was strictly spirit energy. As such it could interact with everything else. When the spirit was encased in matter, however, it became isolated by its own material limitations. But upon our physical death, the spirit energy is released and is free to interact with all other energies in the universe. Is all this pure hogwash or could there be something to it?

3. Why do you think the Early Christians decided that the original sin was of a sexual nature?

4. Why, in your opinion, did issues related to human sexuality matters play such a great role in the development of the Christian doctrine?

5. Even now some people become monks or nuns. Why do you think they do it?

6. Do you know of anyone who has renounced sex? Why did he or she do it?

7. Some people say that homosexuality is genetic, and some that it is environmental. Since young children like to emulate their teachers, are we increasing the population of homosexuals by placing them in authority positions as teachers or pastors? Does it matter?

8. There was a big furor when the Boy Scouts refused to allow male homosexuals to become scout leaders. I say that if Girl Scouts allowed heterosexual men to lead young girls in wilderness trips (which I don't think they do) then it would be consistent to allow gay scout leaders to lead young boys. What do you think?

9. Do you think that homosexuality has a genetic or an environmental cause? Why do you think so? Does it matter?

Selected References

Brown, Peter. *The Body and Society: Men, Women, and Sexual Renunciation in Early Christianity.*

Brueggemann, Walter, William C. Placher, and Brian K. Blount. *Struggling with Scripture.*

Gagnon, Robert A. J. *The Bible and Homosexual Practice: Texts and Hermeneutics.*

Gomes, Peter J. *The Good Book: Reading the Bible with Mind and Heart.*

Kuefler, Mathew. *The Manly Eunuch: Masculinity, Gender Ambiguity, and Christian Ideology in Late Antiquity.*

Scroggs, Robin. *The New Testament and Homosexuality: Contextual Background for Contemporary Debate.*

Siker, Jeffrey S., ed. *Homosexuality in the Church: Both Sides of the Debate.*

CHAPTER 28

REACHING FOR A NEW CHRISTIANITY

We started this book reading about Michael Goulder who, faced with scientific and historical facts that disputed the Bible story, renounced his religion to become an atheist. We are now standing in his shoes. We have seen that the Bible is a work of men, not of God, and much of what it says is incorrect. Worse yet, it was written by people who had their own specific objectives to advance. Then we discovered that the Church beliefs and dogmas were also more politically directed than God-inspired. Should we also give up our religion, as Goulder did, or should we continue going along with what we know are untruths? I believe there is a better option.

We have seen that even though the stories in the Bible may be false, there is a great deal of truth hidden in them. Maybe God did not talk to Moses from a burning bush. Perhaps there was no Moses. But I do believe that God, whoever or whatever he or it may be, is trying to talk to every one of us each day, and that we will hear him if we only try. I believe that there is a spiritual world around us, which we cannot describe any better than the Bible does, but one which we will join some day. I believe that we are not alone in this world and that there is a power which cares for us and often helps us. I believe that we have an immortal soul, despite the fact that we cannot describe it and science cannot locate it. And I believe that in many ways our minds are more delicate instruments than the best machines in the physicists' labs.

I believe that Jesus was a son of God, but I also believe that you and I are God's sons and daughters. And for once I will follow Paul's thinking to assert that if we are God's sons and daughters, then we are also his heirs.

FACING THE TRUTH

One day long ago, my young daughter found some baby teeth inside a box in the dresser. Being a smart girl, she immediately realized that they were the teeth that the Tooth Fairy had taken from beneath her pillow and had replaced with shiny quarters. Ergo, no Tooth Fairy. And, being a smart girl, what did she do? Did she proudly announce to everybody her discovery? Of course not. She was smart enough to know that if there were no Tooth Fairy, there would be no one to place the shiny quarters beneath her pillow. She still sported some baby teeth, so she kept quiet. She did not even tell her sister about it. Let's continue playing the game, she must have thought. Don't rock the boat.

Does anyone broadcast that there is no Santa Claus, or St. Nick, or St. Basil, or whatever he is called in other countries? Have you ever entered a store in December to see a big sign proclaiming "No Christmas toys, decorations, wrapping paper, or cards, because Santa Claus does not exist"? Entire industries flourish to serve the needs of a non-existent person!

So what happens when priests, ministers, and theologians recognize that the Scriptures are false? That God did not create the world in six days? That there was no Eve to share a forbidden apple with Adam and bequeath to humanity an eternal sin? That God did not give Moses the Ten Commandments carved on two flat stones? That Jesus was not born of a virgin? Just like my daughter, the vast majority does nothing! Don't rock the boat. Don't destroy something you cannot rebuild. So we end up with preachers who will talk at length in church about man being created in the image of God, but when you ask them what that means, they will tell you that it is really man who created God in his image. They will then continue, unruffled, with what they had been saying before. They will admit that the story of Adam and Eve is a myth and yet explain at great length how Jesus' death on the cross atoned for the mythical couple's sin. They recognize the problem but, seeing no way out, they ignore it.

Some, like Michael Goulder, decide they have had enough and throw in the towel. If the Scriptures are false, then there is no God, and they join the mounting crowd of atheists. The way they look at it, if their God does not exist, then no God does. And this, despite the fact that they often talk about God being so different from us that we humans can never comprehend what he is like. Imagine that Columbus arrives at the Caribbean islands and he somehow realizes that they are not part of India. Would he turn around and sail home? Is it India or nothing? My God or no God?

Philosopher Theologians

Some theologians try to resolve the problem by donning a philosopher's hat. We can illustrate the difference between a philosopher and a scientist with a story about Aristotle, the great Greek philosopher. It seems that one day while he was sitting under an olive tree—Greece was full of olive trees in those days and Greeks did not have much work to do since their slaves did everything—he used the powers of his reason to deduce that Mrs. Aristotle, and all women in general, had fewer teeth than men. This is what philosophers do: they figure out things logically. But suppose that Aristotle had been a scientist instead. What would he have done then? Simple: he would have asked Mrs. Aristotle to open her mouth so that he could count her teeth.

Perhaps one of the greatest philosopher-theologians of the twentieth century was Paul Tillich. Unfortunately, his writings are not easy to read and are even harder to understand. His solution to the problem of errant Scriptures was to consider them as symbolic representations of the truth. A flag, for instance, is a symbol representing a country. But it is not a country. Burning the flag does not harm the country. Yet because a symbol partakes of what it symbolizes, by burning the flag you insult the country. When a symbol becomes that which it symbolizes, then the symbol becomes an idol. The cross points to

Jesus, to salvation. But if you revere the cross itself, then it becomes an idol.[1] Hence the old arguments against icons or the injunction against making graven images.

Tillich defined God as the ground of being. Something that is perhaps too difficult to explain simply and intelligibly: "Everything that is in the world we encounter rests on the ultimate ground of being."[2] God himself is "the ultimate reality, being itself, ground of being, power of being; he is the highest being in which everything that we have does exist in the most perfect way."[3] Strangely enough, these ideas, though confusing, are not original. Jesus said, "It is I who am the light which is above them all. It is I who am the All. From me did the All come forth, and unto me did the All extend. Split a piece of wood and I am there. Lift up the stone and you will find me there," (Thomas 77).

We already saw that in both the Zoroastrian religion and Plato's philosophy, all things in the world, whether physical entities or mental concepts, were considered to be representations of their perfect originals in the heavenly circle. We encountered a similar concept in Sheldrake's Morphic Resonance hypothesis in chapter 17. Thomas includes a related saying: "Jesus said, 'When you see your likeness you rejoice. But when you see your images which came into being before you, and which neither die nor become manifest, how much you will have to bear'" (Thomas 84).

The problem, however, with defining God as the source of all being, or something similar, is that it simply defines God as Nature. Nature is all, and everything derives from it and exists in it, but it is not a God that meets the requirements we specified in chapter 7. Nature is not aware of us except in a purely mechanistic way, and it will certainly not respond to our prayers.

Unbelieving Believers

Some people try to have their cake and eat it, too. A perfect example would be John Shelby Spong. While a practicing bishop in the Episcopal Church, he published numerous books deriding the veracity of the Scriptures. He rejected the existence of an external God, accepting Tillich's definition of God as the ground of all being. So how did he explain his continuing position as a practicing priest? In fact a bishop, a leader of priests?

> I believe that there is a transcending reality present in the very heart of life. I name that reality God.
>
> I believe in Jesus, called Messiah, or Christ.
>
> I believe that in his life this transcendent reality has been revealed so completely that it caused people to refer to him as God's son, even God's only son. The burning God intensity was so real in him that I look at his life and say, "In you I see the meaning of God, so for me you are both Lord and Christ."
>
> I believe that Jesus was a God presence.
>
> I believe in that gift of the Spirit who was called "the giver of life." Once we located God only externally and called this God the Father Almighty. Next we located this God in Jesus, and we called him the Son Incarnate. Now we locate this God in every person, and we call this God the Holy Spirit.
>
> I believe that this Spirit inevitably creates a community of faith that will come, in time, to open the world to God as the very Ground of its life and Being. We call the community "one" because the source of life is one. We call it "holy" because the holy God is seen through it. We call it "catholic" because it is universal and must embrace the whole creation and the families of faith. We call it "apostolic" because it was recognized as present in Jesus, and it flowed to us through the witness of his apostles and disciples. That is our point of entry. It will not be our conclusion.[4]

Some may find these words deep and meaningful. Others may look at them as clever but meaningless wordsmithing. If you read them enough times you will find that they make some sense, and you may even agree with them. But even so, I find them somehow lacking of real meaning or a connection to the Scriptures. I prefer the approach of United Methodist Bishop Joseph Sprague, who prompts us to look at the Scriptures as metaphorical allegories. We should "unpack this ancient, obtuse language...to make a semblance of finite sense out of the infinite mystery, on behalf of those who find the ancient creedal language confusing or implausible."[5] Although he also follows Tillich's thinking, he concludes:

> God is not a Supreme Being "out there" in the great beyond. Rather the word God is the sound image we humans employ to point to the very Essence of it all that is both in our midst and beyond the boundaries of time and existence. Symbolically, if we employ the spatial metaphor developed by Paul Tillich, God is not a Being "out there" or "up there," but the foundation or Ground of all Being. Not limited by time or space, history, or creation, God has been and ever shall be. God is the Essence of it all and is constantly (preveniently, as John Wesley said) at work creating, loving, doing justice, calling humans and all creation into relationship by forgiving, reconciling, empowering, and transforming so that all human beings and the whole created order might be saved. God as Ground of Being, never quits being God and does not cease from revealing the Essence of it all.
>
> In Jesus, God's Essence found confluence with a human being and the kingdom/reign was incarnated and ushered into being.[6]

Sprague's God appears to be more of a caring entity than a purely philosophical Essence. The Scriptures point to it, and it is a willful, rather than a passive power. I can be almost comfortable with it, but not quite. What bothers me with these philosophical concepts of God is that they still interpose a huge, unexplainable chasm between this power and me.

My God

There are things that we have discussed in this book, which are not reflected in these philosophical constructions. So when one of my manuscript readers asked me, "But what do *you* think?" I decided that in this last section of this last chapter of this book, I would shed my objectivity and present my own personal feelings. As I discussed in chapter 16, our minds are search engines constantly trying to make sense out of the contents of our brains. So what personal reality did my mind construct after adding the information in this book into its existing databanks? I will describe it in the next few pages. Although I may try to justify some of its conclusions, I will not attempt to convince the reader that he should think in a similar manner. In fact, the reader can probably skip the rest of this chapter without losing anything of objective value. Am I personally comfortable with the conclusions my mind has drawn? Frankly, so and so! The energy universe it came up with seems a little too extreme. But for the sake of completeness, of intellectual honesty, and to satisfy the curious, this is what my mind has constructed.

First of all, my God is a personal God. One who answers my prayers. If he did not do so, he would have been completely irrelevant to me. Of intellectual interest, perhaps, but of no practical importance. The reason I believe that he answers my prayers is that he has done so in the past. Often not in the way I would have expected, perhaps not even in a way I would have liked, but always in an effectual manner. "Be careful what you ask for, because you may get it," is a saying that describes well my God's actions. But because the word "God" carries such anthropomorphic connotations, it is perhaps better to refer to him as the Spirit, which in the Greek (*pneuma*) is of neuter gender.

Einstein proposed that matter and energy are interchangeable. And as I argued before, we and everything around us are just energy manifestations which our senses and instruments interpret as solid matter.[7] Think of it as frozen energy. Paul said that in death our natural bodies are replaced by spiritual bodies and that the brightness of the heavenly bodies is different from that of earthly ones.[8] Remember that light was the closest concept to energy that the ancients had. But if our bodies are energy, then so also must be our thoughts. I would expect that in a universe of energy, the Spirit, whatever it is, must also be made of energy, so we are indeed created in its image.

After we die our body molecules disintegrate, but something remains. Our selves, perhaps represented by our thought and memory records, survive at an energy level that we cannot sense at the present time. I believe this, because occasionally the energy presence of a dead person can be perceived. Does the released human spirit join and become one with the God-Spirit in the world, satisfying Jesus' saying, "I am in you, and you are in me"? Is this why Jesus' spirit stayed around for only a few days after his resurrection and then left, promising to send the Holy Spirit to his disciples? Is this why visions of the very recently dead people are far more frequent than those of individuals long dead? Do these energy personalities last forever, or are they attenuated by being merged with others? I cannot guess.[9]

I am inclined to believe that the Spirit of which I am talking is not the creator of the world but one of the creations—formed of all the spirits released throughout Earth's history, including perhaps those of dolphins, chimpanzees, dogs, cats, and other animals down the intelligence scale. This Spirit thus becomes somewhat similar to the sentient Gaia of which many science-fiction authors have written. Is there more than this? Is there a creator God? Perhaps, but I can neither prove it nor disprove it.

Is there a sentient Evil in the world? I believe so. Just as upon their death good people release good spirits into the world, which we call angels, so also evil people release evil spirits, which we call devils. But don't think that we must die before our spirits affect our surroundings. As we move through life we all exude a good or evil aura (that some people claim to be able to recognize). Even those of us who can't see auras can sense goodness or evil in people. This is why Jesus equated a sinful thought with the sin itself.

If all this is true, why should we bother with the Bible, with religion, with the Church? As manmade constructs, what is their value? But consider the Ten Commandments, one of the most important directives in the Judeo-Christian religion. Would they be less important if God did not really give them to Moses on the mountaintop?[10] Would Jesus' words "Love your neighbor like yourself," be of less value if he were not the one and only son of God?

Perhaps the Bible is like a fable. A long fictional story full of morals. When Aesop's tortoise beat the hare in the race, we learned an important lesson. Does it matter that the race never took place and that animals can't talk? We have already discussed the famous fable of Adam and Eve. Just as there was a lesson there, there are lessons in the rest of the Bible. And where can we study them better than in the church?

I believe that the Scriptures attempt to express the Spirit's directives to its people. But we must remember that much must have been lost in their interpretation. Just as sometimes in a telephone conversation there is a lot of noise in the connection, the Scriptures probably also contain a lot of noise. Some people confuse the noise with the message. One can make noise by banging on metal garbage-can lids; in fact there is at least one band in this country, that pretends to make music in this manner (for some obscure reason, the company for which I last worked once hired such a group to perform at a sales meeting), yet don't expect to hear them play Beethoven's Fifth. Furthermore, we only hear what we want to hear. So it is possible that some of the Spirit's directives written down in our Bible represent the wishes and interests of the ancient scribes more than the intentions of the Spirit.

An equally important matter is that the Spirit talks to people in their own time. The Scriptures address the problems of the people who were alive at the time the Scriptures were written. The worst thing that we can do is to try to interpret them literally, in our time. Jesus said, "If your hand or your foot causes you to sin, cut it off and throw it away; it is better for you to enter [eternal] life maimed or lame than to have two hands or two feet and be thrown into everlasting fire" (Mat. 18:8 AMP).[11] So when the Church followed Columbus to the New World, it tried to save the souls of the native Indians by torturing them and burning them alive at the stakes.

Today, the Spirit probably tries to communicate with us through contemporary writings, art, even music. It might be through Tillich, or Gomes, or the writers of the *Touched by an Angel* TV series, or the church preacher on Sunday. I believe, however, that one thing is true: the Spirit is trying to communicate with you and me. But we must keep our ears and minds open to hear its words.[12] So even if for no other reason, this is why one should go to church. It is far more likely to hear the Spirit's words there than in the local bar, although there probably are exceptions. I hope and pray that while writing this book I was successfully listening to its message.

I accept Christianity as the belief in the teachings of Jesus of Nazareth, as transmitted to us by the Evangelists, always keeping in mind that in the process some of them may have been altered or even completely fabricated. I reject the Nicene Creed in its entirety as an unsuccessful attempt by man to define and even describe God.

Some Questions to Consider

1. If the Bible is not a hundred-percent God-inspired book, we should evaluate each one of its commandments separately to determine whether we should follow it. Does this make sense? Is this not the same as accepting what we like from the Bible and rejecting what we dislike?

2. None of the modern theologians explains how it all started. Where does the creation process fit in?

3. Does Jesus' description of himself as a pointer ("I Am the Way") make him a symbol and therefore worshipping him becomes idolatry?

4. When one says that "God is the ground of all being," what does he mean?

5. The author's God appears to be a minor local deity, if anything. He refuses to acknowledge the existence of the true God, the Ruler of the Universe. What do you think? Who would this Ruler of the Universe be?

6. How does the Christian faith differ from the Church faith?

Selected References

Church, F. Forrester. *The Essential Tillich: An Anthology of the Writings of Paul Tillich.*

Spong, John Shelby. *Why Christianity Must Change or Die: A Bishop Speaks to Believers in Exile.*

_________. *A New Christianity for a New World.*

Sprague, C. Joseph. *Affirmations of a Dissenter.*

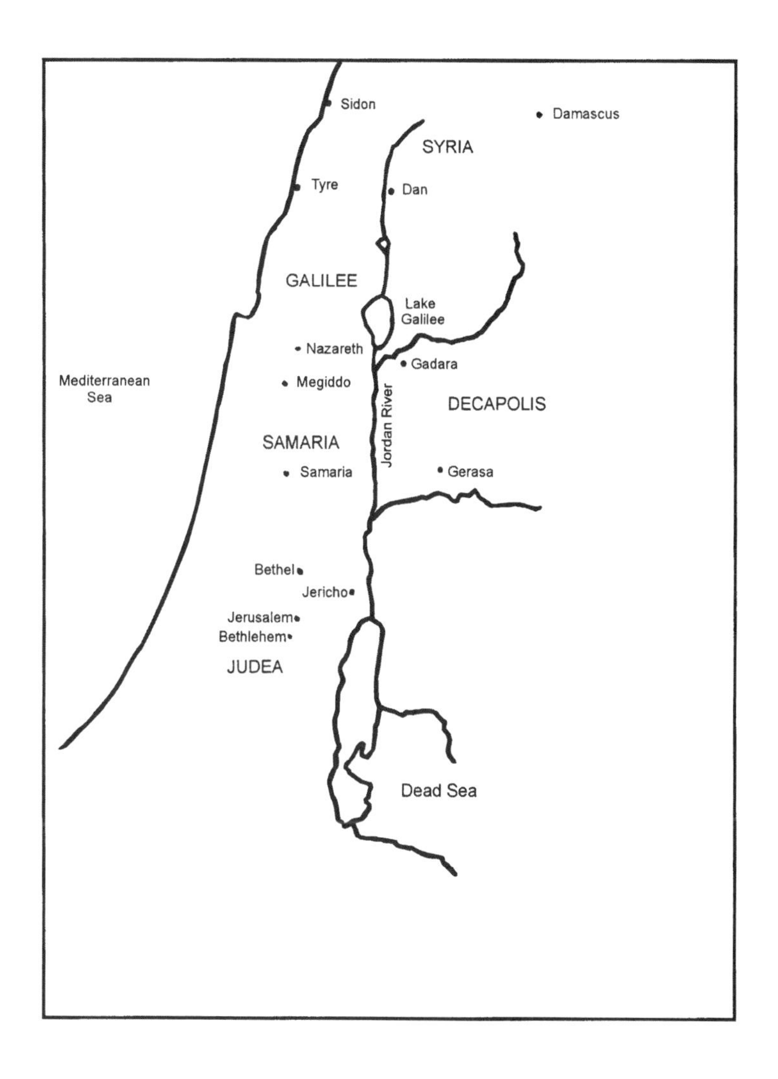

PALESTINE 2000 YEARS AGO

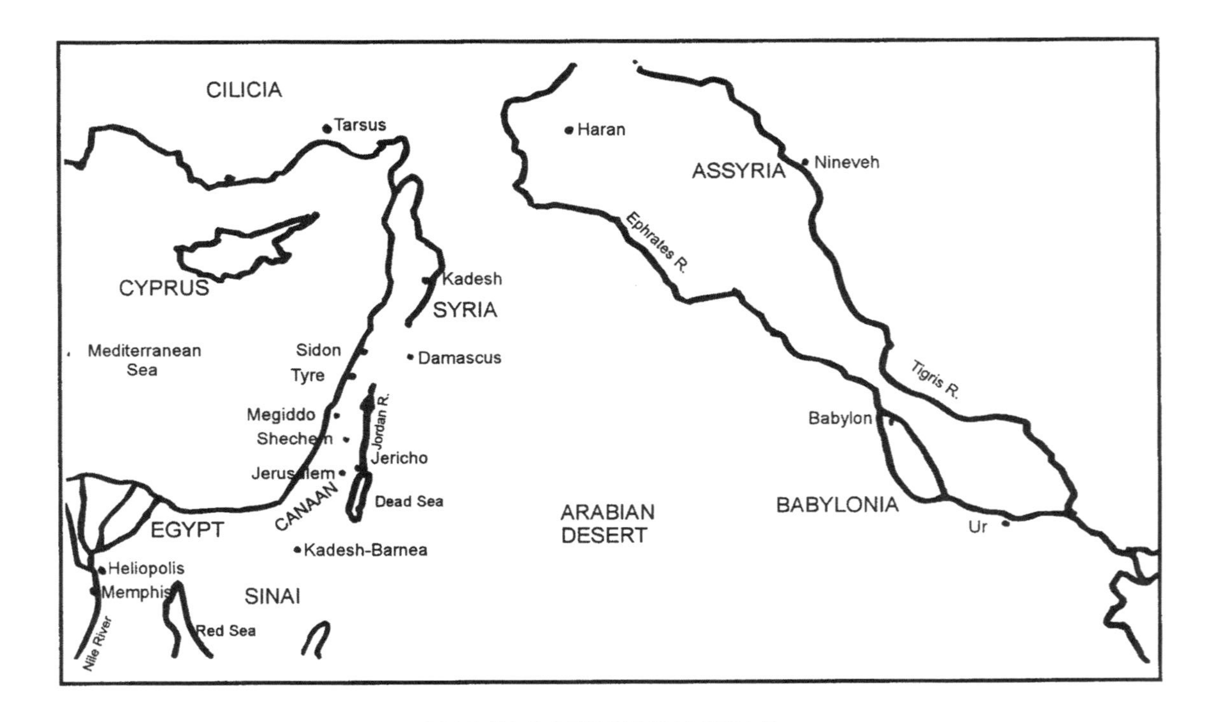

THE OLD HEBREW WORLD

CHAPTER NOTES

1—THE FIRST FIVE BOOKS OF THE BIBLE

[1] The first ten generations starting with Adam up to, but not including Noah, had a lifespan of over 900 years. From Noah to Abraham's father the lifespan declined from 777 to 148 years.

[2] It was Abraham's father who took the family from Ur to Haran. When his father died, Abraham left Haran with his family and his nephew Lot.

[3] After Sarah's death, Abraham remarried and had six additional sons (we don't know about daughters since they were not counted in those days). Although he willed all he had to his son Isaac, he gave his other sons grants and sent them to Kedem, the country of the East. In later days, tribes in the Arabian Desert carried the names of those sons. Here starts the practice in the Bible of naming people of the Genesis generation after nations and tribes of later years.

[4] See Schwartz (1977, 81).

[5] "Angel of God," normally stands for "apparition of God." In Hebrew, one meaning of "Israel," is "a man who wrestles with God."

[6] The Bible reports that the Israelites stayed in Egypt 430 years (Ex. 12:40) but others maintain that this period started from the day Abraham came into Canaan, in which case they stayed only 230 years in Egypt (Josephus 1999, 107).

[7] Before the Hebrews left Egypt, Moses instructed them to borrow from their Egyptian neighbors all the clothing and gold, silver, and brass ornaments they could. Today this is explained as back wages for the work Hebrews had done during their years of slavery.

[8] Note that the original Bible text quotes Aaron telling the Hebrews, "These be thy Gods, O Israel, which brought thee out of the land of Egypt" (Ex. 32:4). Since there was only one bull, however, some recent versions have changed the passage to the singular, "This is your God." For one explanation of the plural see chapter 6, note 7.

[9] According to Exodus, on the first occasion the tablets were prepared by God and written by his own finger (Ex. 31:18); but on the second occasion, they were prepared and written by Moses (Ex. 34:28), despite the fact that originally God had said that he was going to write them (Ex. 34:1). Deuteronomy, however, states that God himself wrote the second set of tablets just as he had written the first (Deut 10:4).

[10] See Kugel (1997, 441) for some explanations provided by later writers.

[11] We will see in Chapter 6, that the introductory sentence varies depending on who was the author of that part of the Bible.

[12] A later census showed that the tribe of Simeon was reduced to less than half its earlier size, indicating, perhaps, that they were the hotheaded ones.

[13] And in a similar manner, later, on the way to Sinai when the Amalek (an aboriginal tribe of the area) attacked, Moses stood on top of a hill overlooking the battle, keeping his hands raised. As long as his hands were up to bring down God's help, the Israelites would win, but when his hands got tired and drooped, the Amalek would get the upper hand. (Finally Aaron and Hur had to stand next to him and support his arms.)

2—IS THE BIBLE HISTORY?

[1] Here the Jewish title *El Shaddai*, where *El* means God, almost certainly identifies one of many gods in a polytheistic pantheon.

[2] Josephus (1999, 48).

[3] Maier (1999, 23) quoting Eusebius.

[4] The story has been quoted from *The Alphabet of Ben Sira* an eleventh-century Arabic writing.

[5] I am following here Friedman, *Who Wrote the Bible?* starting with p. 18.

[6] Friedman (1987, 246-255).

[7] Since the Genesis story was written thousands of years after the events it claims to depict, how reliable is it? Obviously the author was not depending on his memory of the events, and equally obviously, stories that descended down the generations would contain very little reality. Think of the children's game where the first one whispers a sentence into the ear of his neighbor, who then repeats it to his neighbor, until after a few repetitions, the message becomes completely garbled. On the other hand, consider the second question posed at the end of this chapter. We can look at the Bible history at best as inspired writing, with all the reservations that inspiration carries, and at worst as an outright fabrication—fairy tales.

[8] Referencing a one-hour PBS-Nova television documentary titled Evolution: Episode 7: What about God? But see also a response in www.answersingenesis.org.

3—CREATION: SIX DAYS OR A BIG BANG?

[1] Most Bible translations give a specific blessing: "Be fertile, multiply, and fill the water of the seas; and let the birds multiply upon the earth" (Gen. 1:22 JER).

[2] The previous three lines are from the Jerusalem Bible translation (2: 23).

[3] In his book, *In the Beginning*, Walt Brown (2001) has attempted to prove that the story of the Creation and the flood should be accepted as literal truths. The eighth edition of his book is a beautifully printed volume with glossy paper and color pictures throughout, which, at less than twenty dollars per book, is a value only matched by the *National Geographic*; it was probably possible only because it was printed in Hong Kong. Although he tries to prove scientifically the literal truth of the Bible, his arguments are objectively unsound, and unfortunately this beautiful book represents a waste of reading time. While trying to debunk standard scientific knowledge, he states that "evidence must be observable and verifiable" (p. 249), something which most of his own theories lack. Nevertheless, I am sure that this volume has found a prominent place in the homes of most Fundamentalists.

[4] There is a play on words here: in Hebrew, adam means man and adama means ground.

[5] It is not clear what kind of partner man was expected to find among the animals. Certainly sexual relationships were strictly forbidden, as specified later in Leviticus 19:23.

[6] The calculations actually show that the point of origin was reached only eight billion years ago, inconsistent with the 13-billion-year age of some stars. Discrepancies such as this are common in the cosmology field, and are usually resolved later as new data are obtained and theories are modified.

[7] Kelvin degrees are the same as centigrade, but start at absolute zero, -273 C. 10^{28} is equal to 1 followed by 28 zeros, a very large number. 10^{-32} is equal to 1 divided by 10^{32}, a very small number.

[8] *Chicago Tribune,* February 12, 2002. A special telescope will be designed and placed in the Antarctic to investigate this phenomenon (*Chicago Tribune,* August 30, 2002.) The early decrease in the separation rate between galaxies could explain the discrepancy between the 13-billion-year age of the oldest star and the 8-billion-year universe origin that was calculated using a constant separation speed of the galaxies.

[9] For years I had a magazine cutout pasted in front of my desk at work with the following underlined: "People mistake confidence for competence." Since scientists always exude confidence when they speak, we never question their competence. (There is no doubt that I would have been far more successful in my own career had I followed this advice rather than just pinning it on the wall.)

[10] Darwin based his conclusions mostly on the observations he made during his 1831-1836 voyage to the Galápagos Islands, where he studied many birds and animals that had evolved in isolation from the outside world. However he did not publish his results until Alfred Wallace wrote to him in 1858 describing his own, essentially identical, evolutionary theory. Wallace based his theory mainly on the results of a comprehensive analysis of the beetle population in Singapore and the surrounding area. The ideas of the two men were identical, except that Wallace believed that man possessed a divine spirit. Because of the rather extreme views of spirituality that he later developed, Wallace has been forgotten and his body lies in a long-neglected grave. Darwin's follow-up book, *The Descent of Man,* where he argued that man had evolved from apes raised an even greater furor, and even today many Christians try to poke fun at it; nevertheless scientists accept it as an undeniable truth.

[11] There is even disagreement about the exact event that is called the Big Bang and when it started. Some say it refers to the original stable production of the first energy-matter in the universe at time zero; others assign the term to the inflation period starting at about 10^{-35} sec., when it was the size of a proton; others, with whom I agree, maintain that the term should apply to the final outward spreading of the universe starting at 10^{-33} sec., when it had already reached the size of a grapefruit.

[12] Physicists use two theories to explain the universe: the theory of relativity, which deals with the effect of huge masses and speeds, and the quantum theory, which deals with incredibly small things. At the very first instants after its creation, the universe consisted of huge masses that were incredibly small. The physics of this environment, therefore, would require the use of both these theories, but unfortunately they cannot be used si-

multaneously—relativity theory cannot be used when distances are small, and quantum mechanics only works for small masses. A possible solution is promised by the theory of strings or superstrings, which proposes that incredibly short energy strings vibrating in ten dimensions (some of which are curled into essentially zero length) form the basic building blocks of the universe. As Greene (2003) explains, however, it will be decades before any meaningful progress results from this work.

[13] This incredible assumption has led to some even more incredible problems (e.g., the Einstein-Podolsky-Rosen Paradox), whose discussion is outside the scope of this book. It is sufficient to say that their solution requires one of two assumptions: either communication is possible at rates exceeding the speed of light, which the theory of relativity says is impossible, or there exists a field in space which relates together all points in the universe. Such fields are unknown at this time, but various avant-garde physicists have postulated similar ones, and we will discuss them in chapter 17.

[14] To illustrate how little we know of our world, the scientists don't understand yet what gives particles one of their most fundamental property, mass. They have postulated that it is the result of frictional drag sustained by particles as they travel through the postulated Higgs field. Since large particles would sustain greater friction, they would also have greater mass. That of course would indicate that since neutrinos have no mass, they are also dimensionless.

[15] I should remind the reader that everything said in this section is an unproven hypothesis, and thus subject to change.

[16] "It was a mind-boggling conclusion. And it was arrived by an ageing postdoc facing certain unemployment unless he could come up with something important to impress his superiors—and fast. Guth succeeded beyond his wildest dreams." (http://www.newscientist.com/ ns/970510/review.html)

[17] Delseme (1998, 294).

[18] A common joke among physicists is that the universe is the result of a failed experiment by a graduate student in a superior

civilization in another universe (Morris 1999, 75). Why not a successful Ph.D. thesis, I ask.

4—THE GARDEN OF EDEN—ORIGINAL SIN

[1] "There are similarities between the story of Adam and Eve and parts of mankind's earliest epic, the Babylonian epic of *Gilgamesh*. In the *Gilgamesh* epic there is a reference to Eden as the most fertile country in the world. In Eden, the goddess *Aruru* formed a hero named *Engidu* or *Enkidu* out of clay. *Enkidu* lost his immortality after being seduced by a woman. However, *Enkidu* was not the first man in these myths. That was *Adapa* or *Adamu*. In another part of the epic, *Gilgamesh* searches for a plant that will give him immortality. He finds it, but loses it to a serpent, and resigns himself to being mortal" (Elbert 2000, 58).

[2] Nowhere in the Bible does it say that either the animals or Adam and Eve were immortal. It is true that the last two were created in the image of God, but we don't know what that means exactly. Besides, creating an immortal race of beings that can procreate would eventually cause an ecological problem. The world is currently facing overpopulation, and we rarely live over ninety years.

[3] Some modern Christian writers interpret this as "God clothed Adam and Eve in garments of skin, i.e. human bodies" (See Hengel 1997, 295). We will have an opportunity to discuss at a later time the difference between human bodies and spiritual bodies or bodies made of energy.

[4] In the apocalyptic and gnostic texts written around the beginning of the first millennium, the act of creation is not performed by God himself but by a surrogate *demiurgos*, who sometimes is incompetent. Even the Gospel of John hints at this process, but without the incompetence: "In the beginning was the Word.... All things were made by him, and without him nothing was made" (John 1:1, 3). So it was not God himself but the Word, as his tool, that created the universe.

[5] Sociologists tell us that the main reason we work is not to make a living but to structure (fill) our time. Many people who retire discover that they need some work to fill their time and

provide them with relationships to people. Rich people who have no need to work always find some non-paying work to keep them busy; examples are Charles, Prince of Wales, and his architectural pursuits, or the past Japanese emperor Hirohito and his study of horticulture.

[6] God certainly broke his rest to sew clothes for Adam and Eve. It is strange that Jews never mention this break in God's Sabbath, but insist that absolutely no work should be done during the day, except to save a life.

[7] We have already seen that Peter expressed the same opinion. This idea has also given rise to the various end-of-the-world millenarian sects that have expected the world to last six or seven of God's days.

[8] According to the Jewish Talmud, written at around 300-500 CE, the original sin cost mankind its immortality. But to preserve the concept of free will, a man could only be charged for sins that he had personally committed, not any that he might have inherited (Cohen 1976, 96). Of course this does not agree with the Bible quote that the sins of the fathers will be visited upon their children for ten generations.

[9] The first believers in the Church of Thessalonica were surprised and very distraught when some of them died. Paul had to write to them and explain that at the time of the second coming the dead would rise first and then be joined by those still living, among whom he expected to be included (1 Thess. 4:13-18).

[10] There are two references in the Bible from where this forgiveness of sins is derived. In Matthew's gospel, Jesus tells Peter: "I will give you the keys to the kingdom of heaven and whatever you bind on earth shall be bound in heaven; and whatever you shall loose on earth shall be loosed in heaven" (Matthew 16:19). In John's gospel, the resurrected Jesus tells the disciples (missing Judas and Thomas): "Receive the Holy Spirit. If you forgive the sins of anyone, they are forgiven; if you retain the sins of anyone they are retained" (John 20:23 AMP).

[11] Before this absolution concept became accepted, soldiers and others who by profession lived a bloody and sinful life put off baptism until they approached death. A famous example is St. Constantine, the Roman emperor who made Christianity the legal religion of the Roman Empire. Although he espoused all Christian causes, he was not personally baptized until he was on his deathbed.

5—EXODUS—TWO MILLION PEOPLE IN THE DESERT?

[1] The number forty appears very often in the Bible, in both New and Old Testaments. It should not be taken as a quantitative but as a qualitative description of a long period of time.

[2] The name "Moses" is probably derived from an Egyptian word for "has been born" (NAB, note 2, 10 on p. 59). The Levites did not possess a tribe area as the other twelve tribes did; according to the Bible, however, they were to be assigned forty-eight cities and pasture land around them (Numbers 35:2-5).

[3] It is usually assumed that only males died, which is probably true since women did not count in those days, but the writing is not clear.

[4] According to the laws given to the Hebrews during the Exodus, any killing of an ox, sheep, or goat was to be handled as a sacrifice, and had to be offered first to God at his Dwelling.

[5] Finkelstein (2001, 54), Mendenhall (2001, 46).

[6] Finkelstein (2001, 63-64).

[7] A record of Egyptian correspondence has survived that details the pursuit of two runaway slaves. This would indicate that chasing runaways was more a matter of policy than of economic need.

[8] According to Jewish legend, God provided the wandering Hebrews with a fresh-water well on account of Miriam, Moses' sister, who had complained about the bitter-tasting water that was available in the desert. This well followed the Hebrews everywhere they went, and finally dried up when Miriam died.

Paul has called this water-bearing rock the pre-existing Christ, who followed the Hebrews during their journeys (1 Cor. 10:4).

[9] Along the same line, some Egyptian letters have survived to tell of a fifty-soldier unit pacifying the Canaan area (Finkelstein 2001, 60).

[10] Finkelstein (2001, 57).

[11] Friedman (1987, 82).

[12] Since the ancient Egyptians practiced circumcision, it is quite possible that Moses picked the concept there, instead of Abraham making a covenant with God. We must also remember that historians distinguish between Moses' god Yahweh whom he brought from Egypt, and Abraham's god El who had existed in Canaan; furthermore the Canaanites did not practice circumcision.

[13] God asked Abraham: "Why did Sarah laugh?" Sarah replied: "I didn't laugh." But he said: "You did." The woman-haters of later generations pointed out that this is the only occasion in the Old Testament when God spoke to a woman. (This may be wrong, since "the Angel of the Lord," which usually means "the Lord," also spoke to Sarah's slave, Hagar, when she had run away into the desert.)

[14] Miles (1996, 67-76)

[15] It is difficult to believe that any sane god would prescribe to a group of desert wanderers laws about not eating shellfish, or that he would require the complicated and detailed service ceremonies that are described in the Scriptures.

[16] It has been pointed out by many that the Ten Commandments are very similar to the Amorite Babylonian Hamurapi's laws. Although later access to the text of those laws may have affected the exact wording of the Commandments, as they were eventually written down in the Bible, it does not detract from the obvious common sense of these laws designed to help people live peacefully together. Similar common sense is found in the Golden Rule (do unto others as you would like them to do unto

you) that exists, in some form or another, in almost all religions.

[17] Until eventually the people asked the prophet Samuel to anoint a king for them so that they would be like their neighbors (1 Samuel 12:12).

[18] Finkelstein (2001, 119).

[19] There is a possibility that, to my knowledge, nobody has ever suggested: Suppose that we do accept the theory that the Hebrews who entered Canaan were few in number and successfully promoted themselves as the priestly clan for the Canaanites. As the chronology table in the appendix shows, the exact time of Exodus is uncertain. It covers two hundred years that encompass the period of Akhenaten, the pharaoh who tried to impose a monotheistic religion on the Egyptians based on the Sun god. Soon after Akhenaten's death, however, pharaoh Horemheb returned Egypt to its old polytheistic religion and tried to erase all records of Akhenaten's innovations. Is it possible that some of Akhenaten's priests fled Egypt upon his death, and found sanctuary in Canaan as the new priestly clan? This would also explain why the Hebrews and the Egyptian priests had many things in common, such as practicing circumcision and abstaining from eating pork.

6—*WHO WROTE THE OLD TESTAMENT?*

[1] Although they were called judges, they were mostly military leaders. Counting Eli and Samuel at the end, fourteen judges are identified in the Bible, including one woman. The stories usually follow the same pattern: The Hebrews serve other gods in addition to Jehovah; they are then conquered and oppressed by one of their neighboring countries until they ask God for help; he tells them to destroy the false gods; and a leader arises and helps deliver them from their oppressors.

[2] Instead of voting as we do, the Hebrews let decisions be made through casting lots, since in this manner God could reveal his choice. In a similar manner they used divinations to decide whether they should do battle. The selection of Saul, a physically imposing figure, since he was a head taller than most, was

probably an astute political decision. He came from the tribe of Benjamin, the smallest of the twelve, and so his tribe was not in a position to lord it over the others. This qualification did not apply to subsequent kings. Solomon, in particular, who was from Judah, exploited the northern tribes for the benefit of his own.

[3] I am following here the dates given at an Internet source, http://www.hostkingdom.net, which presents data on Egyptian and Middle Eastern leaders. This is why I am saying that David reigned for 49 years despite the fact that the Bible clearly states that it was for 40.5 years (2 Sam. 5:5).

[4] Notice that the suffix in Ishbaal's name relates to the Canaanite god Baal.

[5] This is exactly what David told Solomon: "You and your descendants must always faithfully obey the Lord. If you do, he will keep the solemn promise he made to me that someone from our family will always be king of Israel'" (1 Kings 2:4 CEV).

[6] Solomon made sure that he had no enemies by starting his reign with the murder of his older brother, Adonijah, and the two generals who had opposed David's rise to the kingdom (1 Kings 2:13).

[7] At each of the two cities Jeroboam built a temple with a gold statue of a calf (properly a bull). The two bulls were supposed to support God's throne over the entire kingdom, just like the two cherubs supported his throne in the Jerusalem Temple. It has been suggested that this is the origin of the plural in the sentence "These are the Gods that took you out of Egypt," which Aaron was quoted as having told the Hebrews in the desert. Note also that the bull was the sacred animal of the Canaanite god Baal (see Friedman 1987, 47).

[8] According to Finkelstein (2001, 180-191), Ahab built the city of Samaria and many structures and stables normally attributed to Solomon.

[9] This refers to the ten lost tribes of Israel, who were distributed and assimilated in other populations. According to Assyrian

records, 13,500 people were deported. But the population of Israel at its greatest extent, including areas in the Transjordan, amounted to about 350,000 (Finkelstein 2001, 193). It would thus appear that only a small proportion of the population was actually deported, not everybody as claimed in the Bible. The imported people had their own gods, and John's gospel refers to them as the Samaritan woman's five husbands (John 4:18).

[10] Finkelstein (2001, 243).

[11] As a requirement for their job, many of them had to become eunuchs. One of these, Nehemiah, was eventually sent back to Judea as a governor.

[12] Yamauchi (1996, 424).

[13] Berquist (1995, 25).

[14] During this period, the so-called DeuteroIsaiah was written, which includes the famous passages of the suffering servant. Although Christians later said that these referred to Jesus, Berquist thinks that they refer to Cyrus (Berquist 1995, 41). Others think that they refer to the prophet himself. But in Luke's Acts of the Apostles, the Ethiopian eunuch asks Philip to whom these verses refer, and Philip replies to Jesus (Acts 8:34).

[15] The two prophets of the period, Haggai and Zechariah, helped calm the Judeans from their fear of this huge army that was passing so close to their country.

[16] Prophet Malachi who lived during this period castigated both the priests and the people for trying to cut corners in the religious services by sacrificing animals that had some blemishes (Berquist 1995, 95). The early Christian Fathers, however, interpreted his words as God's rejection of all sacrifices as impure; only the perfect sacrifice of his incarnate son would be acceptable.

[17] We will see later that the main objection of the Romans to Christianity was the refusal of Christians to worship the Roman gods. This was an insult to their gods, who might thus become angry and refuse to help the Romans in their wars.

[18] For consistency, in this section I am following exclusively the arguments of Friedman. Most authorities agree with the general outline of his ideas and any disagreement in details, such as the identification of specific authors, is a matter for experts to resolve. In any case, small differences are immaterial to the purposes of this book.

[19] Nobody, to my knowledge, has explained why the author referred to God in the plural. Perhaps it is something left over from the Canaanite days of multiple gods.

[20] Friedman (1987, 146-149). During the final days of Judah's kingdom, when the Babylonians were preparing to attack it, Jeremiah tried to persuade the people in the city to yield to the invaders; he argued that the Babylonians represented God's punishment of the Judahites for the breaking of the covenant not to worship other gods, and thus in fighting them, they were fighting God himself.

[21] Because the first creation story is very close to the Babylonian myths of creation, many maintain that "P" lived at the time of the Babylonian exile. They may well be right, but it does not alter the fact that somebody apparently wrote the story years after the events, and depended on previous writings to do so.

[22] In the book of Kings, for instance, he often says, "the rest...are they not written in the book of the chronicles of the kings of Israel?" (1 Kings 22:39) The writer of the Chronicles refers to such long-lost sources as "the books of the kings of Israel and Judah," "the history of Nathan the prophet," even "the commentary of the Books of Kings."

[23] Since this event is described in the later book of Nehemiah, perhaps Ezra read his book of law in 445 BCE, years after his arrival in Judea.

[24] Nigosian (1993, 104-113).

7—IS THERE A GOD?

[1] Even though God is an it, in most of the book I am going to use the pronoun "he" to refer to God. This is mostly to accom-

modate the reader who tends to think of God as a creator-father. Unfortunately there are many people, who having been abused by their fathers in their early years, have great difficulty relating to a religion that calls a loving power "Father."

[2] A rabbi, whose name I have forgotten, claims that God answers all prayers unless they fall in one of three categories: they require undoing something that has already taken place, they are injurious to another person, or there are opposing requests on the same matter.

[3] www.progressiveawareness.com "The Active Subconscious: Mind, Body and Memory Dependent Wellness," p. 11 of 34.

[4] The reverse is also true: we often think ourselves sick. Hysterical paralysis, where the patient cannot move despite the fact that there is nothing physically wrong with him, is one obvious example.

[5] Even this will not necessarily prove the existence of an external god, since some can argue that we have an unknown and uncontrollable power to affect our external environment (Tart 1997, 118-127)

[6] Google.com "Catherine Rauch Probing the Power of Prayer."

[7] Tart (1997, 113).

[8] Walsch (2001, 24).

[9] The Episcopalian Bishop John Shelby Spong has written many bestsellers, where, in an effort to be overly inclusive, he belittles the Christian faith. In his *Why Christianity Must Change or Die*, he presents God as an all-encompassing presence, using Tillich's term "the ground of all being," where God is part of us, while we are part of him. I have no objection to this concept. What I violently disagree with, however, is that his god lives an uninterested, passive existence. He asks, "Can God be real if there is no divine entity that can be invoked to come to us in our moments of need?" (p. 59). For me, such a god is a god for philosophers; he is a completely irrelevant god; his existence or nonexistence is completely immaterial to us, since it does not touch us. But most troubling for me is his approach to

prayer. He says that prayers addressed to an external being have little meaning for him. Then he admits that he does not know how to pray, although he has read numerous books on the subject. He says that he fixed up a corner in his study for prayer, with a prayer desk to remind him that this is a prayer place. He printed a cross on his watch to remind him to pray. Despite all his efforts, however, he was unsuccessful because, as he says, the God to whom he had been taught to pray was fading from his view (pp. 136-7). And this man was a practicing Christian bishop, one who was supposed to lead his congregation in prayer! Is it possible for someone to pray when he has no faith? What is astonishing, however, is how tightly closed he kept his mind. He relates how his first wife was diagnosed with a fatal case of cancer. Because of their position in the religious community, prayer groups were mobilized throughout the diocese; her name was added to the prayer lists of all the churches of the diocese, even in ecumenical settings. It turns out that she went into remission and lived for another six and a half years, "beyond anything that the doctors had led us to believe was possible." And what was the good bishop's reaction to all this? He said that if God had responded to all the prayers on behalf of his wife, which resulted from his high religious position, then God must be a capricious being, not affording the same opportunity to the wife of a sanitary worker, who presumably could not put together such an effective prayer chain (pp. 142-3). He refused to believe in such a God, even if he existed! He insisted on defining God the way he wanted him to be.

[10] Pickthall (1999, iv).

[11] Moslem Surah 96:1-5, translated by Pickthall (1999, 723).

[12] Armstrong (1993, 5).

[13] On the other hand, how do we ever know that the products of our imagination are not divinely inspired? In recent years there have been many movies and television shows involving angels, which have the feel of a real inspiration, *Touched by An Angel* being an excellent example.

[14] Morris (1999, 194).

[15] A rather extraordinary example is that of Einstein who in 1905 published seminal papers on the theories of Brownian movement, the thermoelectric Compton effect, and special relativity, plus two others of somewhat lesser importance. It is true, however, that he had the able help of his thirty-year-old wife, Mileva, who may well have been as great a physicist as he was. (Her presence also explains how he succeeded in turning out this large quantity of work while he was fully employed at the Suisse patent office.) In 1912, at the age of thirty-two Einstein published his general theory of relativity. Mileva's contributions to his work had probably stopped a couple of years previously. Thereafter, he made no particularly important contributions to physics, perhaps hampered by the fact that he refused to acknowledge the validity of the new science of quantum mechanics—"God does not play with dice," was how he explained his disagreement.

[16] Farmelo (2002, xi).

[17] See Hoeller's remarkable book for more information. According to Jung's gnostic thinking, at the start was the pleroma, something devoid of everything. Out of this nothing emerged opposites, good and evil, light and dark, god and devil, and so on. Compare this with the ideas physicists developed during the same period: a nothing, out of which emerged the opposite electron and positron pairs. Where did these similar ideas come from?

[18] See Sheldrake (1995) and Talbot (1991).

8—WHY BAD THINGS HAPPEN

[1] I remember once when I wanted to buy one specific stock, and I had problems getting through to the broker. In the middle of April I was getting an automated telephone response telling me that they were closed for the Christmas holidays, and so on. After many unsuccessful attempts, I told myself that somebody was trying to warn me to lay off. But being a scientific person, I ignored my intuition and persevered, eventually getting through and making the purchase. Within one hour some adverse news was announced and the stock dropped ten percent. By the time

I eventually sold it, it had lost another twenty percent. It appears that guardian angels may also be interested in minor things, not just in life-and-death situations.

[2] According to the old Hebrew beliefs, a person was rewarded or punished in this life for his righteousness or sins, not after his death.

[3] 1t has been pointed out to me that Job's woes resulted from the fact that "he was righteous in his own eyes" (Job 32:1), somewhat like the Pharisee in Luke's Gospel who had exalted himself while praying (Luke 18:9-14). But I think that the two cases are different: 1) Job mentioned his own righteousness only while trying to understand the reason for his punishments; 2) God himself had found Job to be righteous.

9—*APOCALYPTIC WRITINGS—PROLOGUE TO CHRISTIANITY*

[1] In an obviously legendary story related by Josephus, the Judeans refused to offer provisions and reinforcements to Alexander because they had given an oath to Darius not to bear arms against him. So Alexander became extremely angry and at the first opportunity marched against Jerusalem. He was greeted outside the city's gates by a fearful high priest dressed in purple and scarlet, and all the inhabitants of Jerusalem dressed in white. And to the bewilderment of all, Alexander saluted the priest, paying homage to God's name written on his hat, for the priest reminded Alexander of a person he had seen in a dream who had told him to go to Persia where he would defeat Darius. Then Alexander proceeded to the Temple and sacrificed to God according to the priests' directions; and the priests showed him the Book of Daniel (11:3) that predicted his arrival (Josephus Ant. XI. 316-337). In addition to the general improbability of such an event, we must remember that only properly purified priests were permitted to perform sacrifices, and that the Book of Daniel had not been written yet.

[2] After many other massacres of unresisting Judeans on the Sabbath, this rule was eventually reinterpreted by the Maccabees to exclude occasions of self-preservation. On a separate sub-

ject, it was said that Ptolemy went to the Temple, and when he entered the Holy of Holies he was greatly surprised to find it completely empty.

[3] It has been suggested that there might have been two Sanhedrin groups: one dealing with theological matters and one with political. Also that in smaller towns there were local twenty-three-member Sanhedrins acting as courts.

[4] The naked participation of Judeans in athletic events led them to accept the painful practice of epispasm that reversed the visible results of circumcision, so that they could blend in better with the Greeks.

[5] A story regarding this translation, which was attributed to someone called Aristeas soon became widespread and has been used widely to claim the inspired nature of the translation. According to the story, while Ptolemy Philadelphus was starting the library in Alexandria, he was told that he should add a translation of the Hebrew sacred books. He agreed and asked Eleazar, the high priest in Jerusalem, to send him a copy of the book, along with seventy-two interpreters, six from every tribe of Israel. The fact that ten of these tribes did not exist any longer at that time points out the fictitious nature of the story. In the wildest variant of the story, the seventy-two interpreters were paired and isolated in separate cells, where they each produced a Greek translation of the Hebrew Scriptures. When compared, the thirty-six copies proved to be identical down to every single word. In practice the Septuagint (meaning seventy) translation, as it is called, is of varying quality, with the Torah being translated the best, and the Book of Daniel the worst. Even so, there are many Hebrew sections that have been omitted, and strangely, some translated parts are in accordance with the Samaritan version of the Torah instead of the Judean, something that has never been satisfactorily explained. During the second century CE, three separate, corrected translations were introduced, which were compared side by side by Origen in the third century. Of these, only Theodotion's story of Daniel was retained and substituted for the Septuagint original.

[6] No reference exists of any overt objections by the Jerusalemites to the transformation of their city into the city of Antioch-in-Jerusalem (Hayes, p. 53).

[7] The Parthians were part of the old Persian empire, and sometime before 250 BCE moved into what we would now call north-eastern Iran. They fought almost continuously with the Seleucids and the Romans. In 53 BCE the Parthians destroyed a forty-thousand-man Roman army, and thirteen years later they conquered areas in Cilicia and Syria. The Romans never conquered the Parthians and eventually a border was established at the Euphrates river, in what is now Iraq.

[8] According to Jewish tradition, there was only one vat of non-defiled oil left in the Temple, enough to burn for only one day. It would have taken the Judeans seven days to prepare more purified oil. Miraculously, however, the menorah (a seven-cup candelabra) burned for eight days with the existing oil. This is why today's Hanukkah menorah has eight candles (plus a ninth, the *Shamash*, used to light the others) and the reason that Jews celebrate the holiday for eight days.

[9] Alexander Jannaeus (103-76 BCE), Simon's grandson, undertook a series of wars to expand his territory, at considerable cost to the Judean population. Once, when presiding at the Feast of the Booths, the people pelted him with lemons, whereupon his soldiers attacked and killed six thousand of them. On a later occasion, when the people arose against him, he fought them for six years, killing another fifty thousand. When the Seleucid king Demetrius responded to the pleas of the Judeans, Alexander was forced to flee after sustaining great losses. But he later recovered, and with the help of the Judeans who now felt sorry for him, chased Demetrius out of the country. Then, turning around, he killed many of those Judeans who had earlier opposed him. Then "as he was feasting with his concubines, in the sight of all the city, he ordered about eight hundred of [the captured Pharisees] to be crucified; and while they were still alive he ordered the throats of their children and wives to be cut before their eyes" (Josephus, Antiquities 13:380).

[10] Around 400 BCE, Joel described the apocalyptic destruction of Jerusalem by locusts, and then, after the repentance of its inhabitants, its rebirth and the destruction of the neighboring kingdoms.

[11] These revelations are always explained by an otherworldly creature and involve both the end of time on earth and another space where life of some type continues.

[12] Ezekiel's death and life speech (33:10-12) probably does not refer to an individual's after-life existence, but to the continuing survival of Israel.

[13] Their passive inaction was justified by the belief that if they were slaughtered unjustly, God would eventually interfere, since according to Deuteronomy 32:43, "He avenges the blood of his servants and purges his people's land." In the apocryphal Testament of Moses, Taxo says to his seven sons: "Behold a second punishment has befallen the people; cruel, impure, going beyond the bounds of mercy.... Never did [our] fathers nor their ancestors tempt God by transgressing his commandments.... Here is what we shall do. We shall fast for a three-day period and on the fourth we shall go into a cave, which is in the open country. There let us die rather than transgress the commands of the Lord.... For if we do this, and do die, our blood will be avenged before the Lord" (Testament of Moses 9, from Charlesworth p. 931).

[14] The ideas of life after death and of the eventual judgment of all people, living or dead, was a Zoroastrian belief that was current among the Persians during the Babylonian exile. Yet no hint of such ideas appear in the final "Law" that was presumably put together during, or immediately after, the captivity. It required an additional two hundred years of life under Persian rule for these ideas to slowly percolate down to the Judean population.

[15] A second prediction of resurrection occurs during a particular gory scene in 2 Maccabees 7:14, where the dying young man says, "God has promised to raise us to life! And so we are willing to die, but you have no hope for life after death (CEV)."

Note that in this case there is no mention of an eternal punishment.

[16] Biblical commentators often propose that the 365 years that Genesis states Enoch spent on earth, are related to the 364 days of the calendar that was suggested in 1 Enoch (Charlesworth 1983, 60). The problem with this reasoning is that the two writings are separated by centuries. It would require for somebody to edit all copies of the Genesis document years after it was written and translated in Greek, something very unlikely.

[17] Prophet (2000, 9).

[18] In 1947, in a cave near Qumran, by the northwest shore of the Dead Sea, Bedouin shepherds discovered a jar containing the disintegrated remains of numerous ancient scrolls. In the following years, additional jars containing the remains of even more scrolls were discovered in ten other nearby caves. Because of the great deterioration of these scrolls, the political strife in the area, and the bickering between competing groups of scholars, we still do not have a complete reading and translation of these documents. It is assumed, however, that they were written during the two hundred fifty years that preceded the Roman army's conquest of Qumran in 68 CE and perhaps stored in the caves to save them from destruction. Just as the Naj Hammadi findings revolutionized our understanding of Christian beginnings, the Dead Sea Scrolls provided a wealth of information about the beliefs of the pre-Christian era.

[19] The Persian Zoroastrians who were probably the originators of this dualism, believed that each side would get to govern for three thousand years, and then they would fight for another three thousand years until one of the two sides won. Man, through his good or bad actions, would influence the final outcome.

10—MESSIAH

[1] Which elicited Jesus' famous response: "You are Peter, and upon this rock (*petra* in Greek) I will build my church, and the gates of hell shall not prevail against it. And I will give you the keys of the kingdom of heaven. And whatever you shall bind on

earth shall be bound in heaven; and whatever you shall loose on earth shall be loosed in heaven" (Matt. 16:18-19). It is upon this statement, not found in any of the other gospels, and upon the widespread belief that Peter had been the first bishop in Rome, that the Roman Catholic Church has claimed primacy over the other Christian communities. This passage, however, is generally considered to have been a later addition; Jesus, after all, was an eschatological preacher who was expecting the advent of God's kingdom, not of man's Church (Kessler p. 38).

[2] After God instructed Samuel that he had selected Saul as the first Hebrew king, "Samuel took a vial of oil and poured it upon [Saul's] head, and kissed him, and said: 'The Lord has anointed you to be captain over his heritage'" (1 Sam. 10:1).

[3] The Gospel of Thomas presents a different exchange: "Jesus said to his disciples, 'Compare me to someone and tell me whom I am like.' Simon Peter said to him, 'You are like a righteous angel.' (That is, the Messiah.) Matthew said to him, 'You are like a wise philosopher.' (That is, a rabbi.) Thomas said to him, 'Master my mouth is wholly incapable of saying whom you are like'" (Thomas 13). Whereupon Jesus took him aside and revealed to him some secrets.

[4] According to Maccoby (1980, 128), Hebrew coronations were performed before representatives of the twelve tribes, and he interprets all four gospels' description of the transfiguration scene as Jesus' coronation; Peter, John, and James were stand-ins for the twelve disciples, who in turn were representing the twelve Hebrew tribes.

[5] Although nowadays we translate "Hosanna" as "praise" or "glory," the old Aramaic meaning was "save us."

[6] It was common in Hebrew poetry to repeat a line using a slightly different wording. In this case, the "ass," and "the colt, the foal of an ass," were one and the same animal. Matthew mindlessly copied the wording, and so described Jesus as "riding on an ass and on a colt, the foal of a beast of burden" (Matthew 21:5). Some scholars consider this error as an indication that

Matthew was a Gentile rather than a Jew, since he was not conversant with the Hebrew way of writing poetry.

[7]" My house shall be called a house of prayer for all people" (Isaiah 56:7)

[8] Mark 14:3-9, Matthew 26:6-13, John 12:2-8. It has also been argued that Jesus was anointed at his baptism when the dove descended on his head.

[9] Perhaps because of the close relationship between the Persians and the Judean high priests whom they appointed, there is no indication that the Judeans ever contemplated a revolt against them.

[10] In the scrolls found in the Qumran Caves, the "end of days" is always associated with the coming of the Messiah (Collins 1995, 61).

[11] Deutero-Isaiah has four sections dealing with an unidentified suffering servant: the first two deal with his selection and his task to bring the Hebrews back to Jerusalem; the other two deal with the opposition that he encounters and his eventual humiliating and painful death. Since the first passage talks about future events, and the last (part of which I quoted in the text) talks about events that have already taken place, one can conclude that the prophet is referring to someone current. For lack of others, it has been proposed that the reference points either to the prophet himself, taunted for unfulfilled predictions, or to King Cyrus who did release the Hebrews to return to Jerusalem and rebuild the Temple (Berquist 1995, 41). Remember that we saw in chapter 6 that the prophet referred to Cyrus as Messiah: "Thus says the Lord to his anointed, to Cyrus whose right hand I have held" (Isaiah 45:1). However as far as we know, Cyrus died as a successful ruler while fighting the nomadic Massagetae, so it is difficult to imagine him as being the suffering servant.

[12] Brenton (1986, 889).

[13]Commentary note, NAB, Acts 3:18 p. 1174.

[14] "When the atonement of the sanctuary, the Tent of Meeting, and the altar is complete, [Aaron] is to bring the other goat that

is still alive. Aaron must lay his hands on its head and confess all the faults of the sons of Israel, all their transgressions and all their sins, and lay them to its charge. Having thus laid them on the goat's head, he shall send it out in the desert led by a man waiting ready, and the goat will bear all their faults away with it into the desert place" (Lev. 16:20-22 JER).

[15] Collins (1995, 126).

11—THE TIME OF JESUS

[1] There was a 10 percent contribution (tithe) to the Temple, and another 10 percent had to be spent in Jerusalem during the high feast days. Every third year another tithe was given to the poor, although it is possible that this might have substituted for the tithe that would have been spent in Jerusalem. Then there was the offering of the first fruits of the field each year, as well as the first male born from each animal (except that donkeys were redeemed by a lamb, and firstborn sons by 5 shekels)—often called heave or wave offering—calculated at about 2 percent. There was a 2-drachma head-tax paid to the Temple by each Jew, which amounted to another 0.4 percent. Then there were the Roman taxes, which amounted to either 1 percent of the value of all the property, or 12.5 percent of the production, depending which modern writer you believe. Finally there were customs payments, tolls and tributes. Fishermen had to pay for fishing rights. And to make things worse, all officials had to be bribed in order to get anything done. For additional discussion see Sanders (1992, 146-169).

[2] According to Deuteronomy 15:1-2, land lost because of non-payment of debts had to be returned during the seventh year. However, the "prosbul" law attributed to Rabbi Hillel overturned that provision, allowing the land to go to creditors through the courts. In this way, the oligarchs amassed even greater wealth.

[3] It has been suggested (Hanson 1998, 126) that Jesus was perhaps referring to these absentee landowners in his parable of the gold coins (see chapter 13), where the noble is accused of reaping where he had not sown (Luke 19:21).

[4] Hanson (1998, 80).

[5] In her recently published book *Honor Lost*, Norma Kouri describes how her Jordanian girlfriend was killed by her father in 1995, because she had a Catholic boyfriend. The killing was pronounced a misdemeanor, and her father was given a three-month suspended sentence. The veracity of this story, however, has been questioned recently in a front page article in the Chicago Tribune (August 1, 2004).

[6] Jeremias (1969, 364).

[7] The gospels rarely mention the Sadducees. They appear in Mark only once (12:18) when they thought they would embarrass Jesus by asking him a question about life after death, a story repeated in both Matthew and Luke. John's gospel does not mention them at all. Matthew lumps Sadducees and Pharisees together a couple of times, presenting them as general opponents to Jesus and John the Baptist. In the Acts they appear three times, in direct opposition to the Pharisees.

[8] The scribes were often Pharisees, and as such Luke sometimes lumps the two groups together. Mark refers to them more often than to the Pharisees but in the corresponding stories, Matthew frequently substitutes Pharisees for scribes. According to Saldarini (1988, 159) Matthew appears to be less adversarial to the scribes than Mark, probably because his church included scribes as members.

12—THE TEACHINGS OF JESUS

[1] In Isaiah 64:7 there is a reference to God as our father: "But now, Oh Lord, you are our father; we are the clay and you our potter; and we are all the work of your hand." But the context refers to a creator rather than to a parent.

[2] The New American Bible notes that despite the difficulty in identifying who were these who acted in an evil manner, "a strong case can be made for the view that in the evangelist's sense the sufferers are Christians, possibly Christian missionaries whose sufferings were brought upon them by their preaching of the gospel" (NAB, commentary on Mat. 25:31-46, 1054). The statements are therefore Matthew's, not those of Jesus.

[3] It can be argued that by exporting jobs overseas, the total wages paid in this country will shrink. But if the overseas products cost less, as they must, then we can afford to buy more of them with less money, so our overall living style can still increase. There are two problems, however, with exporting jobs: first, the income of some specific individuals can drop markedly, or even disappear, causing personal hardships; second, the negative balance of payments increases, and this can be corrected only by eventually selling our country's assets, such as buildings and companies.

[4] The injunction against evil thoughts appears in the Old Testament also: "You shall not covet your neighbor's house, you shall not covet your neighbor's wife, nor his male servant, nor his female servant, nor his ox, nor his ass, nor any thing else that is your neighbor's" (Ex. 20:17).

[5] The last sentence also appears in Mark 7:19 and many translators added the conclusion "Thus he declared all foods clean." This phrase, however, does not appear in the original.

[6] But later on, the Didache, an early Christian teaching, added: "Let your contribution sweat in the palms of your hands until you know to whom you are giving it."

13—PARABLES

[1] "Humans can relate to stories better than they can to pure logic or objective facts. It is simply easier to keep track of a complex argument if it includes people, places, and events rather than propositions, syllogisms, and symbolic logic" (Shermer 2000, 148).

[2] In the parable of the Good Samaritan, the good neighbor is identified as the Samaritan who took care of the hurt person. This looks at the teaching from the wrong viewpoint. We all agree that whoever takes care of us in our need is our good neighbor. The purpose of the story, however, is to teach us that whoever needs our help is our neighbor.

[3] Jeremias (1972, 59) believes that the nobleman in the story refers to Herod's son Archelaus, who went to Rome to get ap-

proval for his kingship. Because the Judeans sent a delegation of fifty people to oppose him, the Roman senate appointed him only an ethnarch instead of a king. So when he returned to Judea, he dealt severely with his accusers.

[4] Renditions of "The Two Sons" parable are often confusing because some ancient texts had transposed the order of the two sons in the story, thus rendering the explanation difficult to justify.

[5] The final admonishment in the parable—to stay awake since one does not know the time of the judgment—makes no sense since all the maidens fell asleep; some of them were just better prepared than others.

[6] Compare with the words of John the Baptist: "Every tree that does not bring forth good fruit is cut down and thrown into the fire" (Mat. 3:10).

[7] Although this parable should belong with the Last Judgment (F) group, Jesus' words specifically place it here.

[8] Calvin's quotation is from Young (1998, 235).

[9] On the other hand, if one believes that Jesus was really a revolutionary, these parables could be explained in the reverse manner: Plan well before you start; your numbers will overwhelm them; let each one use his position to do his share; can't you see all the injustice around you? The last statement can be supported by his other difficult-to-explain saying: "To him who has, more shall be given; and from him who has not, even that which he has shall be taken away" (Mark 4:25).

[10] The parable of the Tenants is based on Isaiah's Vineyard Song (Isa. 5:1-7).

[11] The Church's interpretation is also incorrect on three counts: 1) The Hebrews (tenants) did not kill the son, but the Romans did; 2) The killing was not performed inside Jerusalem (vineyard) but outside its gates; 3) The body was not cast out, but given a burial. This parable also appears in the Gospel of Thomas, but it ends with the killing of the son, not the retaliation against the workers. If that is the original version, then this is

another parable that can only be explained as a call to civil disobedience.

[12] The word parable has other meanings also, and unfortunately "riddle" is one of them. Admittedly, Jesus' parables were less than crystal clear. According to Hultgren (2000, 456) the second-century gnostic *Apocryphon of James* tells that, after his resurrection, Jesus spent eighteen days explaining his parables to his disciples. Similarly, the writings of Irenaeus around 180 CE say that certain Valentinian Gnostics of his day claimed that Jesus conversed with his disciples for eighteen months after his resurrection in order to clarify his teachings. This indicates that the early Christians also felt that the parables lacked clarity.

[13] The CEV bible gives a different translation:

"These people will look and look, but never see.
They will listen and listen, but never understand,
If they did, they would turn to God, and he would forgive them."
(Mark 4:12 CEV)

Although this rendering resolves the problem, it does not correspond to a correct translation of the original Greek text. As discussed in the next note below, however, it does correspond to the Septuagint translation of the Isaiah quotation.

[14] The Septuagint text of Isaiah 6:8-10 goes as follows: "And [God] said, 'Go, and say to this people, "Ye shall hear indeed, but ye shall not understand; and ye shall see indeed, but ye shall not perceive." For the heart of this people has become gross, and their ears are dull of hearing, and their eyes have they closed; lest they should see with their eyes, and hear with their ears, and understand with their heart, and be converted, and I should heal them'" (Brenton 1986, 841). In the Septuagint version it is not God who hardens the hearts of people, but the people themselves. Although Funk (1997, 192) mentions the fact that Matthew used the Septuagint translation for his quote, he makes no other comment about it. Funk's position, instead, is that this part in the gospels should be attributed solely to the evangelists and not to Jesus.

[15] Johnson (1999, 171) has suggested that Jesus spoke in parables to hide the real meaning of his message and protect himself from violent opposition to his teachings. A similar explanation is sometimes given to why he always referred to himself as the "son of man," since that need not mean anything beyond "man."

14—THE REIGN OF GOD

[1] Abstracted from Nigosian (1993, ch 4).

[2] The ritual purity rights of the Zoroastrians were probably stricter than the eventual Judean rules. Unfortunately we cannot tell how much the Judeans copied. The Torah was probably finally edited during the Persian regime, but the Zoroastrian temples and sacred writings were destroyed by Alexander when he burned Persepolis in 330 BCE. Those Zoroastrian records we now have date from the start of the third century CE when efforts were made to gather together the ancient scriptures. So we cannot tell with certainty whether early-day Hebrews borrowed Zoroastrian ideas, or latter-day Zoroastrians copied Hebrew ideas.

[3] The Zurvanites, a later Zoroastrian sect, attempted to return to monotheism by declaring that Ahura Mazda and Ahriman were the twin sons of a single deity. This heresy, however, was quashed by the tenth century CE.

[4] Just like the Zoroastrians, the Judeans in Jesus' days also did not believe that someone was truly dead until three days had elapsed. This is why Jesus waited until Lazarus had been dead for three full days before he went to resurrect him; otherwise he could not have claimed that he had brought him back from the dead. Strangely, Jesus' own resurrection took place less than three days after his own death.

[5] The reconstruction of one's physical body out of its original elements was also an early Christian concept associated with life after death.

[6] "Armageddon: in Hebrew, this means 'Mountain of Megiddo.' Since Megiddo was the scene of many decisive battles in antiquity (Judg. 5: 19-20; 2 Kings 9:27; 2 Chron. 35:20-24) the town

became the symbol of the final disastrous rout of the forces of evil" (NAB, note to Revelations 16:16, 1388). Notice that the Zoroastrian religion was described using a term from later Hebrew ideas, showing how the concepts migrated between the two religions.

[7] There are, however, a few occasions in the Old Testament where people were tempted by God to do evil. When David was king, for instance, "The anger of the Lord once again blazed against the Israelites and he incited David against them. 'Go,' he said 'take a census of Israel and Judah'" (2 Sam. 24:1 JER), an action for which God later punished the Hebrews. Similarly, in 1 Samuel 16:14, "The spirit of the Lord departed from Saul, and an evil spirit from the Lord troubled him." As the two examples show, God was not above performing actions that we would normally attribute to Satan. In fact, many scholars have attributed to God an evil side that prompts him to act in ways that we humans find unjust and excessively severe.

[8] "It was the devil's envy that brought death into the world, as those who are his partners will discover" (Wisdom 2:24 JER).

[9] Zechariah 14, Micah 7, Obediah 17-21, Amos 9:11-15, Joel 13:17-21, Hosea 14:4-9, Daniel 12:1-3, Ezekiel 39:21-48:34, Isaiah 65:17-25.

[10] Ehrman (1999) maintains that Jesus did expect a sudden Armageddon type of ending that would introduce God's kingdom. Afterwards life would continue more or less unchanged, except that there would be no demonic possessions, sickness, poverty, oppression, and violence; the last would be first, and the first last. But the "Son of man" would also come to judge everybody, and Ehrman does not make clear how the final judgment and life continuing as before, except better, could be part of the same event. Jehovah's Witnesses, however, also hold similar expectations: the select few go to heaven, the evil are punished, and the rest live a peaceful, bucolic life on earth.

[11] Luke tried to relax the apocalyptic call when he said that "the kingdom of God is among you," something Mark and Matthew did not. John never mentions the kingdom.

[12] Here we have the description of "Rapture," the sudden change of human beings into spiritual beings. This idea is one of the centerpieces of the *Left Behind* novels, so popular in the early 2000s.

[13] There is one book in the Scriptures that predicts such a clash between good and evil: The Revelation to John, or the Apocalypse as it is often called, the last book in your Bible. Written about 95 CE, perhaps by a follower of John's church, it is full of allegorical language incredibly easy to misunderstand. It is perhaps the perfect example why the Church in the Middle Ages wanted to keep the Bible out of the hands of Christians for fear that they would misinterpret it. It is also the only religious document I am acquainted with that limits the number of the truly saved (to twelve thousand for each of the Hebrew tribes, for a total of 144,000). I am not discussing this writing in this book for two reasons: I am not qualified and I think it is too easy to draw the wrong conclusions from it. Nevertheless, two very important groups have based their existence on John's Revelation: Jehovah's Witnesses, a group that sends a couple of their members over to proselytize me every two years but is otherwise harmless, and the dispensationalists, one of the most dangerous groups in the world today. The dispensationalists believe that they can hasten the Second Coming by helping ignite a war between the Jews in Israel and their Muslim neighbors. This will require the Jews to take over all the old Israel territory and build the third Temple. As a result, some of the most faithful evangelical Christians have become the staunchest helpers of Zionist causes, and by the early 2000s they held considerable influence in Washington. For more information check the Internet including the following: http://www.nationalreview.com/dreher/dreher040502.asp and http://www.natcath.com/NCROnline/archives/101102/101102a.htm.

15—SON OF MAN OR SON OF GOD?

[1] It is not clear whether the term "son of man" should be capitalized or not. In literature and in the Bible it appears both ways. In Daniel, for instance, the NAB and CEV versions say "son of man" but the King James version writes "Son of man."

Good editing practice would require consistency, yet capitalization should depend on the meaning of the term. I decided to capitalize when the term clearly refers to Jesus, and to use lower case when it could refer to someone who looks like a man.

[2] According to Danker, *A Greek-English Dictionary* (2000, 902).

[3] For example: "You have heard that it was said to your ancestors...but I say to you..." (Mat. 5:33); or the more common expression "Amen, I say to you..." Another occasion was when Jesus asked his disciples, "Who do people say that I am?" (Mark 8:27).

[4] "Compare Luke 6:22 with Mat. 5:11; Luke 12:8 with Mat. 10:32; Mat. 16:13 with Mark 8:27 and Luke 9:18" (Stein 1994, 144).

[5] This dating of the Similitudes is based on Charlesworth (1983, 7). Copies of the Similitudes, however, have not been found among the Dead Sea Scrolls, despite the fact that all the other portions of 1 Enoch were represented there. If these writings did not in fact predate Jesus, then the "son of man" statements may have very little apocalyptic basis.

[6] Notice that in contrast to Matthew's gospel (28:9) where the women embraced Jesus' feet, there is no physical contact described in John's gospel. I will discuss a possible meaning for the "don't hold on to me" phrase in chapter 17.

[7] Stein (1994, 88) believes, that although in these words Jesus used the terms "my Father," and "your Father," this is somehow different from saying "our Father." I disagree; I believe that in the English language, "my" + "your" = "our."

16—OUR SELVES AND THE WORLD

[1] These are not static firings, however, but repeated on-off events, occurring at defined frequencies, which depend on the person's state of attention. In deep sleep, the neurons fire at 1-2 cps (cycles per second) known as the delta wave; when we are awake but drowsy, the neurons fire at 4-8 cps, the theta wave; when we are somewhat foggily aware of things, they fire

at 18-30 cps, the beta waves; and finally, when we are fully conscious and paying attention to our surroundings, the neurons fire at 40 cps. It is common practice to refer to the gray matter inside our heads as the "brain," and to its operation as the "mind." Not all of the brain, however, is working at any given moment. Depending on what we are doing and of what we are conscious, some parts will be active, firing at 40 cps, while the rest may be relaxing at the lower 18-30 cps rate (Carter 2002, 121-125) . From a purely physical point of view, we are conscious whenever we have brain neurons firing synchronously at a 40 cps rate. From a somewhat metaphysical point of view, modern Western mystics endeavor to attain pure consciousness by shutting off the outside world and concentrating on their emotions and thoughts. Conversely, Eastern mystics believe that our emotions and thoughts are smokescreens obscuring the real consciousness that can be attained only when there exists a feeling of stillness and freedom from thought, anxiety, and perception; accomplishing this gives one a feeling of being outside his own body, a feeling of oneness with everything else, a feeling of ecstasy (Carter 2002, 278)

[2] Cytowic (1998, 169).

[3] I wonder if all the time variables in this experiment have been adequately controlled. The volunteer was asked to make a decision to move his hand, to make a decision to note the time on the clock, and to actually note the time on the clock. Are all three events transpiring simultaneously? How can we be sure?

[4] Gazzaniga (1992, 37). This is why Jesus' cure of the congenitally blind person in John's gospel (9:1-7) is a physical impossibility.

[5] Cytowic (1998, 54). One such person was the nineteenth-century Russian composer Alexander Scriabin.

[6] ibid p. 65.

[7] Johnston (1999, 6).

[8] Plotinus, a pagan philosopher in the third century CE, believed that the whole material world was created by the soul and had no real existence in itself. When a painter asked him

for permission to paint him, he said, "Why paint an illusion of an illusion?" (Barnstone 1984, 725).

[9] This becomes sadly apparent when an Alzheimer patient loses a block of his memory and then uses what remains to reconstruct a coherent history that can explain his observed environment.

[10] Berne's most popular book, *Games People Play*, explained how most of our interactions with others have as a hidden agenda the acquisition of ego rewards, called "strokes," most of which are bad instead of good. (As wise mothers know, if a child cannot get attention through good behavior, he will misbehave so that he will at least attract attention in the form of a scolding.)

[11] According to Berne, successful communication can take place between two people only when they operate from identical personality levels (child to child, adult to adult and parent to parent) or between complementary states (child to parent paired with parent to child) Some examples would include two people in their parent states complaining about how the new generation will not amount to much, or in their child states enjoying themselves in an uninhibited manner. An example of a complementary exchange would be a person giving advice from his parent persona to someone who willingly accepts it while in his child persona.

[12] See Berne (1972), also Steiner (1984). Although it sounds like a lot of hocus-pocus, check it out by asking some of your friends about their favorite childhood tale and see how it relates to their lifestyle. Be aware, however, that one can identify with anybody in the story, not necessarily the protagonist. In the movie *It's a Wonderful Life*, for instance, Jimmy Stewart's character's favorite childhood story would have been the parable of The Prodigal Son, and he would have identified not with the title character, but with the older brother, who had never left home but had lived an uneventful, seemingly boring life centered around his work and family.

[13] According to one of his students, Eric Berne himself died when he was sixty years old from a broken heart (heart attack), fulfilling his personal script (Steiner 1984, 20-24).

[14] Timothy Wilson (2002) discusses what he calls the adaptive unconscious that not only learns without our awareness but also influences our emotions and decisions while remaining completely hidden.

[15] This is not easy. In his autobiography, Bishop Paul Moore describes how as a young deacon in New Jersey he once returned to the rectory after a rare evening at a New York ballet production, only to find "a dirty, ragged man covered with vomit, lying unconscious on the floor of the porch. I stepped over him, closing my eyes and my nose to his presence. It was too much. Oh, I knew he was more important than the fantasy world I'd come from. I knew Christ dwelt in him, that indeed he was Christ to me. And yet I could not face him, the stench of his vomit, nor my own priesthood, which bound me to him." A friend priest at the rectory said that he would handle the man on the porch, and Moore went up to bed. (From Paul Moore, *Presences: A Bishop's Life in the City* [New York: Farrar, Straus, and Giroux, 1997], 116.)

[16] In computer language, for instance, persistence would be a program like this: (until done: try doing it). Someone who gives up quickly, however, would run a program like this: (do ten times: try doing it; if successful exit).

[17] The concept of ideas having their own independent lives describes what Richard Dawkins has called "memes" (*The Selfish Gene* [Oxford: Oxford University Press, 1976]). Memes are ideas that we can copy, imitate, and in the process alter. As we modify the memes, they undergo an evolution that is similar to the evolution of genes and whose purpose is the survival of the meme and therefore of the majority of those who carry it, even to the relative disadvantage of the specific individuals. According to this theory there is no such thing as a conscious self, but only a conglomeration of memes, things we have learned.

[18] Wilson (2002, 93-115) asserts that our conscious mind does not make any decisions at all, but only justifies the decisions reached by our nonconscious, and that it does even this incorrectly.

17—SURVIVAL OF THE SOUL

[1] See Ferguson (1993, 313-315, 364-370).

[2] In the Old Testament, except for the book of Daniel, there is only one short verse in Isaiah that mentions resurrection: "We have won no victories, and have no descendants to take over the earth. Your people will rise to life! Tell them to leave the graves and celebrate with shouts. You refresh the earth like morning dew; you give life to the dead" (Isaiah 26:18-19 CEV).

[3] To quote Ecclesiastes 9:10, written around the third century BCE: "Anything you can turn your hand to, do with what power you have; for there will be no work, nor reason, nor knowledge, nor wisdom in the netherworld where you are going." The Hebrew people should have encountered the idea of life after death while they were in Egypt, helping to build the pharaohs' pyramids, yet nothing about it is mentioned in the Old Testament.

[4] Notice that Jesus chose to ignore the Genesis story of angels marrying human women.

[5] Today it may seem obvious that mankind will have a different kind of body in the next world, if there is one, but this was not always so. Isaiah describes God creating a new heaven and earth but not a new mankind (Isa. 66:22). Even now, the Jehovah's Witnesses represent the new world with a painting of a typical family playing on rolling green hills. We will see in chapter 20 that for a long time the Church believed that the resurrected body would consist of the exact atoms that were present in the physical earthly body.

[6] As reported by Shermer (2000, 23) quoting Andrew Greeley, a University of Chicago sociologist.

[7] "Between 50 and 75 percent of the population, when confronted with the loss of someone they love, will have an after-death communication from that person" (Morse 2000, 90).

[8] Here I have to add my own witness to this effect. My wife died unexpectedly in 1980, leaving behind two eleven-year-old daughters. Two or three weeks later, as I was falling asleep,

I felt her enter the bedroom and sit on the edge of the bed. She placed her hand on my shoulder and I felt her saying good-bye. I raised myself up and tried to put my arm around her, but it went right through her, and I was suddenly wide awake. One can easily explain away this event, attributing it to my unconscious desire to make up for the missed leave taking. What I cannot explain, however, was the strong feeling of "how sorry I am for what you have to go through" that she transmitted. That was not a feeling that I had felt either before that time or at any time afterwards; it was an idea completely unexpected and foreign to my thoughts.

[9] Following the great success of his first book, Dr. Moody wrote numerous others, of increasingly greater spiritual and metaphysical character. In his *Reunions: Visionary Encounters with Departed Loved Ones* (New York: Ivy Books, 1993) he discusses his investigations of summoning spirits in (or out of) mirrors and other shining surfaces. He describes in great detail his experimental techniques, where after suitable emotional preparation, subjects were placed alone in a dark room with only a mirror, out of which one of their beloved would sometimes emerge. Yet any serious scientist would have placed a hidden TV camera to establish whether the event was actually physical and generally observable or whether it was restricted to the participant's subjective state. I am afraid that his later work may have endangered his credibility. The problem I have with near-death experience book authors is that they keep embellishing the basic experiences to the point of straining the willingness of the reader to keep an open mind. To me one such example consists of stories about prebirth experiences of souls that consciously select among the available genetic material to decide what characteristics and even birth afflictions their bodies will have upon birth (Atwater 2000, 62). I have to admit, however, that I do know a person who maintains that through hypnosis he is able to contact spirits of dead, but still earthbound, people which have expressed this exact concept.

[10] Even the congenital blind described a type of seeing, or visualization, both in this physical world and wherever they went afterward.

[11] Some writers relate cases where the near-death-experience participants made numerous inventions later in their lives based on knowledge they received during their out-of-body experiences. Perhaps they are true, but I will abstain from quoting them.

[12] Again I will add a personal story. My wife asked me one morning if there had been a thunderstorm with lightning the night before. I said no, and asked why. Whereupon she told me that she was awakened during the night by a strong light in the room, and then she somehow left her body and floated to the ceiling from where she could see herself lying on the bed. A voice (from the light?) said, "Come!" But she got scared and jumped back into her body. There was nothing physically wrong with her at that time, but two weeks later she suddenly died from unknown causes, although it might have been perhaps the toxic shock syndrome that was identified some months later. Separately, it is interesting to note that the earliest recorded near-death-experience story is found in Plato's *Republic*: a presumably dead soldier whose body had been placed on the funeral pyre ten days after the battle, returned to consciousness, and related a judgment scene and other sights. But since the soldier's body had not started to deteriorate like those of all the others, we must conclude that he had not been really dead during those ten days but had only failed to show signs of life; assuming, of course, that we decide to believe the story.

[13] This brings to mind the scene in John's gospel where Mary Magdalene sees the resurrected Jesus and calls to him, "Rabbouni." He replies, "Stop holding on to me, for I have not yet ascended to the Father" (John 20:17).

[14] An electrical field describes mathematically the distribution of electrical charges, such as electrons, protons, and ions, in the space around it. A magnetic field describes mathematically the magnetic forces generated by moving charges. Since these two fields are interdependent, they are usually combined and called an electromagnetic field. The information such a field carries is stored in the magnitude of time-related waves of various frequencies that affect preferentially different physical organs or physical instruments. This is a very difficult concept to

explain briefly to a nonphysicist but one can visualize the waves in a field by comparing them to the ripples created when you drop a stone in the water.

[15] Even then the particles follow rules completely opposite to our physical-based understanding: they can be at two places at once, they can move in two opposite directions at the same time; they can exchange information with each other faster than the speed of light. These ideas are disconcerting, but scientists insist that such inconsistencies don't matter because once these small particles combine to form larger ones, their behavior follows sensible rules. Thus the whole is different from its parts. But it seems to me that if the constituent parts of something are mathematical equations, then the whole something must also be an equation.

[16] As discussed in Mindell (2000, 471).

[17] This solves the problem of fundamental particles communicating at instantaneous speeds. No matter how far apart they are, they are also at the same spot.

[18] Rupert Sheldrake carried this thinking further when he suggested that all objects, living or not, all concepts, are copies of prototypical templates. He pointed out that the difficulty of doing something for the first time diminished greatly once somebody did it. Nobody had ever run a four-minute mile, but once someone did it for the first time, many others succeeded soon after. This is not a new idea. Both Plato and the Zoroastrians believed that every object and concept on earth has a perfect prototype in the heavenly sphere.

[19] See, for instance, Morse (2000, 61-75).

[20] *Chakra* in Sanskrit means wheel. There are supposed to be seven chakras, one each over the base of the spine, the genital area, the solar plexus, the heart, the throat, the brow, and the crown of the head (Atwater 2000, 287) . The most noticeable one is presumably the one over the head.

18—*ANGELS, DEVILS, AND MIRACLES*

[1] Other examples of an apparition being called both angel and God can be found in Genesis 16:7 ff, 21:17 ff, 22:11 ff, 31:11 ff, and Judges 6:11 ff, 13:21 ff.

[2] In an effort to escape such an interpretation, Saint Augustine (354-430 CE) and before him Julius Africanus, insisted that by "sons of God" the Bible meant the sons of Seth, and that the daughters of man were really those of Cain. Nevertheless, Enoch clearly described how two hundred angels came down from heaven, married human women, and taught mankind many new arts (1 Enoch 6-16) . They were later punished by God for all these things somewhat reminiscent of Prometheus in Greek mythology, who was punished for stealing fire to bring it to humanity.

[3] In the Zoroastrian religion, which appears to have so many similarities to ours, all inanimate elements and living beings in the universe have their own *Fravashi* in Heaven, their exact but perfect duplicates. In later centuries, the Gnostic Valentinus believed that baptism re-establishes the broken link between the conscious person and his angel, a being that represents his latent truest self.

[4] According to Pseudo-Dionysius, the angels were divided in the following categories, from the highest order to the lowest: Seraphim, Cherubim, Thrones, Dominions, Virtues, Powers, Principalities, Archangels, and Angels (Fox 1996, 51) . Paul names some of them when he talks about the creations of the pre-incarnated Jesus (Colossians 1:16).

[5] Josephus describes one of the exorcisms performed by Eleazar as follows: "He put a ring that had a root of one of those sorts mentioned by Solomon to the nostrils of the demoniac, after which he drew out the demon through his nostrils; and when the man fell down immediately, he renounced him to return to him no more, making still mention of Solomon, and reciting the incantations which he composed. And when Eleazar would persuade and demonstrate to the spectators that he had such power, he set a little way off a cup or basin full of water, and commanded the demon, as he went out of the man, to overturn

it, and thereby to let the spectators know that he had left the man" (Josephus, ant. 8:47-48).

[6] See chapter 16, note 4.

[7] It has been also suggested that perhaps Jesus' powers were finite, and that this is why he was loath to cure Gentiles before Jews, as was the case with the Syrophenician woman (Mark 7:24), or the Canaanite woman in Matthew (15:22). See Theissen (1991, 63).

[8] At one point during their victorious conquest of Canaan, as the Hebrews were approaching an area west of Jericho, Balak, a Canaanite chieftain, became fearful and summoned the seer Balaam to come and curse the invaders. Balaam told the messengers to stay overnight, and he would give them an answer in the morning. In his dreams God told him that he could go, but had to do exactly what God told him to do. So Balaam saddled his donkey and started traveling. But in the inscrutable ways of God in the Old Testament, he became angry with Balaam for going and sent out an angel who "stood in the way as an *adversary* against him" Num. 22:22 AMPL, my emphasis) . The donkey sees the angel who is blocking its way and stops, but Balaam does not see anything so he starts beating the donkey. Whereupon the donkey turns around and asks the man why he is beating him. This is the second of the only two occasions when speaking animals appear in the Bible. The first one, of course, was when the snake spoke to Eve in the Garden of Eden.

[9] We read, for instance, that "The Lord then stirred up an adversary to Solomon" (1 Kings 11:14); the word "adversary" appears as "satan" in both Greek and Hebrew Scriptures. In Psalm 109:6 we read, "Set thou a wicked man over him: and let Satan [*diabolos]* stand at his right hand."

[10] Stoyanov (2000, 59).

[11] Vine (1999, 93).

[12] Although the Chronicles did not reach their final form until 350-300 BCE, they may have been first written as early as 515 BCE (Page, 1995, 34). Others believe that the author of Ezra

and Nehemiah, probably Ezra himself, may have written them around 400 BCE (NAB 347).

[13] The Jubilees attributed to Satan a number of unsavory actions that were previously thought to have been performed by God through the centuries: ordering Abraham to sacrifice his son, attempting to kill Moses when he was on his way to Egypt, killing all the firstborn in Egypt (Stoyanov 2000, 60).

[14] The exorcisms in which Jesus chased away demons should be considered separately from the healing miracles. In the healing miracles it was the faith of the sick persons that helped bring the cure; the exorcisms, however, were accomplished through a straightforward, direct order to the inhabiting evil spirits—the suffering persons were completely passive participants.

[15] I have argued that all properties, not all entities, have opposites. Meek may be the opposite of ferocious, but lamb is not the opposite of lion. So for angels and demons to be opposites they must be properties of the same entity. Christianity's concept of fallen sons of God (angels) would satisfy this requirement.

[16] The Rev. Todd McGregor

[17] See for example M. Scott Peck (1983), Malachi Martin (1992) and Ralph Archie (2001).

[18] In 1977, 24-year-old David Berkowitz killed twelve young girls in New York's lover's lanes. He became known as Son of Sam, after a ranting letter he sent to the police in which he also claimed that he was Beelzebul. After his capture he maintained in his defense that he was possessed by Satan and was following orders sent to him through a neighbor's Labrador retriever. At his conviction, he received twelve 25-year prison terms, one for each offense, to be run simultaneously, so he will be out in 2007. He now says that he is a born-again Christian. Maybe he was crazy and maybe not, but certainly nobody ever considered the possibility that he might have been telling the truth.

19—THE DEATH OF A REVOLUTIONARY

[1] Moses and Elijah greeted Jesus on the mountaintop. But how did Peter recognize them? There were no portraits in the Hebrew world. What really happened on the mountaintop that day? How much of the story do we owe to Peter's imagination, and how much to the evangelist's inspiration? We will never know.

[2] The gospels disagree on who exactly came to arrest Jesus. Mark and Matthew say it was a crowd with swords and staves that had come from the chief priests and elders, and that Jesus addressed the crowd. Luke also writes that a crowd came to arrest him, but in that gospel Jesus addresses "the chief priests and temple guards and elders" who had come to arrest him. John's gospel differs in that Judas leads a cohort (600 Roman soldiers) and also servants of the chief priests and Pharisees, carrying torches, lamps, and weapons. Most Bible translations do not make clear the presence of so many Roman soldiers in the crowd.

[3] In Mark's and Matthew's gospels, the Sanhedrin and all the witnesses meet at night at the house of the chief priest. In Luke, Jesus is just kept at the house overnight, and the Sanhedrin meet the next morning. In John's gospel, Jesus is taken directly to the house of Annas, the father-in-law of the high priest Caiaphas. Annas interrogates Jesus and then sends him to Caiaphas, who apparently does nothing except send Jesus to Pilate in the morning. In this gospel there is no meeting of the Sanhedrin.

[4] "This is the disciple who testifies of these things, and wrote these things; and we know that his testimony is true" (John 21:24).

[5] Mazza (1998, 29-30)

[6] John has no detailed story of the Last Supper, but earlier on, speaks at some length on this subject: "Whoever eats my flesh, and drinks my blood, has eternal life; and I will raise him up at the last day. For my flesh is true food and my blood true drink. He who eats my flesh, and drinks my blood, dwells in me, and

I in him" (John 6:54-56) . But this was too much for many of his followers, and they left him. We have to remember that John's gospel is addressed to the Gentiles, and is at odds with the early Christian Judeans (the Nazorites and Ebionites) who still adhered to the old Hebrew beliefs.

[7] Some people have attributed Judas's sudden betrayal to his disillusionment with Jesus, after he told people to give to Caesar what is Caesar's and to God what is God's, instead of calling for an immediate stop to tax payments.

[8] See more extended discussion in Wilson (1992, chapters 8 and 9).

[9] The NAB notes explain that Jesus was speaking figuratively about a hostile world in the future, and that he cut the conversation short when he was misunderstood. But is this true? Could Jesus have been thinking of a world forty years distant, when he fervently believed that the kingdom of God was at hand? Of course the evangelist himself could have inserted it later, although in that case one would have expected to find it in the more polemic gospels of Matthew and John instead of in Luke. All four gospels, however, show Peter having with him in the garden the sword with which he cut off the ear of the chief priest's servant.

[10] "The branch from the stump of Jesse" (David's father) indicates the coming Messiah.

[11] The last sentence is from Micah 4:4. It appears that the prophets did not mind plagiarizing each other.

[12] This story appears only in Matthew's gospel (26:14-16), and it is probably a fabrication by that evangelist as he continues his efforts to connect Jesus' life to Old Testament predictions. The reference to the potter's field is supposed to relate to Jeremiah 18:2-3 and 32:6-15, but I personally find the association too vague.

[13] According to the gnostic *Gospel of Peter*, which reached its final form around 150 CE, Judas remained a member of the group of disciples: "But we, the twelve disciples of the Lord, were weeping and in sorrow" (XV 59).

[14] This was Maccoby's argument in his first book. (In a more recent publication, he took other, even more extreme positions, with which I do not agree.) The reasonableness of this interpretation, however, is supported by the story of the Egyptian as given by Josephus (Ant. 20:169-172). It appears that a prophet led a multitude to the Mount of Olives telling them that at his command the city walls would come down. Again the Roman soldiers arrived, killing and capturing hundreds. The Egyptian himself escaped, and Paul was later mistaken for him (Acts 21:38).

[15] In Luke's gospel the trials by the Sanhedrin and Pilate, as well as the crucifixion took place the same morning. This is almost certainly impossible. John's gospel that omits the Sanhedrin trial is a little more accommodating.

[16] Matt. 21:18-19. The same story also appears in Mark 11:13-14, but here the lack of figs is explained: "It was not the time for figs." Then why did he look for figs, and why did he curse the tree when he did not find any? Are we to believe that Jesus, who lived all his life among fig trees, did not know when they bear fruit? A possible explanation, however, is given by Josephus (and the information is substantiated by a note in Whiston's translation) who points out that the climate by the lake of Genesareth (Galilee) is so mild that figs hang on the trees for ten months of the year (Josephus 1999, 802-804). Perhaps Jesus was a far less traveled person than we have supposed.

[17] As I discussed in the preface of the book, 'Judeans' were the people who lived in Judea in the days of Jesus (and the previous 600 years); 'Jews' are all the adherents to the Hebrew religion, then and now. So when Matthew has the crowd of "Jews" shouting, "His blood be upon us and upon our children," he leaves Jews throughout all ages open to later revenge by Christian extremists. This has lasted for two millennia and has resulted in the death of untold millions of innocent people.

[18] Only a few of the original texts included Barabbas's first name, and until the NAB edition, it was not quoted in the Bible.

[19] Hengel (1997, 53) points out that by these words Matthew was probably referring to the anti-Jewish pogroms of 66 CE in

Syria and Palestine, during which tens of thousands of Jews were killed.

[20] When, during a famine, two villages were unable to pay the taxes due, Roman soldiers attacked, put the men to the sword, sold the women and children to slavery, and repopulated the land with new inhabitants, who presumably would be more conscientious about their tax-paying duties.

[21] On one occasion, when the Judeans tried to protest against Pilate's robbing of the Temple treasury to pay for an aqueduct that he had built, he had his soldiers disguise themselves and mingle within the group, and upon his signal they attacked, beating and killing many of the unarmed protesters. There was another time when a Samaritan so-called prophet aroused his people, asking them to follow him up Mount Gerizim where he would show them where Moses had hidden the sacred vessels. When Pilate heard about the gathering, he dispatched his soldiers, who chased and killed many of the people. But even Rome could not stand for this needless action, and finally recalled him from his post.

[22] The Hebrew high priest and the Judean aristocracy were allowed to keep their positions because the Romans expected them to maintain order. They were held personally responsible for any uprising or other disorder that occurred.

[23] It seems doubtful that the 600 Roman soldiers, having captured Jesus, would have delivered him to the house of the high priest. It is more reasonable to assume that they would have taken him directly to prison until Pilate had time to examine the matter at his leisure. In that case, the entire trial by the Jewish authorities may well have been a fabrication of the first three evangelists, designed to direct the blame away from the Romans and towards the Jews.

[24] This killing outraged the prominent Judeans, who then complained to King Agrippa II, who had oversight of the Temple. He immediately replaced the chief priest, who had been just appointed in his absence and who thus lasted barely three months on the job.

20—CRUCIFIXION AND RESURRECTION

[1] This explains why Jesus was unable to carry the cross beam of his cross the full 650 yards to Golgotha, and why Simon the Cyrene was enlisted to do it instead. It is also the probable reason why he died so soon after being crucified. Incidentally, the condemned man did not carry the entire cross to his execution place, as is usually pictured, but only the cross beam. Even so, it is estimated that the cross beam weighed between 80 and 110 pounds.

[2] Hengel (1977, 37).

[3] The crucifixion scene, however, must have appeared on smaller objects earlier than this. Ehrman (2000, 73) shows a picture of a fourth-century carving of Jesus' crucifixion on a miniature ivory panel. The scene also contains a Roman soldier, Mary, and John, and also Judas hanging from a tree.

[4] "It was another, their father, who drank the gall and vinegar; it was not I. They struck me with the reed; it was another, Simon who bore the Cross on his shoulder. It was another upon whom they placed the crown of thorns. But I was rejoicing in the height over all the wealth of the archons and the offspring of their error, of their empty glory. And I was laughing at their ignorance" (Robinson 1978, 365 [The Nag Harumadi Library, The Second Treatise of the Great Seth VII, 2; 56:5]). "The Savior said to me. 'He whom you saw on the tree, glad and laughing, this is the living Jesus. But the one into whose hands and feet they drive the nails in his fleshy part, which is the substitute being put to shame, the one who came into being in his likeness. But look at him and me'" (Robinson 1978, 377 [The Nag Hammadi Library, Apocalypse of Peter, VII 3, 81:15]). I have to admit, however, that the thought of Jesus sitting on a tree and laughing at somebody being crucified in his place does not ring true.

[5] "And because of their saying: We slew the Messiah, Jesus son of Mary, Allah's messenger—They slew him not, nor crucified, but it appeared so unto them" (The *Qur'an*, Surah 4, 157).

[6] See Ehrman (2000, 324-330) for a detailed analysis.

[7] According to Romans 6:3-4. See Ehrman (2000, 327) for these arguments.

[8] Not everyone agrees. Marguerite Shuster prefers to stay with the Job philosophy: "Historical evils have been so overwhelming that contemporary theologians of many stripes have bolstered their confession of the all-loving God by suggesting that constraints of one kind or another prevent God from eliminating these evils, which he surely would at least mitigated if only he could. The short-term psychological gain from this procedure, however, obviously brings heavy long-term loss: If God can do nothing about historical evils except, perhaps, be present with sufferers, *then it is hard to conceive the hypothesis that he can ever give them meaning by bringing them to serve his sovereign purposes for good; and it is equally hard to believe that he has truly conquered that final evil, death*. Hence we are left to our own likewise impotent devices. If, however, we believe firmly in the finality of God's power as evidenced in the resurrection, we may have to hold in abeyance, not our moral horror at historical evil, but our tendency to use that horror to undo our confidence in the only one who can deliver us from it." (From *The Resurrection*, edited by Davis et al., emphasis added.)

[9] *Cephas* is the Aramaic word for "Peter," both words meaning stone, rock. As a result they are used interchangeably in the gospels. *The Amplified Bible*, which lists all possible meanings of a translated word, uses "Peter (Cephas)," throughout. In Matthew 16:18, Jesus tells his disciple Simon: "You are Peter [(Petros)], and upon this rock [(petra)] I will build my church." Ehrman (2000, 307), however, believes that there might have been two distinct persons, one called Peter and one called Cephas. See note 4 in Chapter 21 for a further discussion of this.

[10] Only John's gospel welcomes Gentiles in the Christian community, and only that gospel reports baptisms by Jesus. Regarding the baptismal command, the NAB notes suggest that it might have been the baptismal formula of Matthew's church (Mat. 28 n.19 NAB). There is much evidence to support that the Early Christians baptized only in the name of Jesus. In the Acts, for

instance, Peter instructs the people after his Pentecost speech to "Repent and be baptized, everyone of you in the name of Jesus Christ for the remissions of sins, and you shall receive the gift of the Holy Ghost" (Acts 2:38). Later on in the same work, however, we are told that because the converts in Samaria had been baptized in the name of Jesus only, they had not received the Holy Spirit (Acts 8:16). So Peter and John were sent to re-baptize the Samaritan believers.

[11] The entire sentence "saying, 'The Lord is risen indeed and has appeared to Simon'" sounds like a later addition to the flow of the story.

[12] The gospels tend to identify Peter as the informal leader of the disciples, and someone to whom, on occasion, Jesus showed some partiality. This has led the later Christian writers and the Roman Catholic Church to expect that the risen Christ would have shown him preferential consideration, perhaps appearing to him first. Many theories have been proposed to explain when this occurred. One is that the ascension scene witnessed by Peter was really a resurrection scene, and was misplaced in the gospels. Another deals with the mostly lost Gospel of Peter, which continues with the Markan story saying that the disciples returned to their fishing business in Galilee. It ends abruptly; perhaps the missing segment described an appearance to Peter along the lines of the last story in John's gospel. We should keep in mind, however, that during his life Jesus never acted as we would have expected him to act. He neither sought nor used for his purposes the strong and powerful, but the weak and powerless. Would it not have been consistent if after his death he first appeared to the weakest and most powerless, the women, rather than to the head of his disciples?

[13] "And when the [risen] Lord had given the linen cloth to the servant of the priest, he went to James and appeared to him. For James had sworn that he would not eat bread from that hour in which he had drunk from the cup of the Lord until he would see him risen from among those that sleep. And shortly thereafter the Lord said: Bring a table and bread! And immediately it is added: he took the bread, blessed it and brake it and gave it to James the Just and said to him: My brother, eat your bread, for

the Son of man is risen from among them that sleep" [The Gospel of the Hebrews, as it appears in Barnstone (1984, 335)].

[14] But as I pointed out in the previous chapter, the Gospel of Peter did refer to the "twelve disciples of the Lord, weeping in sorrow."

[15] See Kessler (1988, 66).

[16] One of the gnostic writings recognized and dealt with the difficulty that the resurrected Jesus would have in entering through closed doors. The Apocryphon of James that was written sometime between the first and third centuries was based on a collection of independent sayings that was contemporary with other early Christian works. It describes Jesus telling his disciples: "For I have descended to dwell with you in order that you may dwell with me. And when I found that your houses had no ceilings over them, I dwelt in houses which would have been able to receive me when I descended" (Barnstone 1984, 347).

[17] This belief in the resurrection of the actual atoms of the dead person was based entirely on Job's words: "In my flesh I shall see God" (Job 19:26). Yet we have already discussed that the story of Job was an ancient novel that examined the relationship between God and evil; it was not an inspired writing of truths on which to base one's theology. Furthermore it is obvious that Job did not believe in life after death. He said: "So man lies down and rises not again. Till the heavens are no more, they shall not awake, nor be raised out of their sleep" (Job 14:12).

[18] Matthew's gospel says that the tomb was guarded by soldiers. If true, it would have been difficult to steal the body. Paying off the soldiers to say that they fell asleep and that somebody stole the body they were guarding, as suggested by Matthew, is completely unrealistic, since the soldiers would almost certainly have been executed for dereliction of duty. In any case, I believe that this explanation could have no historical basis because before the first Easter Sunday nobody expected Jesus to rise from the dead, certainly not his disciples. Some experts have proposed that the Romans threw Jesus' body in a common grave pit with the other executed persons, and that either wild

animals carried it away, or Jesus' family retrieved it for proper burial.

21—THE FIRST CHRISTIANS

[1] As you would expect, by selling their farms and spending the proceeds, the community eventually lost much income-generating capability and by the forties CE was forced to seek aid from the Christians in Syria and Greece.

[2] Note that the Hellenists in the Christian society of this chapter were probably Gentiles who had converted to Judaism and thus were a different group from the pagan Hellenists of chapter 20 who believed in Plato's theories of the soul.

[3] The word "church" corresponds to the Greek word ecclesia, the term used for political assemblies of citizens in ancient Greek cities. When capitalized, "Church" denotes the organization comprised of the believers and their leadership, and includes their agreed upon beliefs.

[4] Baptism is a subject much neglected in the gospels, though often discussed in the Acts. The gospels do say that John the Baptist "came into all the country about Jordan, preaching the baptism of repentance for the remissions of sins" (Luke 3:3). And this introduces the scene of Jesus' baptism with the Holy Spirit descending upon him in the form of a dove. But except for these instances, there are only two other references to baptism in the gospels, which we discussed in chapter 20. And of these two, the important one, Jesus' command to baptize in the name of the Father, the Son, and the Holy Spirit, does not correspond to practices in Jesus' days. Hebrews, however, did baptize new proselytes to their religion and the early Christians probably followed that tradition. But it was Paul, working among the Gentiles, who first preached that baptism bestows eternal life and who considered it a substitute for the Hebrew circumcision rite. (See also question #4.)

[5] Almost all writers, ancient and contemporary, tell us that the Ebionites were not the first group of Jesus' followers, but that they split away from the Nazarenes at the end of the first century; I disagree. The beliefs of the Ebionites were more restricted

than those of the Nazarenes, and a splinter group always adds to the beliefs it inherits, not subtracts—something computer programmers call inheritance. I think that their beliefs clearly identify the Ebionites as being the Samaritan converts of Philip, since they include ideas in which only the Samaritans believed. Remember that the Samaritans were those Hebrews from the nation of Israel who remained behind after the Assyrians conquered their land and carried away most of the inhabitants. Then people from five other countries were brought in by the conquerors to fill the emptied land. Since the Babylonian exiles returned to Jerusalem, not Samaria, its inhabitants never received the newer books written by the Hebrews during the exile. Furthermore, they never sacrificed at the Jerusalem Temple, but only in their own temple at Mount Gerizim that was later destroyed by the Maccabees. It is quite possible, therefore, that after its destruction they became vegetarians, since the Torah specifies that one can kill an animal only at the Temple, while sacrificing to God. Finally, since as we saw earlier it was Peter who brought them the Holy Spirit to complete their Christianizing process (see Acts 8:14), it would make sense for them to venerate him rather than James. The Ebionites followed a book called the Gospel of the Hebrews, but it was probably different from the identically named book used by the Nazarenes.

[6] When Paul was brought to trial in front of Felix, the governor at Caesarea, he was asked if he was the ringleader of the "sect of the Nazoreans." He did not deny it, but replied that he was a follower of the Way, "which they call a sect." We should not confuse the term Nazarene or Nazorean, with Nazirite. This last refers to a Hebrew's firstborn son who is consecrated to God, and who never cuts his hair or drinks wine. John the Baptist was such a Nazirite. We don't know whether Jesus cut his hair or not, but it is well known that he did drink wine. Another possible source for the Nazorean appellation may relate to Isaiah 11:1-10, where he says that a branch (*nezer*) will sprout from Jesse's (David's father) stump.

[7] When Peter left the city in 44 CE to avoid persecution by Herod, he told the disciples to "inform James and the brothers." This is hardly a declaration about passing the baton. Even

before this, the apostles in Jerusalem "sent Peter and John" to Samaria, indicating that Peter was not the leader of the group.

[8] Some have questioned this, because around 66 CE the gentile inhabitants of Caesarea killed about twenty thousand Jews, whereupon the Jews retaliated by sacking various Syrian cities, including Pella. It has been argued, therefore, that the Jews would not have been welcomed there just a few years later, although perhaps gentile Christians may have helped them. Some of the old writings say that Jesus' followers received a divine revelation in 67 CE instructing them to leave the city, but we can take that with a grain of salt.

[9] See Maier (1999, 116-117). Eusebius refers to the Nazarenes as a second type of Ebionites, and argues that they were named poor because of the poverty of their intellect. I should point out that only John's gospel mentions Jesus' pre-existence as the Logos. No such reference is made in the first three gospels.

[10] The *Birkat ha-Minin* was only introduced in Palestine, where Jewish-Christians (converted Jews) lived. It was not practiced in western synagogues, in Greece and Rome, where Paul's gentile Christians were concentrated.

[11] Although the four gospels don't admit it, it appears that John the Baptist left an equal or even greater impression on his contemporaries than Jesus did. The Jewish historian Josephus mentions, at considerable length, the Baptist, his teachings, his baptisms, and his death (Ant. 18:116-119) . Although he also refers to Jesus' teachings, crucifixion, and resurrection (Ant. 18:63-64), this second account shows substantial evidence of having being tampered with by later Christians. The presence of this story, however, is also the probable reason why Josephus' extensive work has survived almost intact through all these centuries.

[12] All inhabitants of the Roman Empire were required to sacrifice to the statues of the State gods, which often included those of the emperors themselves. Only the Jews were exempted from this obligation because their old religion predated all the others. Since at first the Christians were thought to be Jews, they were also exempted; but then the Jews publicly disowned them

and started pointing them out and exposing them to the authorities.

[13] For many years it had been assumed that the "Beloved Disciple" was John Zebedee, the presumed writer of the gospel, who had modestly concealed his name. We now know that that John was not the author of the gospel, and most people doubt that he was the Beloved Disciple. Many other possibilities have been proposed, including even concepts such as the Church. The only person whom the gospels say that Jesus had loved was Lazarus, but he was not a disciple. I will present my own theory near the end of the book. It is interesting to note, however, that in addition to the Beloved Disciple, Jesus' mother is also never mentioned by name in John's gospel.

[14] Deuteronomy called for a punishment of forty lashes for religious apostasy. But only thirty-nine were usually given, just in case someone miscounted. That would have resulted in more lashes than the prescribed number, which would have been a disgrace for meting out an improperly severe punishment.

22—*THE INVENTION OF CHRISTIANITY*

[1] In those days Tarsus was known for its many schools of rhetoric, and many scholars believe that Paul's writing style followed the accepted rules of the day. Tarsus was also known as the center of Mithraism and many writers have commented on the similarity between the bread and wine in Christian Eucharist and the bread and water in Mithraism's secret ceremonies.

[2] This event is described three times in the Acts (9:2, 22:5, 26:12), yet some details are unclear. Damascus was in Syria, outside Judea and thus presumably out of the control of the Temple's priests. Paul refers to his "companions" on the trip; was he leading a posse? Could he get away with such actions in the Roman Empire? Finally, he is described as telling King Agrippa II that he had tried to get Jesus' followers to blaspheme in the synagogues, so that they would reveal themselves and be punished. This practice, however, did not originate until 80 CE, long after his death and the destruction of Jerusalem. This is

another example of the careless way in which Luke handles dates while inventing things to improve his story.

[3] Another, perhaps less believable, explanation has also been proposed. The lost *Ascension of James—an* Ebionite writing dated about 150 CE and only known from Epiphanius' *Heresies*—relates how Paul, a Gentile (according to the writing), went to Jerusalem and fell in love with the high priest's daughter. Although he had himself circumcised in order to marry her, he was unsuccessful in winning her, whereupon he became furious and started writing against circumcision.

[4] Paul quotes only two minor teachings of Jesus: 1) "Let not the wife depart from her husband" (1 Cor. 7:10), corresponding to "Whoever shall put away his wife...causes her to commit adultery, and whoever marries a divorced woman, commits adultery" (Mat. 5:32); 2) "The Lord ordained that they who preach the gospel should live by the gospel" (1 Cor. 9:14), compared to "For the laborer is worthy of his hire" (Luke 10:7) told to the seventy-two disciples when Jesus sent them out to spread the gospel.

[5] Griffith-Jones (2000) has made an interesting speculation. We know that the gentile Antioch church was started when "Some...men of Cyprus and Cyrene...came to Antioch, [and] spoke to the Greeks preaching about the Lord Jesus" (Acts 11:20). Later in his letter to the Romans, Paul sends greetings to Rufus and his mother, whom he calls his own mother also. Could this Rufus be the son of Simon the Cyrene, the father of Alexander and Rufus, who Mark says carried Jesus' cross (15:21)? Could it be that Paul received some of his early gentile-oriented information about Jesus here in Antioch? I find it an interesting but probably farfetched speculation.

[6] The only other Bible use of the word "Christians" is in 1 Peter 4:16, "If any man suffers as a Christian, let him not be ashamed...."

[7] In 57 CE the Jerusalem Christians even forced Paul to purify himself and appear at the Temple to make an offering (Acts 21:26).

[8] Most, if not all, Bible versions use the names Cephas and Peter interchangeably. The Amplified Bible (AMP) even uses it in a combined manner Peter/Cephas. However, Ehrman (2000, 307) believes that the Cephas mentioned here might not be Peter, but some new member of the community. He points out that Paul wrote, "On the contrary, when they saw that the gospel for the uncircumcised was committed unto me, as the gospel of the circumcision was unto *Peter* and when James, *Cephas*, and John, who seemed to be pillars, perceived the grace that was given unto me, they gave me and Barnabas the right hands of fellowship; that we should go to the heathen, and they to the circumcised" (Gal. 2:7,9, emphasis added) . This reading appears to clearly differentiate Peter from Cephas, and it is the only place where Paul's letters in the original Greek use the name Cephas. In addition, we know that Peter and many other Christians had been forced to leave Jerusalem in 43 CE, to escape the persecution of Agrippa I; it is possible, of course, that Peter could have returned after Agrippa's death. Eusebius also agrees with this reasoning, for he quotes Clement as saying that the Cephas to whom Paul refers was one of the seventy disciples who had the same name as the apostle Peter (Maier 1999, 47).

[9] Paul believed that the Second Coming would occur only after all the people of the world had been given a chance to hear the gospel and be saved (see Mark 13:10). Since Spain was the end of the known world, Paul felt that he had to go there and preach in order to help bring this about. (Today many far-right evangelicals believe that the Second Coming will take place when Israel becomes a nation again and all Jews return to it.)

[10] Or "in him" as he says in Galatians 1:16.

[11] Remember that God changed Abram's name to Abraham when he made with him the covenant of circumcision as described in Genesis 17.

[12] This sentence has also been given a different interpretation. The Contemporary English Version translates it as follows: "If we can be acceptable to God by observing the Law, it was useless for Christ to die." Notice that this says that even if we

did follow the Law, God would not be satisfied. It does not say that we are incapable of following the Law; only that it is not good enough. Whatever the interpretation, humanity's salvation through Jesus' sacrifice is one of the main points in John's gospel also, which carries out Paul's theme that "Christ our Passover [lamb] is sacrificed for us" (1 Cor. 5:7).

23—THE GOSPELS

[1] The Immaculate Conception stories in Matthew and Luke are discounted by scholars as embellishments made by the evangelists to emulate the usual pagan claims that important people were sons of gods—even Zoroaster was supposed to have been born by a fifteen-year-old virgin. Perhaps it is more significant that Mithras, the god of one of the most serious religions competing with Christianity, was also born by an immaculate virgin. The concept of immaculate conception, of course, is not by itself unscientific. It is true that parthenogenesis has not been properly documented in the human species (although it apparently has in turkeys) and in any case it could only result in a clone of the mother, thus never in a boy. In our era, however, immaculate conception is very common, but known by another name: artificial insemination. The problem, of course, is that such techniques did not exist two thousand years ago, unless one belongs to one of the UFO cults and points to the angel Gabriel who brought the news to Mary. There is another remote possibility of course: I had an aunt who used to insist that a young relative of hers became pregnant while taking a bath, but then no one ever accused my aunt of being overly bright. (See, however, the end of note 3 in chapter 25.)

[2] According to the Acts (20:7), the early Christians gathered together at sunset after Sabbath, on the first day of the week, to celebrate the Eucharist. When Emperor Trajan (98-117 CE), who persecuted Christians, forbade evening assemblies, the Christians gathered for Eucharist on Sunday morning before dawn. Sunday, the day of the Sun, was also the day when the followers of Mithras celebrated their god, so it has been argued that in later years Sunday celebrations helped to increase the similarity between the two religions even further; another example of at-

tempting to make the two religions appear similar was the setting of Jesus' birth date on December 25, the feast day of Mithraism.

[3] This was taken to the extreme by the self-appointed scholars of the so-called Jesus Seminar, an organization with a goal of separating the true words of Jesus from those attributed to him by the gospel writers. One of the criteria the scholars used was that Jesus' sayings had to be unique. Thus they never gave him credit for saying anything that might have been previously uttered by someone else, even if he would have agreed with it. If they are right and Jesus only spoke original thoughts, can you imagine how hard it must have been for him to go through life expressing solely unique ideas?

[4] Translating the Old Testament must have been even harder because only the consonants were used to write Hebrew words; the vowels had to be deduced. To appreciate how difficult this can be, pick an English word, remove all the vowels, and see how many different words you can construct by selecting various vowels. For example 'mtr' could be meter, motor, miter, mater (English for muter), meteor.

[5] *Synoptic* is the Greek word for "seeing together," thus presenting one view. Since Matthew's and Luke's gospels were in large part based on Mark's gospel, all three agree in their historical representation of Jesus and his mission. This contrasts with John's gospel, which is mainly a theological treatise, and which presents an entirely different history of Jesus' travels and actions.

[6] The statement "thus he declared all foods to be clean" is found in most Bible translations enclosed in parentheses, and is a conclusion reached by the translators. This sentence does not exist in the original Greek writings. What the original Greek and the King James Version say is that the digestion process purifies all food. This sentence is not found in Matthew's gospel.

[7] Although all the gospels report Peter's denial of Jesus, indicating that the event was probably true, Luke takes another stab at Peter's reliability. In his Acts he relates how King Agrippa had

both Peter and James Zebedee arrested, and how after James's execution Peter managed to escape. So twice Peter escaped while a friend of his was executed. This story is continued in *The Acts of Peter*, an Eastern Christian apocryphal story dating from the second century. There Peter is described as leaving Rome, trying to escape Nero's persecution of the Christians, when he meets at the gates Jesus entering the city. "Lord, where are you going?" he asks. And Jesus replies: "I am going to Rome to be crucified." And Peter says to him: "Lord, are you being crucified again?" And Jesus replies: "Yes, I am being crucified again." And Peter sees Jesus ascending into heaven and realizes that it is he himself who is going to be crucified in Jesus' place, and he returns to Rome. So finally, on the third occasion, Peter redeems himself and dies for his Lord.

[8] Most people believe that Mark's gospel was written around 68 CE, after the first Jewish revolt began but before the fall of Jerusalem in 70 CE. A few scholars, however, maintain that it was written one or two years after the city's destruction.

[9] The information we have about Mark and his gospel derives from the writings of Eusebius, who in turn quotes a certain presbyter named Papias who lived in the early second century (Maier 1999, 129) . Much of Eusebius's writings, however, are of dubious accuracy. In this particular case his conclusions are based on the greeting at the end of Peter's first letter: "The church at Babylon...salutes you; and so does Marcus my son" (1 Peter 5:13), where Babylon is understood to mean Rome. This same sentence, incidentally, has also been used to place Peter in Rome, and it remains the only justification for such a belief. Many scholars, however, question Peter's authorship of this letter: it sounds a great deal like Paul, and it is addressed to some churches established by Paul, something unlikely for Peter to have done while Paul was still alive. Returning to Mark, the letter that Paul wrote to Philemon from Rome about 61-63 CE also includes greetings from Mark, possibly indicating that Mark had actually made up with Paul and was in Rome with him, not Peter. On the other hand, there is no reason to believe that all existing references to Mark relate to the same person. But since Mark's gospel is more pro-Paulist than pro-Petrine, I have diffi-

culties with the concept that he was Peter's disciple rather than Paul's.

[10] Mark exhibits an astonishing ignorance of Palestinian geography. At one point he tells of Jesus going from Tyre to Galilee, forty miles southeast, "by way of Sidon," a city located more than 20 miles north of Tyre (Mark 7:31). At another point he described Jesus and his disciples as crossing the Sea of Galilee and coming "to the other side of the sea, to the territory of the Gerasenes" (Mark 5:1). Yet the town of Gerasa is a full thirty-five miles away from the sea; the town of Gedara is far closer, perhaps seven miles away. In all fairness to Mark, however, I should point out that considerable confusion arises from inconsistencies in the original manuscripts that identify the city variously as Gadara, Gergesa, or Gerasa. (See NAB note Matt. 8:28, p. 1021 for discussion of this point.) In any case some people have used these inconsistencies to argue that Mark was not a native of Palestine and perhaps he was not the person he is assumed to have been.

[11] These words were used by the Gnostics to justify their belief that Jesus was really a man, who had been briefly taken over by the divine spirit. According to them, the spirit left him at the time of his crucifixion, leaving the human Jesus on the cross to suddenly and painfully realize that he was alone.

[12] The fact that Mark never referred to Jesus' teachings is another reason for believing that he was a disciple of Paul and not of Peter. There is nothing to indicate that Paul had any knowledge of Jesus' teachings, but obviously Peter did, since he was with him during his entire ministry.

[13] One strange and unexplained aspect of Matthew's gospel is that he changes many of Mark's events into doublets: Mark's blind man in Bethsaida (Mark 8:22) becomes two blind men in Matthew (Mat. 9:27); Mark's Gadarene demoniac (Mark 5:1) is turned by Matthew into two Gergesene demoniacs (Mat. 8:28), where I have used the King James version nomenclatures. On the way to Jerusalem outside of Jericho Mark's Jesus meets the blind Bartimaeus who greets him as the son of David (Mark 10:46); in Matthew, Jesus encounters two blind men who call

him "Son of David" (Mat. 20:30). It all reaches the height of absurdity, when Matthew has Jesus enter Jerusalem "riding on an ass *and* on a colt" (Mat. 21:5). (Try to visualize that.) For this last error, however, Matthew can blame the Septuagint translation he was quoting.

[14] Scholars agree that Matthew's gospel was originally written in Greek, because it improved on Mark's Greek when retelling his stories (Johnson 1999, 188). Nevertheless, the Church historian Eusebius quotes Pappias as saying that "Matthew compiled the sayings (logia of Christ) in the Hebrew language (probably Aramaic), and each interpreted them as best as he could" (Eusebius, Missions and Persecutions 39, as quoted by Maier 1999, 130). Matthew, however, was almost certainly more comfortable with Greek than Hebrew. We have already noted that he used the Septuagint translation of Isaiah when trying to explain why Jesus spoke in parables, and he apparently also quoted Zechariah from it.

[15] In Luke's gospel Jesus advises his followers in this manner: "When they take you before synagogues and magistrates and authorities, do not worry about how to defend yourselves, or what to say, because when the time comes the Holy Ghost will teach you what you must say" (Luke 12:11-12 JER).

[16] It has been suggested that Luke made an anachronistic error when he wrote that Mary Magdalene, Joanna, Chuza, Susanna, and other women followed and helped Jesus through his travels. He probably confused them with those women who helped the Christian cause after Jesus' death.

[17] Seventy-two was the number of members in the Jewish Sanhedrin. Maccoby (1980, 129) has proposed that this was an indication that Jesus was preparing to establish his political kingdom on earth.

[18] As we saw in chapter 22, however, Paul did ascribe to Christ both an existence before his earthly life and a hand in the creation of the world (1 Cor. 8:6). These points were amplified in Col. 1:15-16, written by one of his followers.

[19] The first verse in John sounds a little different in the original Greek, because of the selected use and disuse of articles: "And the Word was with *the* God, and the Word was [a] God. This was first noted by Origen, who claimed for Jesus a derived divinity (Kelly 1978, 113) Nevertheless, the first mention of Christ as the Creator appears in 1 Cor. 8:6, and later in two inauthentic letters of Paul. In the Epistle to the Hebrews that was probably written between 70 and 90 CE, we read: "In these last days [God] spoke to us through his son, whom he has appointed heir of all things and through whom he also made the worlds" (Heb. 1:2). Similarly, in the letter to the Colossians we read: "For by him [God's Son] all things were created, those in heaven and on earth, visible and invisible, whether they be thrones, or dominions, or principalities, or powers [i.e. various orders of angels]; all things were created by him, and for him" (Col. 1:16).

[20] It appears that the identification of Christ the *Logos*, with Wisdom of the Old Testament predates John's gospel. The concept of the "Logos" (Word) originated with Thales of Miletus, who first proposed around 600 BCE that the world does not work in an erratic manner in response to actions of gods and demons as everyone thought at the time, but follows certain immutable laws of nature, which man is capable of discovering. Around 500 BCE, his follower Heraclitus of Ephesus named the rational principle according to which the world was created "logos," and perhaps the Hebrew "Wisdom" originated from here. Later on, the Gnostics conceived the world as consisting of opposites and forever-warring principles: good and evil, body and soul, spirit and matter. They came to believe that since God was pure spirit, he was diametrically opposed to the base material of creation, and so he could not have been its creator. The world was created instead by an evil *demiurgos* (i.e., creator) god. According to some of the early Christian Gnostics, the Hebrew God Yahweh of the Old Testament was this evil *demiurgos*. Jesus, on the other hand, was thought to have been the *Logos*, God's son, who tried to bring to selected humans the wisdom (*gnosis*) necessary to allow their souls to escape this evil world and return to the spirit world from where they had

come. See Isaac Asimov, *Guide to the Bible*, vol. 2 (New York: Avon Books, 1971), pp. 298-302.

[21] One of the main reasons for dating Thomas' gospel so early is that it contains many parables in a more primitive form than do the others. Gregory Riley, in his *Resurrection Reconsidered*, argues that John presented such a negative picture of Thomas in his gospel in order to oppose his followers. Not only did Thomas not believe in Jesus' resurrection as the other disciples did, but he was also absent when Jesus breathed the Holy Spirit on the other disciples, so he did not become empowered to continue Jesus' mission. If he is right, the Gospel of Thomas must have predated that of John's.

[22] Much of the pagan literature was also destroyed in 391 CE, when the Sarapeum (the temple to Isis in Alexandria) and its huge library were burned down under orders of the Christian Emperor Theodosius.

[23] One of Jesus' four brothers was named Judas—the others were James, Joses, and Simon. Of course the idea that Jesus had a twin brother would play havoc with the entire concept of Immaculate Conception introduced in Matthew's and Luke's gospels.

[24] Interestingly this is the only time in the Gospel of Thomas that Jesus used the term "Kingdom of Heaven," so common in the synoptic gospels. On one other occasion (Thomas 20) the disciples use the term. Throughout the rest of the gospel, Jesus speaks of the "Kingdom of the Father" or just plainly of "the Kingdom."

24—TRINITY: POLYTHEISM, SPLIT PERSONALITY, OR LEGISLATED MYSTERY?

[1] The following was the creed developed in Nicaea:

We believe in one God, the Father almighty, maker of all things visible and invisible; And in one Lord Jesus Christ, the Son of God, begotten from the Father, only begotten, that is, from the substance of the Father, God from God, light from light, true God from true God, begotten not made, of one substance with the Father, through Whom all things came into being, things in heaven and things on earth. Who because of us men and because of our salvation, came down and became incarnate, becoming man, suffered and rose again on the third day, ascended to the heavens, and will come to judge the living and the dead; And in the Holy Spirit. But as for those who say, There was when He was not, and, Before being born He was not, and that he came into existence out of nothing, or who assert that the Son of God is from a different hypostasis or substance, or is created, or is subject to alteration or change—these the Catholic Church anathematizes. (Kelly 1978, 232)

[2] 1n addition to reaffirming the Nicene Creed, the Council at Chalcedon adopted the following:

In agreement, therefore, with the holy fathers, we all unanimously teach that we should confess that our Lord Jesus Christ is one and the same Son, the same perfect in Godhead and the same perfect in manhood, truly God and truly man, the same of a rational body and soul, consubstantial with the Father in Godhead, and the same consubstantial with us in manhood, like us in all things except sin; begotten from the Father before the ages as regards his Godhead, and in the last days, the same, because of us and because of our salvation begotten from the Virgin Mary, the *Theotokos*, as regards His manhood; One and the same Christ, Son, Lord, only-begotten, made known in two natures without confusion, without change, without division, without separation, the difference of the natures being by no means removed because of the union, but the property of each nation being preserved and coalescing in one prosopon and one hypostasis—not parted or divided in two prosopa, but one and the same Son, only-begotten, divine Word, the Lord Jesus Christ, as the prophets of old and Jesus Christ himself have taught us about Him and the creed of our fathers has handed down. (Kelly 1978, 339-340).

[3] Is our immortality after we die the same as the immortality that Adam had when he was created? If we will not become immortal in a flesh-and-blood way, what was Adam's immortal state? Some gnostic writers believed that Adam had been a spiritual being, and only upon his fall he became flesh and blood.

Consider the passage in Genesis just before their expulsion from the Garden: "For Adam and also for his wife the Lord God made garments of skin, and clothed them" (Gen. 3:21) . These garments the Gnostics interpreted as our physical bodies. You can point out that God had already used clay to make the human body, but that of course was in the second creation story, not the first. The problem with all this kind of reasoning is that by treating an accepted myth as history, we just create more myths.

[4] Ehrman (1993, 187-194) maintains that Luke's scene of Jesus praying in the garden in such mortal anguish that he sweated blood (22:44) was a later addition to strengthen the idea of Jesus' humanity.

[5] Something similar happened in the Judaic history after the Temple was built and the Laws codified: "There was general agreement that prophecy, in its special connotation, ceased with the overthrow of the first Temple, although it lingered with a few men during the exile "When the later prophets Haggai, Zechariah, and Malachi died, the Holy Spirit departed from Israel (Sanh. lla)" (Cohen 1975, 124).

[6] The construction of the canon, accomplished two things. First it stopped the proliferation of sacred books. But second, and equally important, it specified which books would be considered sacred. The gnostic Marcion, for instance, who had preceded Montanus, had repudiated the entire Old Testament, and only accepted Luke's gospel and Paul's writings.

[7] The gnostic writings describe the dove that descended upon Jesus during his baptism as actually entering into him, making the adoption image even stronger.

[8] Translators have taken some liberty with this passage: is it one transgression, or one's transgression? is it one righteousness, or one's righteousness? And if it is one's righteousness, is this one a person, that is a man, as some translations say?

[9] What we have here is a child's riddle: who came to be first, the father or the son? Obviously neither, since they both came to be at the same time. There cannot be a son without a father,

and there cannot be a father without a son (or daughter). But a person can exist without being a father. The *Logos* may be defined by his sonhood, but God is not defined by his fatherhood. So philosophical arguments notwithstanding, we can visualize God without the *Logos*, but not the reverse.

[10] See Hughes, http://www.christusrex.org/www1/CDHN/coun2.html, p. 1.

[11] You can read about the Council of Chalcedon, and the other nineteen General Councils, in Philip Hughes, *The Church in Crisis: A History of the General Councils: 325-1870*, posted on the internet (see bibliography).

[12] Everybody did not agree. The Docetic schism of monophysitism came into being and still exists fifteen hundred years later.

25—THE EUCHARIST

[1] In Robert Heinlein's *Stranger in a Strange Land*, we also have a prophet who starts a religion based on love. When he dies, his disciples make a broth of him (it is not clear how, since they later bury his body) and drink it. Less gory perhaps, but the same idea.

[2] The rest of the quote, which is almost identical with Paul's words, is missing from one old text, which may indicate that it might have been a later addition inserted by copying scribes.

[3] Many authors have pointed out the remarkable similarities in the rituals of the Christian church, and of Mithraism, a religion that had originated in Persia. This was first pointed out in 140 CE by Justin Martyr, one of the early Church Fathers, while discussing the origin of the Eucharist:

> The apostles, in the memoirs they composed called "Gospels," have passed on to us what was given to them: that Jesus took bread, and when he had given thanks, said, "Do this in memory of me. This is my body." In the same way, after taking the cup and giving thanks, he said, "This is my blood," and he gave it to them alone. This the wicked devils have imitated,

commanding the same thing to be done in the mysteries of Mithras. There, in the mystic rites of initiation, bread and a cup of water are placed amid certain incantations. This you already know or can find out. (Aquilina 2001, 85-86)

During the first three centuries, Mithraism was one of the strongest competitors of the early Christian religion. The two shared many of the same concepts, and even language. But Mithraism in prior forms had existed from much earlier.

Around 1500 to 2000 BCE the Aryan tribes descended from the Russian steppes to India, bringing with them a god called Mithra or Mitra. From there the god Mitra reached Iran, where the Zoroastrian god Ahura Mazda declared Mithra to be as worthy of worship as himself. In 67 BCE, some captured Cilician pirates brought to the Roman Empire a slightly different form of the religion, with a god named Mithras. It is not clear how popular this new religion became. On one hand, most discovered places of worship could only hold a few dozen people at most; on the other hand, it attracted powerful people in Roman society, such as the emperors Trajan, the first persecutor of Christianity empire-wide, and Commodus. (Interestingly, in the sect's murals, emperors were pictured with halos around their heads, like Christian saints.) Although the sect did not admit women, it was open to slaves and soldiers, who could often advance in the religion's military structure to higher positions than the richer and more powerful members. But it was not an easy religion to follow, as it made great demands on its members. They had to be moral, abstain from sex and divorce, and be brave, fearless, truthful, and scrupulously honest.

Mithraism had many outward similarities with the rising Christianity. Their religious ceremonies included a communion of little bread loaves with an inscribed cross, and water (which was miraculously transformed into wine?). Their full meal was partaken only by the higher grades, perhaps once a month, and included cooked meat, fruit, and wine. The members of the community called each other brother; their leader was called father and wore a ring, a pointed hat (like the Christian miter), and carried a hooked sword (Constantine's cross was also in the shape of a sword with lateral member). The father of fathers who lived in Rome governed their priests. They celebrated the birth of their god on December 25. Their places of worship (Mithraeums) were built underground, like the catacombs in which the Roman Christians hid. Although Mithraism did not accept women, it was closely affiliated with the religion of Isis (another very strong contender against Christianity with women members and priestesses) and Mithraist priests often served there.

Mithraism was closely related to astronomical constellations and it is postulated that the sudden growth of the religion was related to the discovery of the precession of equinoxes. According to the original myth Mithras was born from a rock, naked but holding a knife in his right hand and various other things in his left. Murals of the event show many attendants, which often include shepherds with sheep and goats. According to the religion, the great creation event was Mithras's slaying of the sacred bull. As the bull died, the world came into being and time began (thus Mithras was born before time, and so had always existed)

In the Persian myth, Mithras was incarnated by Anahita, an immaculate virgin once worshipped as a goddess of fertility, who was impregnated from the seed of Zarathustra which had survived in a Persian lake. Interestingly, Paul was born and spent many years in Tarsus, in Cilicia, an area known as a center for Mithraism. But despite occasional allegations, nobody has ever proven that it affected his theology in any way. The similarities, however, between Christian communion and the equivalent Mithraic ceremony remain puzzling.

[4] I am following here Bouyer (1968), starting with p. 79.

[5] Compare with Deut. 8:10: "When you have eaten and are full, then you shall bless the Lord your God for the good country he has given you."

[6] http://www.globalprojects.org/equine/Didache.htm or Holmes (1999, 261).

[7] See note 4 in chapter 22 for Paul's only two references to Jesus words.

[8] The letter of Ignatius to the Smyrnaeans, 6, from Holmes (1999, 189).

26—*WOMEN*

[1] Nobody ever pointed out that Eve was fooled by a cunning snake, perhaps by Satan himself, whereas Adam was fooled by a simple, guileless woman. Who showed less reasoning power?

[2] Later rabbinic law forbade women from having intercourse during the first seven days after a period of ritual uncleanness.

[3] The *Shema* was the basic Hebrew statement of belief proclaimed twice every day: "Hear [*Shema*] O Israel, the Lord our God, the Lord is one, and you shall love the Lord your God with all your heart, and with all your soul, and with all your might." The passage then continues with the importance of the Ten Commandments (Sanders 1992, 195-196).

[4] *Phylacteries* (Greek for talisman, or *teffilin* in Hebrew) are the devices used to strap key portions of the Bible on the left arm and forehead of a person during prayer, as well as on the doorposts of the house. They are called for in Exodus 13:16; and Deuteronomy 6:8 and 11: 18, 20.

[5] For seven days in autumn, during the festival of the Booths, all Hebrews were required to live outside (or on top) of their houses, in areas covered by tree branches, to commemorate the travels of their ancestors through the desert. An additional eighth day was declared a festival day during which no work was done. These requirements are called for in Leviticus 23:33-36 and explained in Nehemiah 8:13-16.

[6] The woman of the household was required to light the candle at the Sabbath dinner. This, however, was considered to be one of the penances for Eve's disobedience.

[7] Spending their time spinning appears to have been something that all women were supposed to be doing during the old days. Even the richest Roman matron, whose slaves did everything, was required to spend her time spinning, just like the king's daughter in the Sleeping Beauty tale.

[8] It was later argued that this direct heredity transmission from father to son was the way that Adam's original sin was passed on to all future generations.

[9] Brown (1998, 10) quoting Aretaeus (a second-century CE physician), *Causes and Symptoms of Chronic Diseases*.

[10] This concept is still used today to oppress women. There are unofficial Islamic courts in London which use the *Shariah* (Islamic law) to settle affairs among Muslims. In cases of divorce the father takes custody of the children after they reach the age of seven years. One of the judges explained the reasoning: "A child comes from the seed of a man. The woman is the soil in which the seed is planted. A man is fully entitled to the fruit of his seed" *Chicago Tribune*, October 1, 2003.

[11] *Patria Potestas* was the power a Roman father had over the persons of his children, grandchildren, and other descendants, by virtue of his paternity.

[12] But see note 16 in chapter 23 for an opposite view.

[13] Luke's gospel implies that women were among Jesus' followers even while he was in Galilee. After his resurrection, the angel at the tomb told the women, "He is not here, but he has been raised. Remember what he said *to you* while he was still in Galilee?" (Luke 24:6, emphasis added).

[14] You will find this position explained at much greater length by Ramon K. Jusino in *Mary Magdalene: Author of the Fourth Gospel*, at www.BelovedDisciple.org

[15] The editing of the Bible continues to this day. Most newer versions, for instance, describe the disciple as "reclining at Jesus' side" (John 13:23) . This removes the intimacy from the words of the gospel and substitutes the normal position assumed in those days by people eating for the more intimate "reclining in the bosom of Jesus" position described in the original. (The King James version says "leaning on Jesus' bosom.")

[16] In the Gospel of Philip we read: "And the companion of the Savior is Mary Magdalene. But Christ loved her more than all the disciples, and used to kiss her often on her mouth. The rest

of the disciples were offended by it, and expressed disapproval. They said to him: 'Why do you love her more than all of us?' The Savior answered and said to them, 'Why do I not love you like her? When a blind man and one who sees are both together in darkness, they are no different from one another. When the light comes, then he who sees will see the light, and he who is blind will remain in darkness.'" Quoted in Barnstone (1984, 92).

In the Gospel of Mary we read: "Peter said to Mary, 'Sister, we know that the Savior loved you more than the rest of the women. Tell us the words of the Savior which you remember—which you know (but) we do not, nor have we heard them.' Mary answered and said, 'What is hidden from you I will proclaim to you.'...Levi answered and said to Peter, 'Peter, you have always been hot-tempered. Now I see you contending against the woman like the adversaries. But if the Savior made her worthy, who are you to reject her? Surely the Savior knows her very well. That is why he loved her more than us.'" Quoted from Robinson (1990, 525-527).

[17] According to legend, Mary Magdalene appeared at the court of Emperor Tiberius preaching about Jesus' death and resurrection. When the emperor challenged her by claiming that one could not rise from the dead any more than an egg could turn red, Mary picked up an egg and it immediately turned red in her hands. To this day Orthodox Christians die their eggs red for Easter.

[18] From *Who Framed Mary Magdalene?* by Heidi Schlumpf, at www.uscatholic.org/2000/04/cov0004.tm

[19] Some people believe that this is not an original Pauline statement but that it was inserted later to combat the Montanists, who believed that the Holy Spirit spoke through both men and women. Montanism arose a hundred years after Paul's letter so, if the assertion were true, it would indicate that our records could have been seamlessly corrupted long after they were first written.

[20] In Corinthians Paul again differentiates between men and women, and adds a very confusing instruction regarding women's

hair: "A man should not cover his head because he is the image and glory of God; but the woman is the glory of man. For man did not come from the woman, but woman from man; neither was man created for woman; but the woman for man; for this reason the woman should have power on her head because of the angels" (1 Cor. 11:7-10).

27—SEX

[1] From the Babylonian Talmud, quoted by Brown (1988, 63).

[2] Armstrong (1993, 77) . This interpretation of the ritual bath is perhaps too romanticized. According to Hebrew Law, the bath would have been necessary in any case in order to wash away the ritual pollution incurred during menstruation. Furthermore, the sex act itself causes additional ritual pollution since it involves fluids exiting the body. As we saw in a previous chapter, the high priest had to be kept awake during the night before his performance at the Temple to prevent nocturnal emissions that would render him unclean.

[3] In John's gospel, Jesus performs his first miracle while attending a wedding feast with his mother and disciples. It is probable, however, that this story was contrived by John to show how Jesus' teachings supplanted those of the Old Testament: his new wine filled the jars that had been used for Hebrew purification rites.

[4] Most modern Bibles translate this a little softer: "Others stay single for the sake of the kingdom of heaven" (CEV) . It appears that many men, endeavoring to become holy, castrated themselves on the basis of this saying.

[5] Many pagan religions, Isis being the most famous one, involved eunuch priests. For whatever reasons, forceful and voluntary castrations in the Roman Empire were sufficiently common for the government to promulgate strict laws against the practice (Kuefler 2001, 32) Interestingly, however, eunuchs rose to many high political positions in the Byzantine Empire, probably because, not having any offspring with which to start their own dynasties, they were judged to be more trustworthy.

[6] Much of what follows in this section comes from Peter Brown's *The Body and Society*.

[7] A common gnostic belief was that God had originally created a combined man and woman: "God created man in his own image, in the image of God he created him; male and female he created them" (Gen. 1:27). The Fall rendered them asunder. Somehow the soul and the spirit separated and became housed in matter. The goal is to allow the soul and spirit to become one again, discarding the matter. We thus read in gnostic writings: "When you make the male and the female one and the same, so that the male not be male, nor the female female…then you will enter the kingdom of heaven" (Gospel of Thomas 22). Also: "When Eve was still in Adam, death did not exist. When she was separated from him, death came into being. If he again becomes complete and attains his former self, death will be no more" (Gospel of Philip).

[8] The Christians were not alone in looking askance at the enjoyment of sexual intercourse. "Already at the end of the second century, Marcus Aurelius had written that 'sins of desire,' in which pleasure predominates, indicates a more self-indulgent and womanish disposition" (Kuefler 2001, 95).

[9] In 385 CE Pope Siriclus informed the bishop of Tarragona that he considered it lechery, indeed a crime, for priests to have intercourse with their wives and beget children after ordination (by Uta Ranke-Heinmann in *Return to Genesis of Eden*, at http://www.dhushara. tripod. com/book/rebirth/comment/eunuchs.htm).

[10] Manliness had always been considered a prime virtue in the Roman Empire. But by the third century the manly Roman soldier, who was clearly superior to all others, was quickly disappearing. To replace the soldier's manliness "Latin Christian writers argued that the renunciation of sex and marriage was a sign of perfection and therefore a sign of manliness" (Keufler 2001, 161).

[11] Many similar stories appeared in later centuries. In all of them, female pursuit of holiness was attained by giving up the feminine identity, renouncing sex and marriage, and dressing in manly clothing.

[12] The exact percentage depends on interpretation. See http://www.chimpanzoo.org/african notecards/chapter 16.html. Also http://www.primate.com/chimp/genushomo.html.

[13] "Terry Gray, an elder in the Orthodox Presbyterian Church, [was charged with heresy] because he taught evolution as part of his duties as a professor of biochemistry at Calvin College in Michigan. 'I was charged with holding the view that Adam had primate ancestors,' Gray said" (*Chicago Tribune*, May 12, 2003).

[14] From Kelly (1978, 361).

[15] The Bible (Gen. 9:20-27) describes how after the flood Noah planted a vineyard, made some wine, and got drunk. He fell asleep naked, but his son Ham saw him and told his two brothers, who walked into the tent backwards carrying their father's coat and covered him. "When Noah woke up from his drunkenness and learned what his youngest son had done to him" he cursed him, saying that Ham's son Canaan would become a slave to the others. This rather strange story is usually interpreted as Ham raping Noah to prove his superiority and so become the head of the family [see Gagnon (2001, 63-71)] . But did this really happen? Of course not, since there was no historical Noah and his three sons were imaginary persons named after the nationalities of the people who lived in the same geographical area as the Hebrews.

[16] The priests and priestesses in the pagan temples were surrogates for the deities. So when men went to the temple and paid money to have intercourse with them, they placed the gods, through their representatives, in a sexually submissive position and the visitors could thus expect to have their requests satisfied. It is worth noting that the Septuagint and the King James Version use the word "sodomite" alone. Later Bible versions substituted "cult prostitutes" indicating the translators' opinions that sodomy was related to religious, not social practices. Only the amplified Bible gives both expressions.

[17] One possible exception is the story between David and Saul's son Jonathan: "The soul of Jonathan was knit with the soul of David, and Jonathan loved him as his own soul" (1 Sam. 18:1).

Here there is no mention of sex but of intense bonding between the two boys and nothing derogatory was said about it.

[18] There is one exception, discussed by Clement of Alexandria, who wrote a letter regarding a so-called Secret Gospel of Mark, describing a possibly compromising situation involving Jesus. Clement accepted some statements in this gospel and rejected others. Some scholars, however, have doubted the existence of this version of Mark's gospel. The interested reader can find more information on the Internet by doing a Google search for "secret Gospel of Mark." The discovery and later disappearance of this letter is also discussed at some length in Ehran (2003, 67-89).

[19] Anyone who has dogs, however, can attest to the fact that they often try to mount each other in an effort to establish their relative status. This is completely unrelated to their sex or to whether they have been neutered or not; it is not a sex act but one of domination.

[20] Philo considered that a man who has sex with a barren woman is committing an act against nature because he wastes his seed: "Those persons who make an art of quenching the life of the seed as it drops, stand confessed as the enemies of nature" [from his Special Laws 111.36 as quoted by Scruggs (1983, 89)].

[21] In 1973 the American Psychiatric Association removed homosexuality from its official *Diagnostics and Statistical Manual*, signifying the end of homosexuality's official status as a disease (www.theatlantic.com/issues/junbuvr2.htm).

28—REACHING FOR A NEW CHRISTIANITY

[1] Nehushtan, the bronze serpent that Moses built to protect the Hebrews from snake venom in the desert, was a symbol, a pointer to God's healing power. In later years the Hebrews attributed the healing power to the construct itself, thus turning it into an idol.

[2] Church (1999, 50).

[3] Church (1999, 51).

[4] Spong (1998, 220-224).

[5] Sprague (2002, 37).

[6] Sprague (2002, 40).

[7] We, as well as the world around us, consist of molecules, which consist of atoms. They in turn are composed of electrons, protons and neutrons. Protons and neutrons, in turn, are made up of quarks. So everything is constructed out of electrons and quarks, which are not little balls of matter, but instead packets of energy. Which means that we, as well as everything around us, are really made up of energy. Compare with Thomas 50 on page 279.

[8] In 1 Cor. 15:40, 44.

[9] Jung believed that the goal of each psyche is to remain differentiated in the existing world. But a psyche could be unsuccessful and dissolve in the nothingness of the Pleroma. (See Hoeller p. 70.)

[10] When a Federal judge recently displayed the Ten Commandments in his courtroom, the American Civil Liberties Union forced him to take them down because they were considered to be a religious (and thus forbidden) teaching, rather than a universal truth.

[11] The Moslem Law implements this exact command by ordering that a thief's hand should be cut off.

[12] There is a story I like that illustrates this point. Joe was a fervent God-fearing farmer who lived by the bank of a river down South. One day the ranger drove by to tell him that the dam upstream had broken and he should leave because the area was going to flood. "Don't worry," Joe said, "God will take care of me." Sure enough, the river left its banks, and the first floor of Joe's house was flooded. While Joe was watching from his second-floor window, the emergency response boat came by. "Jump in," the guy told him, "there is no stopping this water." "Don't worry," Joe replied, "God will take care of me." The waters rose up higher and Joe had to find refuge by climb-

ing onto the roof of his house. A helicopter came by and dropped a ladder. "Climb up," yelled the pilot, "the waters are still rising." "Don't worry," shouted Joe, "God will save me." The helicopter flew away, the waters rose even more, and Joe drowned. He went to heaven, and upon seeing God he started to complain. "I don't understand! I have believed and trusted you in all my life, and when I really needed you, you let me down!" "Let you down," answered the Lord, "I sent you a ranger, I sent you a boat, I sent you a helicopter. You were just too blind to see me, and too deaf to hear me."

ANNOTATED BIBLIOGRAPHY

BIBLES

AMP - *The Amplified Bible*. Grand Rapids, Mich.: Zondervan Publishing House, 1987. The unique feature of this Bible is that it lists all possible translation of doubtful Jewish or Greek words.

CEV - *Holy Bible with Deuterocanonicals/Apocryphals*. Contemporary English Version. New York: American Bible Society, 1975. Very readable edition, but in my opinion the translation often destroys the real meaning and sense.

JER - *The Jerusalem Bible: Reader's Edition*. Garden City, N.Y.: Doubleday & Company, Inc. A very readable and accurate translation. Has been recently replaced by a revised edition.

KJ - *The Holy Bible, King James Version, Translated out of the Original Tongues*. New York: American Bible Society. This had been the standard version for the last few centuries. Unless otherwise noted in the text, all quotations are from this book, but with modernized pronouns, grammatical structure, and occasional words.

NAB - *The New American Bible*. Grand Rapids, Mich.: Catholic World Press, 1987. Approved by the Catholic Church, this is my favorite Bible partly because of the superior translation, but more important because of the excellent notes that accompany the readings.

Berry, George Ricker. *Greek to English Interlinear New Testament, King James Version*. Grand Rapids, Mich.: World Publishing, 1981. Presents the original Greek text, in which all New Testament books have been written, above the English translation.

Brenton, Lancelot C.L., ed. *The Septuagint with Apocrypha: Greek and English*. Peabody, Mass.: Hendrickson Publishers. It is generally accepted now that the original Septuagint Greek translation of the Hebrew writings contains errors of both translation and omission, but it had been used by the Early Christian Fathers and Paul, and remains the approved version in the Orthodox Church.

DICTIONARIES

A Dictionary of Early Christian Beliefs. Berkot, David W., ed. Peabody, Mass.: Hendrickson Publishers, Inc., 1998. Discusses more than 700 subjects and people. Nice to have around for answering questions when they come up.

A Greek-English Dictionary, 3d. edition. Revised and edited by Frederick William Danker. Chicago: University of Chicago Press, 2000. Very helpful to those who want to read the original Greek in the scriptures and compare it with the various translations.

The Anchor Bible Dictionary. Freedman, David Noel, editor in chief. New York: Doubleday, 1992. A huge work that covers everything.

Vine, W. E. *Vine's Concise Dictionary of Bible Words*. Nashville, Tenn.: Thomas Nelson, Inc., 1999. Discusses the origin of basic words used in Bible translations. Interesting.

BOOKS

Aquilina, Michael. *The Mass of the Early Christians*. Huntington, Ind.: Our Sunday Visitor, Inc., 2001. A history of mass and the Eucharist during the first centuries. Small, readable, good.

Armstrong, Karen. *A History of God: The 4000-Year Quest of Judaism, Christianity, and Islam.* New York: Ballantine Books, 1993. A bestseller, but I prefer the one written by Miles.

_________. *Islam: A Short History.* New York: Modern Library, 2000. A brief history of Islam, from its birth to the present. Worthwhile, although its brevity and foreign words render it somewhat difficult to read. Some people have accused her of being overly pro-Islam.

Atwater, P.M.H. with David H. Morgan. *The Complete Idiot's Guide to Near Death Experiences.* Indianapolis, Ind.: Macmillan USA, 2000. The book starts reasonably, but as it progresses, the reader's credibility becomes overstretched.

Barnstone, Willis, ed. *The Other Bible.* San Francisco: HarperCollins Publishers, 1984. A reference-type book, containing numerous Hebrew, and Gnostic writings you will not find anywhere else. For the serious student or someone who wants to impress his friends. Includes the Gospel of Thomas.

Bechtel, William, and Adele Abrahamsen. *Connectionism and the Mind.* 2d ed. Malden, Mass.: Blackwell Publishers, 2002. How computer scientists have simulated the workings of the mind. For the very technically inclined.

Berne, Eric. *Transactional Analysis in Psychotherapy.* New York: Grove Press, 1961. The classic work on the subject. Moderately difficult.

_________. *What Do You Say After You Say Hello? The Psychology of Human Destiny.* New York: Bantam Books, 1972. Easy. Proposes that we arrange to live our lives to fulfill the script we wrote in our early childhood.

Berquist, John L. *Judaism in Persia's Shadow: A Social and Historical Approach.* Minneapolis: Fortress Press, 1995.

The historically minded reader will find it interesting and informative.

Blenkisopp, Joseph. *The Pentateuch: An Introduction to the First Five Books of the Bible*. New York: Doubleday, 1992. Good.

Bouyer, Louis. *Eucharist: Theology and Spirituality of the Eucharistic Prayer.* Translated by Charles Underhill Quinn. London: University of Notre Dame Press, 1968. The dependence of the Eucharist and the mass on the Jewish Berakoth, and how it developed throughout the Roman Empire through the centuries. Very good, but taxing.

Brooten, Bernadette J. *Love Between Women: Early Christian Responses to Female Homoeroticism*. Chicago: University of Chicago Press, 1996. Some good sections, but many boring and, I thought, irrelevant. Long.

Brown, Harold O. J. *Heresies: Heresy and Orthodoxy in the History of the Church*. Grand Rapids, Mich.: Baker Books, 1984; Peabody, Mass.: Hendrickson Publishers, 1988. An excellent, though very lengthy discussion of Christian thinking through the early centuries. Most people will use it to look up items, rather than read it through.

Brown, Peter. *The Body and Society: Men, Women, and Sexual Renunciation in Early Christianity.* New York: Columbia University Press, 1988. An excellent discussion of the early Church's attitude towards sex, and its results. Very lengthy, but good.

Brown, Raymond E. *The Community of the Beloved Disciple*. New York: Paulist Press, 1979. An excellent little book on John's Church and gospel.

Brown, Walter. *In the Beginning: Compelling Evidence for Creation and the Flood*. 7th ed. Phoenix: Center for Scientific Creation, 2001. Beautiful work by a creationist to try to prove that world was indeed created in seven days. His pseudo-scientific explanations are pure drivel.

Brueggemann, Walter, William C. Placher, and Brian K. Blount. *Struggling with Scripture*. Louisville, Ky.: Westminster John Knox Press, 2002. Three short essays originally presented in a conference titled Biblical Authority and the Church, question the literal interpretation of the Bible and its injunctions against homosexuality. Good.

Bynum, Caroline Walker. *The Resurrection of the Body in Western Christianity, 200-1336*. New York: Columbia University Press, 1995. The best (perhaps the only) book in the field describing what people thought throughout the centuries regarding the physical resurrection of the body at the second coming. For the student.

Cahill, Thomas. *Desire of the Everlasting Hills: The World Before and After Jesus*. New York: Random House, 1999. A bestseller history of Jesus and his followers as shown in the gospels. Easy, pleasant read.

Cantarella, Eva. *Pandora's Daughters: The Role and Status of Women in Greek and Roman Antiquity*. Baltimore: Johns Hopkins University Press, 1987. The position of women in ancient Greece and Rome.

Carter, Rita. *Exploring Consciousness*. Los Angeles: University of California Press, 2002. Extremely interesting for those who care about these things. What our senses tell us does not correspond to reality, and what this means. Moderately difficult.

Charlesworth James H., ed. *The Old Testament Pseudepigrapha: Apocalyptic Literature and Testaments*. New York: Doubleday, 1983. A reference book for the scholar. Includes Enoch.

Church, F. Forrester, *The Essential Tillich: An Anthology of the Writings of Paul Tillich*. Chicago: The Chicago University Press, 1999. One of the first Christian philosophers to become widely quoted by clergy. The average person will find that his writings are difficult to read and impossible to understand.

Cohen, Abraham. *Everyman's Talmud: The Major Teachings of the Rabbinic Sages*. New York: Schocken Books, 1975. A reference book.

Collins, John J. *The Apocalyptic Imagination: An Introduction to Jewish Apocalyptic Literature*. 2d ed. Grand Rapids, Mich.: William Eerdmans Publishing Co, 1998. One of the few studies in this field. It is moderately difficult and sometimes drags, tiring the reader.

Collins, John J. *The Scepter and the Star: The Messiahs of the Dead Sea Scrolls and Other Ancient Literature*. New York: Doubleday, 1995. For the student interested in this restricted subject.

Cooper, D. Jason. *Mithras: Mysteries and Initiation Rediscovered*. York Beach, Maine: Samuel Weiser, 1996. A description of one of Christianity's strongest competitor.

Cytowic, Richard E. *The Man Who Tasted Shapes*. New York: G.P. Putnam's Sons, 1993; Cambridge, Mass.: MIT Press, 1998. Astonishing facts about the human mind, brain, and limbic system, and how these affect our conception of reality. Only a little difficult.

Davis, Kenneth C. *Don't Know Much About the Bible*. New York: Eagle Brook/Morrow, 1998. An extremely easy-to-read book, which looks at some easy-to-miss stories in the Bible. Lots of fun.

Davis, Stephen T., Kendall, Daniel SJ, O'Collins, Gerald SJ, eds. *The Resurrection: An Interdisciplinary Symposium on the Resurrection of Jesus*. Oxford: University Press, 1998. A series of talks on the subject. Interesting.

Delsemme, Armand. *Our Cosmic Origins: From the Big Bang to the Emergence of Life and Intelligence*. Cambridge: University Press, 1998. Tries to explain creation in simple terms.

Dunn, James D. G., ed. *Jews and Christians: The Parting of the Ways AD 70 to 135*. Grand Rapids, Mich.: William B.

Eerdmans Publishing Company, 1999. A series of papers first presented in 1989 on early Christianity. A serious student will find them very interesting, and in general they are not too difficult too read.

_________. *Paul and the Mosaic Law.* Grand Rapids, Mich.: William B. Eerdmans Publishing Co., 2001. Detailed study of some points in Paul's theology. For scholars.

Ehrman, Bart D. *The Orthodox Corruption of Scripture.* New York: Oxford University Press, 1993. Instances where Bible statements were altered in order to support specific beliefs. For the serious student.

_________. *Jesus: Apocalyptic Prophet of the New Millennium.* New York: Oxford University Press, 1999. A short, authoritative, yet very readable discussion of the history of the gospels, leading to the author's belief that Jesus was a mystic who expected that the rules governing the existing world would end within his lifetime, to be replaced by the God-ordained rules that he himself espoused.

_________. *The New Testament: A Historic Introduction to the Early Christian Writings.* 2d ed. New York: Oxford University Press, 2000. An outstanding, in-depth study of the New Testament writings. Excellently written; will not bore the reader, but it is a very long textbook. The best choice for the seriously interested reader, but not for everybody.

_________. *Lost Christianities: The Battle for Scripture and the Faiths We Never Knew.* New York: Oxford University Press, 2003. A vivid description of the battles among the adversarial faiths of the early years, but not much regarding them, since little is known. Some fascinating chapters regarding the way that the Scriptures were manipulated to present specific viewpoints.

Elbert, Jerome E. *Are Souls Real?* Amherst, Mass.: Prometheus Books, 2000. The author concludes that science says no.

Farmelo, Graham, ed. *It Must be Beautiful: Great Equations of Modern Science.* London: Granta Books, 2002. Extraordinary scientific discoveries reveal an unexpectedly incredible world. For the scientifically inclined. Difficult.

Ferguson, Everett. *Backgrounds of Early Christianity.* 2d ed. Grand Rapids, Mich.: William B. Eerdmans Publishing, 1993. An excellent treatise on the social world of the day. A must-read for the very serious student, but a long textbook.

Finegan, Jack. *Myth and Mystery: An Introduction to the Pagan Religions of the Biblical World.* Grand Rapids, Mich.: Baker Books, 1989. A brief discussion of some of the Judeo-Christian religion's competitors.

Finkelstein, Israel, and Neil Asher Silberman. *The Bible Unearthed: Archaeology's New Vision of Ancient Israel and The Origin of Its Sacred Texts.* New York: The Free Press, 2001. One of the most fascinating, but sometimes dragging, archaeological books you may read. Proposes a completely different time scale for the Solomonic era. If he is right, a part of the Bible needs to be rewritten.

Fowler, Robert M., *Let the Reader Understand: Reader-Response Criticism and the Gospel of Mark.* Harrisburg, PA.: Trinity Press International, 1996. Presents Mark as an erudite author, but is he right? Moderately difficult.

Fox, Matthew, and Rupert Sheldrake. *The Physics of Angels: Exploring the Realm Where Science and Spirit Meet.* New York: HarperCollins, 1996. An avant-garde physicist and a theologian write a book about their own interpretations of the subject. I did not finish it, and probably you will not either.

Friedman, Richard Elliott. *Who Wrote the Bible?* New York: HarpersCollins Publishers, 1987. My favorite detective story on the writers of the Old Testament. Very readable. A must.

———. *The Hidden Face of God.* New York: HarpersCollins Publishers, 1995. Why God's appearances decrease as we move further away from Genesis. Read this instead of a novel.

Funk, Robert W., Hoover, Roy W., and The Jesus Seminar. *The Five Gospels: The Search for the Authentic Words of Jesus*. New York: HarperCollins Publishers, 1993. The famous (or infamous) work of the Jesus Seminar. Some self-selected scholars vote on what Jesus said and did not say.

Gagnon, Robert A. J. *The Bible and Homosexual Practice: Texts and Hermeneutics*. Nashville, Tenn.: Abingdon Press, 2001. An in-depth reading of the Bible (with preconceived opinions) convinces the author that homosexuality is immoral.

Gazzaniga, Michael S. *Nature's Mind: The Biological Roots of Thinking, Emotions, Sexuality, Language, and Intelligence*. New York: Basic Books, 1992. For the very scientifically inclined.

Goldstein, Jonathan. *Peoples of an Almighty God: Competing Religions in the Ancient World*. New York: Doubleday, 2002. Nations' misfortunes are punishment from God for their misdeeds. Daniel was written in Babylon and adopted by the Judeans. For the expert.

Gomes, Peter J. *The Good Book: Reading the Bible with Mind and Heart*. New York: William Morrow and Company, 1996. Doing so, says the author, changes the Bible from exclusive to inclusive. Tries to justify homosexuality. Interesting.

Goulder, Michael. *St. Paul versus St. Peter: A Tale of Two Missions*. Louisville, Ky.: Westminster John Knox Press, 1994. An excellent and very interesting study of the early Christian thinking, and the different paths followed. Easy to read.

Greene, Brian. *The Elegant Universe*. New York: Vintage Books, 2003. A good, brief description of relativity and quantum mechanics is followed by a lengthy description of how extremely short energy strings, vibrating in ten dimensions, may be the basic building blocks of the matter. The very difficult ideas thwart the author's best efforts to render it understandable. Only for the scientifically well-prepared reader.

Griffith-Jones, Robin. *The Four Witnesses: The Rebel, the Rabbi, the Chronicler, and the Mystic*. San Francisco: HarperSanFrancisco, 2000. A fairly good description of the various Christian communities and beliefs as evidenced through the four gospels. Easy but long.

Guggenheim, Bill, and Judy Guggenheim. *Hello from Heaven*. New York: Bantam Books, 1997. Some results from a seven-year project investigating people who had seen visions, or otherwise communicated with dead people. Too many people were involved over too long a time for everything to have been made up. Inspiring.

Guth, Alan H. *The Inflationary Universe: The Quest for a New Theory of Cosmic Origins*. Cambridge: Perseus Books, 1997. The latest theory regarding the Big Bang. For the very scientifically inclined.

Hanson, K. C., and Douglas E. Oakman. *Palestine in the Time of Jesus*. Minneapolis: Fortress Press, 1998. Presents a lot of interesting information. For the historically minded reader.

Harris, Thomas A. *I'm OK—You're OK*. New York: Avon Books, 1973. Another popular book based on transactional analysis. Good.

Hayes, John H. and Sara R. Mandell. *The Jewish People in Classical Antiquity: From Alexander to Bar Kochba*. Louisville, Ky.: Westminster John Knox Press, 1998. The historically minded reader will find it enthralling.

Helminiak, Daniel A. *What the Bible Really Says about Homosexuality*. Millennium Edition, updated and expanded. San Francisco: Alamo Square Press, 2000. A small booklet presents the standard refutations of all anti-homosexual references in the bible.

Hengel, Martin. *Crucifixion: In the Ancient World and the Folly of the Message of the Cross*. Philadelphia: Fortress Press, 1977. An interesting, very small book covering more than you ever wanted to know about this subject.

Hengel, Martin and Anna Maria Schwemer. *Paul Between Damascus and Antioch: The Unknown Years*. Louisville, Ky.: Westminster John Knox Press, 1997. An illuminating story of Paul's early Christian years. Moderately difficult.

Hoeller, Stephan A. *The Gnostic Jung and the Seven Sermons to the Dead*. Wheaton, Ill., Quest Books, 1982. An extraordinary book. Moderately difficult.

Hoffman, Lawrence A. *Covenant of Blood: Circumcision and Gender in Rabbinic Judaism*. Chicago: University of Chicago Press, 1996. Interesting information on circumcision, blessings, wine, and women.

Holmes, Michael W., ed. *The Apostolic Fathers*. rev. Grand Rapids, Mich.: Baker Books, 1999. Texts of some of the Church Fathers' writings, also in Greek. Includes Didache, Ignatius, Pappias, Clement.

Horsley, Richard A., with John S. Hanson. *Bandits, Prophets, and Messiahs*. Harrisburg, Pa.: Trinity Press International, 1999. A lot of interesting in-depth historical information. Moderate difficulty.

Hughes, Philip. *The Church in Crisis: A History of the General Councils: AD 325 through AD 1870*. http://www.christusrex.org/www1/CDHN/coun0.html. An excellent, very detailed description of the 22 councils, what they accomplished, and how.

Hultgren, Arland J. *The Parables of Jesus: A Commentary*. Grand Rapids, Mich.: William B. Eerdmans Publishing Company, 2000. The best of the three parable-related texts listed in this bibliography.

Ilan, Tal. *Jewish Women in Greco-Roman Palestine*. Peabody, Mass.: Hendrickson Publishers, 1996. All the roles the women played during that period. Very thorough.

Jeffers, James S. *The Greco-Roman World of the New Testament Era: Exploring the Backgrounds of Early Christianity*. Downers Grove, Ill.: InterVarsity Press, 1999. A social study of the period. Moderate difficulty.

Jeremias, Joachim. *The Parables of Jesus*. New York: Charles Scribner's Sons, 1954. A classic on the subject. Too professorial. For the serious student.

________. *Jerusalem in the Time of Jesus*. Philadelphia: Fortress Press, 1969. A classic study of the political, social, and religious practices of the Judeans. Contains a great deal of information that is quoted by later writers. Average difficulty.

Johnson, Luke Timothy. *The Writings of the New Testament: An Interpretation*. Revised edition. Minneapolis: Fortress Press, 1999. Includes 350 pages on Paul's and the other letters in the Bible. For the dedicated student.

Johnston, Victor S. *Why We Feel: The Science of Human Emotions*. Oxford: Perseus Books, 1999. Because it helps our survival. For the scientifically inclined.

Josephus. *The Complete Works of Josephus,* revised and expanded, translated by William Whiston, commentary by Paul L. Maier. Grand Rapids, Mich.: Kregel Publications, 1999. The world from creation to the fall of Jerusalem as seen through the eyes of a turncoat Pharisee. A reference book.

Kelly, J. N. D. *Early Christian Doctrines,* rev. ed. San Francisco: HarperSanFrancisco, 1978. An excellent description of the differences in Christian thinking during the first few centuries. For the serious student.

Kessler, William Thomas. *Peter as the First Witness of the Risen Lord*. Rome: Gregorian University Press, 1988. An attempt to explain how Peter could have been the first person to see the risen Jesus.

Kovach, Sue. *Hidden Files: Law Enforcement's True Case Stories of the Unexplained and Paranormal*. Chicago: Contemporary Books, 1976. Remarkable stories, but are they true?

Kraemer, Ross Sheppard. *Her Share of the Blessings: Women's Religion among Pagans, Jews, and Christians in the Greco-Roman World*. New York: Oxford University Press, 1992. Very good. Presents some new information.

Kuefler, Mathew. *The Manly Eunuch: Masculinity, Gender Ambiguity, and Christian Ideology in Late Antiquity*. Chicago: University of Chicago Press, 2001. How eunuchs came to represent the Christian ideal. The book is better than its title would suggest. But long.

Kugel, James L. *The Bible As It Was*. Cambridge: Harvard University Press, 1997. Presents the interpretations of the Old Testament by latter-day Hebrews. Very interesting, for those interested.

LeDoux, Joseph. *The Emotional Brain: The Mysterious Underpinnings of Emotional Life*. New York: Simon & Schuster, 1996. How emotions, compared to feelings, affect our actions. For the scientifically inclined.

Maccoby, Hyam. *Revolution in Judea: Jesus and the Jewish Resistance*. New York: Taplinger Publishing Co., 1980. Proposes that Jesus was following the Book of Zechariah during his last week in Jerusalem, and expected that God would intervene to bring in his Kingdom. Very interesting, but out of print.

MacMullen, Ramsay, *Christianizing the Roman Empire AD 100-400*. New Haven: Yale University Press, 1984. A look at the growing Christian movement through the eyes of the pagans, and why they converted. For the interested student of the period.

Maier, Paul L. *Eusebius: The Church History*. Grand Rapids, Mich.: Kregel Publications, 1999. A beautifully illustrated book of the writings of the first historian of the Church. For reference.

Martin, Malachi. *Hostage to the Devil: The Possession and Exorcism of Five Contemporary Americans*. New York: Reader's Digest, 1976; San Francisco: HarperSanFrancisco, 1992. Interesting, but is it true?

Mazza, Enrico. *The Celebration of the Eucharist: The Origin of the Rite and the Development of Its Interpretation*. Translated by Mathew J. O'Connell. Collegeville, Minn.: The Liturgical Press, 1998. More detailed information about the Eucharistic mass than the average non-seminarian would care to learn. Strictly the Catholic view. Read Bouyer instead.

Meeks, Wayne A. *The First Urban Christians*. New Haven: Yale University Press, 1983. Presents many interesting points, but will probably tire the average reader.

Mendenhall, George E. *Ancient Israel's Faith and History: An Introduction to the Bible in Context*. Edited by Gary A. Herion. Louisville, Ky.: Westminster John Knox Press, 2001. A somewhat iconoclastic description of the period. The book is easy to read, reasonable in size, and I recommend it.

Metzger, Bruce M. 3d ed., *The Text of the New Testament: Its Transmission, Corruption, and Restoration*. New York: Oxford University Press, 1992. The process whereby Bible writings were corrupted and sometimes restored. For the very serious student.

Miles, Jack. *God: A Biography*. New York: Vintage Books, 1996. A bestseller: looks at God as an actor in the book called Bible. A lot of fun to read.

Mindell, Arnold. *Quantum Mind: The Edge Between Physics and Psychology*. Portland: Lao Tse Press, 2000; distributed by Words Distributing Company, Oakland, Calif. Attempts to relate physics and psyche. Not for the average reader.

Moody, Raymond A., Jr. *Life After Life*. New York: HarperSanFrancisco, 2001. The original work describing the near-death experience.

Morris, Richard. *The Universe, The Eleventh Dimension, and Everything: What We Know and How We Know It*. New York: Four Walls Eight Windows, 1999. Creation theories. Slightly difficult.

Morse, Melvin with Paul Perry. *Where God Lives: The Science of Paranormal and How our Brains are Linked to the Universe*. New York: HarperCollins Publishers, 2000. An effort to scientifically analyze miracles and other paranormal experiences. Very readable, but are they stretching the truth?

Mortley, Raoul. *Womanhood: The Feminine in Ancient Hellenism, Gnosticism, Christianity, and Islam*. Sydney: Delacroix Press, 1981. A very small book that belongs in the library of all feminists. Good.

Nigosian, S. A. *The Zoroastrian Faith, Tradition and Modern Research*. Montreal: McGill-Queen's University Press, 1993. A religion that may have contributed many basic concepts to the Judeo-Christian religion. A small and not-so-interesting book.

Page, Sydney U. T. *Powers of Evil: A Biblical Study of Satan and Demons*. Grand Rapids, Mich.: Baker Books, 1995. The appearances of Satan in the Old and New Testament. Interesting.

Pagels, Elaine. *The Gnostic Gospels*. New York: Vintage Books, 1989. Easy and enjoyable reading on the Gnostic writings.

_________. *Beyond Belief: The Secret Gospel of Thomas*. New York: Random House, 2003. Another book on gnostic teaching. Includes a comparison of John's and Thomas' gospels. Easy read.

Peck, M. Scott. *People of the Lie: The Hope for Healing Human Evil.* New York: Simon and Schuster, 1983. A psychiatrist recognizes personalized evil (Satan) in his patients and others.

Pickthall, Muhammad M., transl. *The Glorious Qur'an.* 2d ed. Elmhurst, New York: Tahrike Tarsile Qur'an, 1999. They say that the Qur'an cannot be properly translated, but here is one translation. Includes the Arabic text for those who can read it.

Polkinghorne, John. *Belief in God in an Age of Science.* New Haven: Yale University Press, 1998. Moderately difficult.

Pomeroy, Sarah B. *Godesses, Whores, Wives, and Slaves: Women in Classical Antiquity.* New York: Schocken Books, 1995. The position of women in the societies of Greece from the bronze age to the Christian age. One of the "women's must-read books."

Pritz, Ray A. *Nazarene Jewish Christianity: From the End of the New Testament Until its Disappearance in the Fourth Century.* Jerusalem: The Hebrew University Magnes Press, 1988. Perhaps the only monograph on the subject. For the interested reader only.

Prophet, Elizabeth Clare. *Fallen Angels and the Origin of Sin: Why the Church Fathers Suppressed the Book of Enoch and its Startling Revelations.* Corwin Springs, Mont.: Summit University Press, 2000. The book of Enoch.

Reese, Martin. *Just Six Numbers: The Deep Forces That Shape the Universe.* New York: Basic Books, 2000. How our universe's existence depends on the specific value of six universal parameters. A small book, but only for the technically minded reader.

Reiser, Marius. *Jesus and Judgment: The Eschatological Proclamation in Its Jewish Context,* translated by Linda M.

Maloney. Minneapolis: Fortress Press, 1997. Focuses on the positions regarding final judgment expressed in Old and New Testament writings. Interesting but somewhat heavy.

Ring, Kenneth and Sharon Cooper. *Mindsight: Near-Death and Out-of-Body Experiences in the Blind*. Kearney, Neb.: Morris Publishing, 1999. Do blind people see during their near-death experiences? The authors say yes, after a fashion.

Robinson, James M., ed. *The Nag Hammadi Library*. 3d revised edition. New York: HarperCollins, 1977. The standard translation of the discovered writings, with placeholders for the missing fragments. Contains the Gospel of Thomas. A reference book.

Rusch, William G. *The Trinitarian Controversy*. Philadelphia: Fortress Press, 1980. The first 62 pages contain a brief but concise history, and some letters of the participants. The rest is junk.

Saldarini, Anthony J. *Pharisees, Scribes, and Sadducees in Palestinian Society*. Edinburgh: T&T Clark, 1988. For the inspired student.

_________. *Matthew's Christian-Jewish Community*. Chicago: University of Chicago Press, 1994. An excellent study of Matthew's community. Interesting but moderately difficult.

Sanders. E.P. *Judaism: Practice and belief 63BCE-66CE*. Philadelphia: Trinity Press International, 1992. Good, but moderately difficult.

Sarchie, Ralph, and Lisa Collier Cool. *Beware the Night: A New York Cop Investigates the Supernatural*. New York: St. Martin's Press, 2001. Interesting, but is it true?

Schacter, Daniel L. and Elaine Scarry, eds. *Memory, Brain and Belief*. Cambridge: Harvard University Press, 2000. A series of papers on belief, memory, and identity. Of medium difficulty.

Schwartz, Regina M. *The Curse of Cain: The Violent Legacy of Monotheism*. Chicago: University of Chicago Press, 1977. An interesting theory that restricting what is desirable results in people's aggressive attempts to seek it and not share it.

Scroggs, Robin. *The New Testament and Homosexuality: Contextual Background for Contemporary Debate*. Philadelphia: Fortress Press, 1983. An attempt at an objective analysis of the Bible readings, it concludes that they are not relevant and do not apply to today's problem. Small, readable, and clear.

Seifrid, Mark A. *Christ, our Righteousness: Paul's Theology of Justification*. Downers Grove, Ill.: Intervarsity Press, 2000. An in-depth analysis of Paul's writings; for the scholar.

Sheldrake, Rupert. *The Rebirth of Nature: Science and God*. Rochester, Vt.: Park Street Press, 1991. For the mystically inclined reader. Most people will find it drags.

_________. *A New Science of Life: The Hypothesis of Morphic Resonance*. Rochester, Vt.: Park Street Press, 1995. A unique theory of creation and of the real world around us.

Shermer, Michael. *How We Believe: The Search for God in an Age of Science*. New York: W. H. Freeman and Co., 2000. The director of the Skeptics Society tries to explain how we fool ourselves to believe what we do. Interesting, but not always correct.

Siker, Jeffrey S., editor. *Homosexuality in the Church: Both Sides of the Debate*. Louisville, Ky.: Westminster Knox Press, 1994. A collection of essays looking at various aspects of the subject. Good.

Spong, John Shelby. *Why Christianity Must Change or Die: A Bishop Speaks to Believers in Exile*. New York: HarperCollins, 1998. An Episcopalian bishop tries to explain how he can remain a Christian while he rejects all Christian beliefs. Very popular in liberal church circles, but I find it hypocritical and I personally detest it.

_________. *A New Christianity for a New World.* New York: HarperCollins, 2001. A more tolerable version of his previous book, written after his retirement.

Sprague, C. Joseph, *Affirmations of a Dissenter.* Nashville, Tenn.: Abingdon Press, 2002. The non-Christian philosophy of a Methodist bishop. A small booklet.

Stegemann, Ekkehard W., and Wolfgang Stegemann. *The Jesus Movement: A Social History of the First Century.* Minneapolis: Fortress Press, 1995. Excellent professional treatment, but its moderate difficulty and great length will bore the average reader.

Stein, Robert H. *The Method and Message of Jesus' Teachings,* rev. ed. Louisville, Ky.: Westminster John Knox Press, 1994. Excellent. For more serious students.

Steiner, Claude M. *Scripts People Play: Transactional Analysis of Life Scripts.* New York: Bantam Books, 1974. Examples of people living scripts created when they were young. Interesting.

Stoyanov, Yuri. *The Other God: Dualist Religions from Antiquity to the Cathar Heresy.* New Haven: Yale Nota Bene, 2000. The history of God-and-Devil dualist thought from antiquity to modern days. Good, but the reader will eventually get tired.

Talbot, Michael. *The Holographic Universe.* New York: HarperCollins Publishers, 1991. A unique theory about how everything and everybody are interconnected. Not too difficult.

Tart, Charles T., ed. *Body Mind Spirit: Exploring the Parapsychology of Spirituality.* Charlottesville, Va.: Hampton Roads Publishing, 1997. Some essays for those willing to examine unconventional ideas. Difficult reading.

Theissen, Gerd. *The Gospels in Context: Social and Political History in the Synoptic Tradition.* Translated by Linda M. Maloney. Minneapolis: Fortress Press, 1991. A discussion of how various socio-political events are represented in the gospels. Interesting, but can become tedious.

________. *The Religion of the Earliest Churches*. Minneapolis, Fortress Press, 1999. For the scholar.

Tomlin, Graham. *The Power of the Cross: Theology and the Death of Christ in Paul, Luther and Pascal*. Carlislc, Cumbria, U.K.: Paternoster Press, 1999. The early Christian Church had no theology regarding the Cross. Paul used it to the Corinthians to correct their behavior, not their theology. Medium heavy.

VandenKamp, James C. *An Introduction to Early Judaism*. Grand Rapids, Mich.: William B. Eerdmans Publishing Company, 2001. Jewish literature between 200 BCE and 100 CE.

Vermes, Geza. *The Changing Faces of Jesus*. New York: Penguin Compass, 2002. A Jew, turned Catholic, turned Jew, examines how the various New Testament Books describe Jesus, who in his opinion was just a charismatic Hassid. Very readable, with much worthwhile content.

Vincent, Ken R. *Visions of God from the Near Death Experience*. Burdett, New York: Larson Publications, 1994. A very small booklet with quotations from people who have undergone near-death experiences.

Walsch, Neale Donald. *Moments of Grace*. Charlottesville, Va.: Hampton Roads Publishing Company, 2001. Actual stories of miracles happening in everyday life. True?

Wilson, A. N. *Jesus: A Life*. New York: Ballantine Books, 1992. One of the better so-called historical studies of Jesus. Very good, but presents many assumptions as probable facts. Easy and pleasant read.

________. *Paul: The Mind of the Apostle*. New York: W. W. Norton and Company, 1997. An excellent biography of the inventor of Christianity.

Wilson, Colin. *After Life: Survival of the Soul.* 2d ed. St. Paul, Minn.: Llewelyn Publications, 2000. Remarkable stories, but are they true?

Wilson, Timothy D. *Strangers to Ourselves: Discovering the Adaptive Unconscious.* Cambridge, Mass.: The Belknap Press of Harvard University Press, 2002. Most of our learning during life takes place unconsciously, and although we use it during our decision process it remains consciously inaccessible. A small, very interesting book.

Witherington III, Ben. *Women and Genesis of Christianity.* Cambridge: Cambridge University Press, 1990. Examines all the references to women in the New Testament. Good.

Witt, R. E. *Isis in the Ancient World.* Baltimore: Johns Hopkins University Press, 1971. One of the strongest competitors to the Christian religion until the days of Constantine.

Yamauchi, Edwin M. *Persia and the Bible.* Grand Rapids, Mich.: Baker Books, 1996. Interesting history, but eventually the reader may get bored.

Young, Brad H. *The Parables: Jewish Tradition and Christian Interpretation.* Peabody, Mass.: Hendrickson Publishers, 1998. Compares the Christian parables to Jewish lore in midrash and halakic texts. Interesting.

INDEX

INDEX OF BIBLICAL QUOTATIONS